AF880286

Andre Bationo • Boaz Waswa • Jeremiah M. Okeyo
Fredah Maina • Job Kihara • Uzo Mokwunye
Editors

Fighting Poverty in Sub-Saharan Africa: The Multiple Roles of Legumes in Integrated Soil Fertility Management

Springer

Editors
Andre Bationo
Alliance for a Green Revolution
in Africa
Soil Health Program
Accra
PMB KIA 114, Airport-Accra
Ghana
abationo@agra-alliance.org

Jeremiah M. Okeyo
Tropical Soil Biology & Fertility (TSBF)
International Centre for Tropical
Agriculture (CIAT)
P.O. Box 30677
00100 Nairobi
Kenya
jmosioma@gmail.com

Job Kihara
Tropical Soil Biology & Fertility (TSBF)
International Centre for Tropical
Agriculture (CIAT)
P.O. Box 30677
00100 Nairobi
Kenya
j.kihara@cgiar.org

Boaz Waswa
Tropical Soil Biology & Fertility (TSBF)
International Centre for Tropical
Agriculture (CIAT)
P.O. Box 30677
00100 Nairobi
Kenya
bswaswa@yahoo.com

Fredah Maina
Kenya Agricultural Research Institute
Socio-economics and Biometrics
P.O. Box 14733
00800 Nairobi
Kenya
fredah.maina@yahoo.com

Uzo Mokwunye
Mokwunye Consulting
Accra
Ghana
uzo.mokwunye@alumni.illinois.edu

ISBN 978-94-007-1535-6 e-ISBN 978-94-007-1536-3
DOI 10.1007/978-94-007-1536-3
Springer Dordrecht Heidelberg London New York

Library of Congress Control Number: 2011935025

© Springer Science+Business Media B.V. 2011
No part of this work may be reproduced, stored in a retrieval system, or transmitted in any form or by any means, electronic, mechanical, photocopying, microfilming, recording or otherwise, without written permission from the Publisher, with the exception of any material supplied specifically for the purpose of being entered and executed on a computer system, for exclusive use by the purchaser of the work.

Printed on acid-free paper

Springer is part of Springer Science+Business Media (www.springer.com)

Fighting Poverty in Sub-Saharan Africa: The Multiple Roles of Legumes in Integrated Soil Fertility Management

Preface

A substantial amount of research has been undertaken on legume crops in Sub-Saharan Africa (SSA), yet their role in fighting poverty has not been sufficiently highlighted in the literature. In the mixed cropping systems that are widely practised in the continent, legume crops constitute an important component. The multiple uses of legume crops are quite evident. First, they are an important source of nutrition not only to humans but also to livestock. They provide the much needed protein; minerals; fibre and vitamins. Secondly, farm households generate income through the sale of legumes seed, leaves and fibre. Finally, the legume crops that are able to biologically fix nitrogen in the soils play an important function for soil fertility management. In addition to increasing inorganic nitrogen in the soil, they improve other soil physical properties, provide ground cover and reduce soil erosion, increase soil organic matter, cation exchange capacity, microbial activity, lower soil temperatures and suppress weeds and pests.

In spite of the huge potential of legume crops, farm households in SSA face enormous challenges in producing these legumes. These include low and declining soil fertility, low soil pH, high salinity, drought and flooding, poor access to improved germsplasm, diseases, pest and weeds. The nature of these challenges calls for a comprehensive integrated approach for research on legume crops.

The chapters in this book address the issues using a multi-disciplinary lens incorporating not only the biophysical but also the socio-economic dimensions of the research questions. Chapter 1 examines the agro-ecological distribution of legumes. It provides insights into the biophysical niches for legumes growth. The information should be useful to researchers and development workers. Chapter 2 tackles the contribution of legumes to the livelihoods in SSA, hence addressing the key consideration of how legumes can fight poverty. The material in this chapter is complemented by Chap. 5 that examines scaling out strategies for legumes-based practices. In Chap. 3, the inter- and intra- specific variation of legumes. The relationship between legumes and soil biodiversity, soil-borne pests and disease dynamics are examined. This chapter sheds light on how legumes affect soil physical properties. Chapters 6–8 address the current and potential role of legumes in integrated soil fertility management through a comparative analysis. Chapter 6 presents the case of West and Central Africa; Chap. 7 examines the East Africa situation; and Chap. 8 looks at the Southern Africa case. These different case studies are truly illuminating. Chapter 9 looks at options for improving plant nutrition to increase common bean productivity in Africa.

There is no doubt that this book makes a very useful contribution to our knowledge on the roles of legumes not only in soil fertility management but also in fighting poverty. The two outcomes are clearly linked. Improvements in soil productivity have direct positive consequences on agricultural growth, food security and the livelihoods of the poor. Hence, when there is growth in the agricultural sector, the poor are most likely to benefit. This is especially true in a continent where over 75% depend on agriculture, which holds a high share in gross domestic product. The United Nations University Institute for Natural Resources in Africa is pleased to be associated with the publication of this book. The Institute places very high importance on research aimed at harnessing the continent's land resources.

Elias T. Ayuk
Director, UNU-INRA

Contents

Chapter 1
Agro-Ecological Distribution of Legumes in Farming Systems and Identification of Biophysical Niches for Legumes Growth

S.M. Nandwa, S.N. Obanyi, and P.L. Mafongoya

Abstract Legumes have potential of enhancing sustainable productivity in crop and forestry systems especially in developing countries where agricultural productivity is on the decline. However, adoption and integration of legumes in farming systems has been difficult to achieve because of biophysical and socio-economic constraints faced by landusers who are predominantly smallholder farmers. There is need therefore, to develop appropriate agroecological and social niches within which different legume crops can be fully realized, be they multipurpose in nature or of specific benefits. Potential niches for the "best bet" legumes fall into temporal niches which are defined by sequential or simultaneous occurrence of food crops and legumes; and spatial niches, defined by the best place on the farm to plant different legumes, depending on farmers priorities and preferences, soil fertility, farm size, climate and other factors. This chapter explores different efforts that have been undertaken to suite legume crops to different agroecological and farmer settings in sub-Saharan Africa.

1.1 Introduction

Although increase in population has been associated with activities and practices that contribute to climate change, it is imperative to emphasize that the consequences of high population are likely to remain of ecological and socio-economic importance

S.M. Nandwa
Kenya Agricultural Research Institute – National Agricultural Research Laboratories (KARI-NARL), Nairobi, Kenya (Deceased)

S.N. Obanyi
Kenya Agricultural Research Institute – National Agricultural Research Laboratories (KARI-NARL), Nairobi Kenya

P.L. Mafongoya (✉)
Department of Soil Science and Agricultural Engineering, University of Zimbabwe, MP 167, Harare, Zimbabwe
e-mail: mfongoya@yahoo.com

A. Bationo et al. (eds.), *Fighting Poverty in Sub-Saharan Africa: The Multiple Roles of Legumes in Integrated Soil Fertility Management*,
DOI 10.1007/978-94-007-1536-3_1, © Springer Science+Business Media B.V. 2011

Table 1.1 Rural poverty by region, 1975–1980

Region	Rural population in poverty (%)
Sub-Saharan Africa	65
Latin America	53
Asia	50
Near East and North Africa	32

Source: World Bank (1986)

irrespective of climate change. The problem at the apex of this subject is that of achieving sustainable food security in decades ahead. This is particularly an arduous task in the developing world where some 800 million people live and where one out of six people are food insecure; 200 million children are malnourished (56% of the deaths of children below 5 years are due to malnutrition, and thus 1,700 children die unnecessarily every hour) (Howard et al. 2001; IPPF et al. 1998; NRSP 2001). Reports from Sub-Saharan Africa (SSA) indicate that out of 1000 children born alive, 112 and 48 children die before their 1st and 5th birthday, respectively; while the remainder suffer from underweight. These circumstances have contributed towards the recognition of the importance of the concept of sustainable cropping and forestry management, to help provide secure foods, fibre, timber and fuel supplies for the rapidly growing population.

Next and related to the state of food insecurity, is the problem of poverty. It is estimated that approximately 1.3 billion people in the world live on less than a dollar per day; and that by 2050, some 24% of the total population in developing world will be living below the poverty line (one dollar per day). Reducing (rural) poverty is one of the most pressing economic and social issues in developing countries. According to World Bank estimates, around 780 million people live in absolute poverty (income of less than 1 US $ per day). In Africa, some 250 million people subsist below this *per capita* income. The incidence of poverty is highest in Sub-Saharan Africa and lowest in the near East and North Africa (Table 1.1) (World Bank 1986).

Poverty is reported to be the major cause of food insecurity and deaths of many people in the developing world. Thus to alleviate this widespread problem of poverty and food insecurity and provide economic empowerment to the impoverished populations, agricultural researchers are looked upon as the core leaders to improve and sustain agricultural productivity under the above-mentioned prevailing global changes and global challenges.

One of the major ways of solving the poverty problem is by promoting agricultural production. However, agricultural growth alone will not be enough except where it is augmented with technological, institutional and political changes (World Bank 1996, 1997). In small-holder farming systems, new technologies can directly benefit the poor producers through raising the productivity of staple crops and consequent greater availability of food (Beets 1990). The management of the available land so as to produce adequate amounts of food then becomes of major importance. The management aspect is therefore a factor which is determined by the farming system being practiced in the area. Farming systems typologies are dictated by climate, production goals and culture. Sustainable productivity of cropping and forestry systems implies continuous and increased yields of agricultural (crops and livestock) as well as forestry products, in the short and long-term. Even without climate change, for the cropping system this is not an issue in the developed countries because most of them are producing surplus crop and livestock products. It is however critical in developing countries, especially Sub-Saharan African countries where sustainable agricultural productivity has become an elusive issue. Similarly, sustainable productivity of forestry systems even without the climate change is not a major issue in developed countries because there is very little natural forestry systems, which are endangered. Most of the forestry systems are conservation and managed systems. It is in the developing countries that forestry systems are endangered not only by climate change but also by human activity. This chapter therefore focuses on the role of legumes for sustainable productivity in crop and forestry systems especially in the developing countries, in the context of declining agricultural productivity in SSA. Developing adoptable legume-based technologies could remarkably improve livelihoods of rural poor (KARI 2000). But this requires judicious management of factors that determine their integration into different land types, farming systems and niches in order to fully realize the multiple purpose and benefits of legumes. For example, in the miombo woodlands, lack of dry season fodder and fuelwood, provides opportunity for introduction of legumes.

In a recent literature review of legume germplasm carried out under the auspices of the African Highlands Initiative (AHI 1997), information was provided on 106 legume species conserved in Kenya and Ethiopia, many of which have high potential as nitrogen fixers, green manure, fodder and cover crops, yet have been under-exploited. A related study targeting on-farm surveys of nutrient dynamics and organic resources from two niches namely, farm hedges and boundaries, revealed 26 species which may be under-exploited, besides lantana, green manure legumes and tithonia.

In Kenya (1995–1996), a study to evaluate suitability of exotic and traditional legume species on low/medium fertility soils and several agroecological zones (mainly midlands and upperlands) for green manures and human and livestock consumption showed that while some species performed well across all sites, a few were confined to certain agroecological zones (Gachene et al. 1999). Out of 40 traditional legumes evaluated, only 11 turned up to be "best bet" for the East African region: *Canavalia ensiformis, Crotalaria grahamiana, C. juncea, C. ochroleuca, Desmodium intortum, D. uncinatum, Lablab purpurens, Mucuna pruriens, Tephrosia vogelii, Vicia bengholensis* and *V. dasycarpa*. This was based on the criteria of early maturity; high biomass production; high BNF capacity; tolerance to pests, diseases and drought; rapid ground cover and weed suppression; ease of establishment and possible alternative uses. A majority of these were suitable for growing in the lower/upper midlands, and a few in the lower highlands. Similar "best bet" studies are underway in Southern Africa (Waddington and Snapp 1999). These studies indicate that despite the use of Legume Cover Crops (LCCs) in the first half of the century, it was/is difficult to find them in farming systems of East Africa. Other studies have shown that in 1980s and 1990s, extensive efforts were made to research and promote *Mucuna pruriens* as a green manure or cover crop in the tropical regions, but despite its good biophysical performance, it was poorly adopted by farmers (Eilitta and Carsky 2002). This was attributed to perceived ambiquity on the state of knowledge e.g. poor suitability for use as food and feed (associated with the presence of toxic and anti-nutritional factors) and lack of assessment and consensus on research and development as a food and feed crop. From the foregoing, it would appear that efforts should be directed in assessing trade off between human food and livestock needs and soil fertility improvement. Thus, the fate of anti-nutritional factors down the ruminant digestive system needs to be determined. Also, there is need for information on level of supplementation of livestock diets and determination of dairy meal equivalencies. The mucuna problem reported above underpins the need for careful assessment of green manure legumes for human consumption.

Although the potential niches for legume growth in different farming systems has been identified and characterized, further research is required on their socio-economic and biophysical characterization. A major impact of the research is likely to come from work on adapting legume germplasm to different agroecological niches in farming systems. Also of importance is research to overcome environmental stresses and constraints to legume integration e.g. drought, soil acidity, nutrient deficiency, aluminium (Al) and manganese (Mn) toxicity and soil salinity (Pessarakli 1999). To what extent can they help preserve biodiversity? Further work needs to be undertaken on how the systems can be modified to increase the size and number of potential niches. This type of research requires a farming systems (Sect. 1.2) and agro-ecological (Sect. 1.3) approach.

1.2 Farming Systems: Major Associated Changes, Degradation and Need for Legume Integration

A farming system is an ecosystem in which all the components – land, operators, hired labour, crops and cropping systems, animals and machinery are considered together to produce goods to meet the production requirements such as food, clothing, shelter etc or to exchange for goods to meet part or all of those needs. In some regions of Sub-Saharan Africa only one farming system can be found i.e. only shifting cultivation in large forest areas. In other areas, several systems can be found in an intricate pattern i.e. a mixture of subsistence farms, small-holder coffee farms, and large scale plantations as found in some parts of Kenya. The term system is therefore used at two different levels, at the farm enterprise level and at the regional or area level or recommendation domain. More recently, Scherr (1997, 1999) defined farming systems in terms of six land types and associated farming practices/systems (Table 1.2). The classification includes the seven major crop- based tropical farming systems described by Beets (1990) namely: shifting cultivation; lowland rice-based farming systems (FS); upland cereal-based systems; smallholder mixed FS; irrigated smallhold FS; smallhold farming with plantation crops; and agroforestry systems. The major feature of both approaches is that recent changes in the farming system are described and related on-site and other resource degradation are listed. There is degradation in all farming systems including: nutrient mining, loss in biodiversity, acidification, salinization, waterlogging, pollution, compaction etc. The irrigated lands encompass lowland rice-based farming systems (phreatic) upland rice-based as well as commercial enterprise type of farming system. A major change includes the

Table 1.2 Major pathways of change in agricultural land use and farming systems in developing countries and associated degradation problems

Land type	Farming systems	Main changes	On-site soil degradation	Other resource degradation
1. Irrigated lands	(a) Commercial enterprise type of technology e.g. drier areas of Kenya, Sudan, Egypt, Chad etc. (b) Traditional techniques of dryland or upland system of irrigation e.g. upland Rice-Based Systems (wetlands with soil flooded to 5–10 cm through rainwater throughout the growing season) (c) Phreatic upland or ground water system depends on both rain and ground water e.g. lowland rice-based systems (swamp land and lower parts of catena)	With the exception of a few countries e.g. Egypt where 40% of all arable land was under irrigation (1994) only 4% of Africa's arable land is under irrigation thus making the system to be unimportant in many of the continent's countries (World Bank 1997). Recently there are increasing number of plans for lowland rice-based development of hitherto unused lowlands in Africa especially in West Africa (WARDA, 1990). This implies intensification and diversification of the system towards the Asia type of green revolution (e.g. use of HYVS, [high yielding varieties] agrochemicals, mechanization, etc)	(a) Salinization- there is need for integration of technical solutions that work under prevailing biophysical and Socio-economic conditions to counteract salinity–related problems. (b) Water logging (c) Nutrient constraints under multiple cropping associ-ated with imperfect marketing infrastructure and supply of production inputs (d) Biological degradation due to blanket excessive use of agro-chemicals (e) Nutrient leaching and soil physical degradation (hardpans etc.)	(a) Lowering of water quality associated with nutrient pollution in ground/surface water (b) Pesticide pollution (c) Water-borne diseases (d) Water use conflicts

(continued)

Table 1.2 (continued)

Land type	Farming systems	Main changes	On-site soil degradation	Other resource degradation
2. High-quality rainfed lands: in areas with naturally deep fertile and less weathered soils e.g. the East African Highlands, Tropical areas with vertisols and alfisols e.g. West African Savannahs.	(a) Upland Cereal-based systems of the East African Rift Valley with soils of high to medium potential, whereby cereals e.g. maize, wheat, barley, triticale are commercially and predominantly produced as cash crops (in cereal mono-cropping farms but sometimes mixed with livestock)	• Transition from predominantly crop-livestock based systems to upland cereal based • Transition from short fallow to continuous cropping with mechanization • Evolution of present system comes naturally via shifting cultivation (e.g. coffee in Ethiopia) or via arable-cultivation fallow farming towards a sequence of arable cropping-perennial crops feeding to stationary system.	(a) Nutrient imbalance due to non-balanced fertilization (b) Nutrient mining due to depletion (c) Increased use of low quality planting materials, fertilizers and pest control contributes to low productivity and low income	• Pesticide pollution • Loss in bio-diversity
	(b) Plantation Crop-Based systems found in Eastern and Western African Highlands with farms dominated with one or more cash or industrial tree or shrub crops (e.g. coffee, tea, cocoa, oil palm, rubber etc.) or non-woody perennials (sugarcane and bananas)	• Exploitation by middlemen leads to a vicious cycle of low income, little plantation maintenance low productivity and low income • Less commercial-oriented farms may revert to multi-storey cropping to improve productivity • Future sustenance will depend on less reliance on world market prices which require policy on the establishment of critical hectarage		

3.	High-medium quality rainfed intensively managed lands in the Eastern and Southern African Highlands. These are areas which are well watered but with considerable poor soil fertility expanse e.g. in Zambia	(a) Small-holder mixed farming systems (e.g. in the highlands of Kenya, Uganda, Rwanda, Cameroon, Ethiopia) with complete crop-livestock integration. Tends to be largely subsistence (finger-millet, maize, potatoes, vegetables, rice, wheat, teff, fruit trees, enset and bananas) sometimes with cash crops (coffee, tea, enset, bananas, chat etc.) (b) Small-holder farming with plantation crops (c) Integration of agro-forestry practices ("bush fallow", home gardens, taungya afforestation, agro-silviculture, traditional agriculture etc.)	• Transition from shifting cultivation to continuous cultivation with rotation/mixed cropping/multistoreys which tends to make the system sound and sustainable. • Transition from traditional practices to HYVS agro-technical sophistication and fine tuning the system with crop intensification and diversification	(a) Nutrient and SOC mining though depletion. (b) Nutrient imbalance due to non-balanced fertilization (c) Acidification (d) Soil erosion (e) Biological degradation with agro-chemicals (f) Loss in biodiversity resulting in environmental deterioration (g) Destruction of soil physical properties thus reducing fertility	• Pesticide pollution • Loss in biodiversity • Poor control of resistant pests and diseases leading to low production of some crops/trees

(continued)

Table 1.2 (continued)

	Land type	Farming systems	Main changes	On-site soil degradation	Other resource degradation
4.	Densely populated, semi-intensively managed marginal lands e.g. hill masses of Machakos District, Kenya	Small-holder crop livestock system of the arid and semi-aridd lands in the Eastern, Southern and Western Africa. The system is common in the seasonally semi-arid tropics with high-textured easily erodible soils and the seasonally humid parts of Eastern and Southern African and the marginal zones of the Sahel and Guinea Savanna parts; with pockets found on the deep vertisols e.g. Kenya, Zimbabwe and Nigeria. The system is based on a self sufficient semi-sedentary form of agriculture primarily founded on grain crops (millet and sorghum etc. in drier areas, maize etc in wetter areas) and cash crops such as cassava, sunflower, forage crops, cotton, groundnuts and sesame.	• Transition from long to short fallows or continuos cropping • Transition from extensive grazing and cropping only in traditional landscapes to cropping in new landscape niches e.g. valley bottoms, high fertility niche crops, inclusion of various low fertility adapted crops	(a) Soil erosion (b) Soil fertility depletion (c) Removal of natural vegetation perennials from landscape (d) Soil compaction, physical degradation from over cultivation (e) Acidification	• Loss of biodiversity • Watershed degradation

5.	Extensively managed marginal lands	The farming systems of the Savanna Margins and Pioneer Forests	• Immigration and land clearing for low-external input agriculture	(a) Soil erosion from land clearing (b) Soil erosion from crop/ livestock production (c) Soil nutrient depletion (d) Weed infestation (e) Biological degradation from top soil removal	• Deforestation • Loss of biodiversity • Watershed degradation
6.	Urban and peri-urban agricultural lands (intensive and semi-intensive system)	(a) Subsistence-oriented, compost-based urban agriculture (b) Urban sewage farming (c) Urban ornamental farming (d) Commercial-oriented peri-urban agriculture	• Rapid urbanization • Diversification of urban food market • Rise in urban poverty	(a) Soil erosion from poor agricultural practices. (b) Soil contamination from urban pollutants (c) Overgrazing and compaction	• Water pollution. • Air pollution • Human disease vectors

expansion of lowland rice cultivation especially in West Africa. A major degradation is salinization and lowering of water quality associated with nutrient and pesticide population resulting in water-bone diseases. Legume integration should strive to mitigate nutrient constraints under multiple cropping with imperfect marketing infrastructure and supply of inputs.

Beside upland cereal-based system, the high quality rainfed lands with naturally deep fertile soils, also consist the plantation crop-based systems (e.g. sugarcane, bananas, coffee, tea, coca, rubber, oil-palm etc.). Major degradation include, nutrient imbalance due to non-balanced fertilization and mining due to depletion and associated loss in biodiversity and pesticide pollution. Due to high rainfall and fertile soils, there are high opportunities for integration of all classes of legumes (grain, forage, green manure and trees) with minimum direct policy intervention (Table 1.3).

The smallholder mixed farming system and smallholder farming with plantation crops (and integration of agroforestry practices) are the main farming systems in the high-medium quality rainfed intensively managed lands. The systems are rapidly changing to continuous cultivation and introduction of high yield varieties agro-technical sophistication, crop intensification and diversification. Besides nutrient imbalance and mining (pollution and biodiversity loss), other degradation aspects include acidification and soil erosion. Legume options for integration are similar to those for high quality rainfed lands.

The main farming system in the densely populated, semi-intensively managed marginal lands, is the smallholder crop-livestock system of the arid and semi-arid lands in SSA. A major change in the system includes reduction in extensive grazing associated with removal of natural vegetation perennials from landscape (loss of biodiversity, watershed degradation, soil erosion, fertility depletion, soil acidification and compaction). There is high potential and contribution of legume integration e.g. introduction of forage crops, groundnuts, sesame, cotton, cassava, sunflower etc in drier areas where millet and sorghum perform poorly.

The farming systems of the savanna margins and pioneer forests are found in the extensively managed marginal lands. The major change includes immigration and land clearing for low external input agriculture. Major degradation includes soil erosion from land clearing and from crop-livestock production, mining from depletion, weed infestation and loss in biodiversity associated with deforestation. Research to identify alternatives to slash and burn has identified suitable legumes for integration into the farming system (Denich et al. 2002).

The urban and peri-urban agriculture is a rapidly evolving phenomena associated with rapid urbanization, diversification of urban food markets and rise in urban poverty. The main farming systems include subsistence-oriented, compost-based urban agriculture or urban sewage farming. Major degradation include soil erosion from poor agricultural practices, soil contamination from urban pollutants (water and air pollution and associated human disease vectors). The next two (Sects. 1.3 and 1.4) deal with the question of; to what extent can the legumes mitigate these constraints and in what land types and farming systems should the research focus be?

A recent study by IFPRI involving areas of policy interventions identified two priority farming systems, namely densely populated, semi-intensively managed marginal lands and irrigated land (land type 4 and 1 in Table 1.2, respectively). This prioritization is based on the severity of the problems faced in these farming systems, and their importance for food security and economic development (Table 1.3) (Hilhorst and Toulmin 2000).

1.3 Agro-Ecological Distribution of Legumes in Farming Systems

As already stated, there are six land types each associated with several major farming systems that have been identified in Sub-Saharan Africa. These cover the four large climatic classifications in Sub-Saharan Africa. These climatic zones are rainy climates, seasonal climates, dry climates and deserts. Out of these four, the seasonal climates and dry climates form the bulk of the agricultural zones. In the seasonal climates, wet and dry climate occupy about 50% of the tropics, mostly around the equatorial belt. In Africa, they include most of the continent between Sahara and Kalahari deserts. On an annual basis precipitation (P) exceeds potential evapotranspiration (PET) in all or most months. In the rainy climates, the main areas include the Congo basin and P exceeds PET in all or most months. The dry climates cover about 15% of the tropics. The largest areas are the Sahel, located between the Savannah belt and the Sahara desert and Kalahari desert. Here, the rainy season is short and the PET exceeds P most of the months. The tropical deserts are extremely dry areas and of little significance in agriculture. We will therefore discuss the distribution of legumes across different farming systems.

1.3.1 Shifting Cultivation Systems

Shifting cultivation is defined by FAO (1982) as "a farming system in which relatively short periods of cultivation are followed by relatively long periods of fallow." The system thus involves the clearing of a plot of land, usually under some sort of forest cover; its use for a few years and then, as soil fertility declines, its abandonment in favour of another newly-cleared plot. About 40% of the total agricultural production (3–500 million people) in the third world depend on this system for their daily livelihood (Lanly 1985; Stocking 1984)

Slash-and-burn cultivation is a common practice and long fallow periods are more sustainable than short fallow periods. Unfortunately land shortage has become a problem and traditional shifting cultivation can hardly be practiced anymore. High erosivity of the soils due to the high rainfall has a pronounced negative effect. To tackle the declining fertility and high erosion multipurpose tree legumes have been and can be used. This technique allows the fallow to accumulate the necessary nutrients and recycle them during the cropping period. For example, in West African

Table 1.3 Importance of soil degradation in different agricultural systems

Farming systems	Anticipated impact of soil degradation					
	Food security and poverty	Market supply of produce	Economic development	National wealth of resources	Overall severity of degradation	Opportunity for effective direct policy intervention
1. Irrigated lands	**	***	***	***	**	**
2/3. High/medium quality rainfed lands	**	***	***	***	*	*
4. Densely populated marginal lands	***	**	***	**	***	***
5. Extensively managed agricultural lands	**	*	*	*	*	*
6. Urban and peri-urban agricultural lands	**	*	*	*	*	*

Source: Adapted from IFPRI, Scherr (1999), p 43
*low
**medium
***high

countries, deforestation, annual bush burning, overgrazing and intensive cultivation have turned many tropical forests to savanna. This change in vegetation has led to a decrease in precipitation, increased erosion problems, decreased soil fertility and soils have become compacted. Experiments have been conducted to restore these ecosystems. Some 30 multipurpose tree species were used whose role was to provide shade, preserve water, reduce the susceptibility to erosion and pump nutrients into the system through nitrogen fixation and recycling from lower soil layers (Badeyo 1998).

In the Ethiopian highlands, wood lots have been established where *Acacia cyanophyda* is used for erosion control and soil fertility improvement. Other included legumes in the farming system are *Eucapyptus globulus* and *E. camaldulensis* which are used for fuel.

1.3.2 Small-Holder Mixed Farming Systems

Complete integration of animal and crop components is rare in Africa, hence the system has more of the character of an upland cereal-based system, and tends to be basically subsistence, but is ecologically stable because of its diversity.

As livestock cannot graze the fields when crops are in critical stage, this requires substitution by crop residues or within/outside farm of fodder trees planted in the homestead, in alleys along borders or erosion control bunds grown in some agro-forestry arrangement (esp. fodder and fuel-wood trees). Particular leguminous fodder trees integrated in the system include *Leucaena calliandra* and *Mimoa species* while fodder crops may include napier, *Leucaena* etc. Very intensive fodder "cut and carry" systems with dairy cattle are found in the limited parcels upland areas of Africa practicing partial or full animal confinement (zero grazing units). The mixing of legume trees with animals often results in beneficial interactions between the different components of the system e.g. trees act as windbreak, provide nutrients partly from biological nitrogen fixation to the annual crops and shade and shelter for the compound. The deep-rooted trees recycle nutritive elements.

1.3.3 Upland Cereal-based Systems

This is a system based on a semi-sedentary form of agriculture (with shorter fallow periods) in the sub-humid and semi-arid tropics (savanna zones), often referred to as "seed agricultural complex" (Okigbo 1979); based on millet and sorghum (drier areas) and maize (wetter parts). There is little integration between crop and livestock activities in this farming system.

Cereal production for self-sufficiency forms the backbone of the system, especially in the Sahel and parts of Eastern and Southern Africa. The objective is sustained through yearly production, at minimal risk, of reasonable levels of a range of products to satisfy both subsistence and commercial (groundnuts, cotton, cassava, tobacco and chilies) needs. The system is therefore characterized by high yield gaps, as the objective is not to maximize production of any one commodity. Output of the system often only provides a farm income below subsistence requirements (especially with small farm sizes). Stable upland cereal-based systems are increasingly becoming rare, most are in transition towards monocropping e.g. maize-maize or cassava-cassava or towards "small-scale mixed farming system integrating agroforestry and green manuring." Natural fallow is most common, often grazed. There is great potential for regulated fallow and growing of green manure crops (especially legumes). Without external inputs, productivity of the system tends to decline to a point where it is not sustainable biologically, with yields declining to low levels, but costs of inputs increasing faster than the market value of the crops, which is financially not viable, and sites may be abandoned for 2–5 years.

1.3.4 Irrigated Smallholder Farming Systems

The system found in drier areas (agriculture impossible without irrigation) is similar to lowland rice-based system but avoids heavy reliance on one crop and the specific conditions of the flooded rice-land regime. The degree of commercialization tends to be quite high, e.g. cotton in Sudan and Egypt. The system has several activities e.g. several crops and often resembles a mixed farm. Most traditional ones are usually found near small rivers while government-constructed schemes are found in large plains (predominantly surface ridge and furrow irrigation). Crop varieties tend to be a mixture of traditional and modern ones which perform sub-optimally. There is often little integration of legumes in this farming system, in spite of their potential benefits.

1.3.5 Smallholder Farming with Plantation Crops

The system is dominated by one or more cash or industrial crops most of which are mostly trees or shrubs, such as rubber, coconut, cashew, vanilla, cocoa, oil-palm, coffee, tea, but may include sugarcane and banana (non-woody perennials). Often the system is found in association with large-scale commercial estates since some of the crops cannot easily be processed by smallholders e.g. tea and coffee in East African Highlands; sugarcane in Kenya and Nigeria, cocoa in Ghana etc. The system may evolve from "arable cultivation-fallow" to arable cropping perennial crops. This is often in crop mixtures, particularly in multi-storey cropping with perennial and annual food crops. Niches for cultivation of plantation crops include tea in hilly areas, rubber in flat areas etc. Poor husbandry in some plantation farming system is attributed to a lag in "cultural adaptation." There is potential for integration of legumes as part of integrated soil fertility management (ISFM).

1.3.6 Lowland Rice-based Farming Systems

Rice is important in small parts of the African lowland. Until recently, most SSA governments paid relatively little attention to developing unused lowland rice areas. The farming system is typical smallholder activity. In most systems, the practice of high output monoculture often over the years exhaust the soil, and to maintain fertility requires integrated soil fertility management strategies. There is need for intensification and diversification of the system if possible towards the green revolution attained in Asia.

1.3.7 Agroforestry System

Agroforestry is strictly a component of a farming system (FS) rather than being an FS itself. It is a landuse in which trees are grown on the same land as crops and/or animals, either in a spatial arrangement or in a time sequence, and in which there are both ecological and economic interactions between the tree and non-tree components. This includes "bush fallow," "home gardens," taungya afforestation, traditional agriculture, aquaculture and fisheries under mangroves, agro-silviculture, silvo-pastoral.

The success of the components depends on the potential of trees to "pump-up" and recycle nutrients, fix nitrogen and minimize nutrient losses via leaching, erosion, etc. A number of different systems are practiced e.g. Shade Tree-Based System, Fodder Tree-Based System and Dispersed Tree in Annual Crop Fields. In Shade Tree-Based System, plantation crops such as coffee and cocoa are often found planted with a variety of leguminous shade trees e.g. *Leucaena leucocephala, Cordia alliodora, Erythrina poeppigiana,* which help produce wood, fix nitrogen, create conductive micro-climate, etc.

In Fodder Tree-Based System in semi-arid regions of Africa, livestock has been reported to rely in the protein-rich "aerial pastures" of woody shrubs and trees which are the most nutritious forage available in the dry season. In general, the success of productive system is often attributed to the association of trees and shrubs with either improved pastures or natural range lands. The strategy of the systems is to let the bulk of nutrients come from the trees and from herbaceous legumes whilst only the most important deficiencies are corrected with external inputs. Similarly, dispersed trees in annual crop fields provide poles, fuel-wood, fodder, edible fruits, nuts and leaves, medicines, and a range of other products without unduly competing with associated annual crops. In East Africa, such trees include local species *Markhamia platycalyx* and exotic *Grevillia robusta* valued for their wood and soil-enriching litter fall. More modern agroforestry components include "alley cropping" and systems with fodder trees which tend to produce more surpluses and are thus more commercialized.

In a study to delineate the agro-ecological suitability of green manure, herbaceous legumes and legume cover crops in Kenya (under optimum conditions) in acrisols, luvisols, nitisols and ferralsols, Gachene et al. (1999) found most legumes to perform well in upper midlands (1,300–1,900 masl) e.g. *mucuna, canavalia, crotolaria, lablab, desmodium,* but very few in the lower highlands (1,900–2,300 masl) e.g. *Vicia, C. grahamiana* and *tephrosia* and also few in the lower midlands (800–1,300 masl). Thus, they concluded that while some species perform well across most sites few tend to be confined to certain agro-ecological zones.

1.4 Identification of Biophysical Niches for Legume Growth

The widespread adoption and integration of legumes in the farming systems has been difficult to achieve because of biophysical and socio-economic constraints in the tropical region e.g. sacrifice of time and space normally used for food crop production. Therefore, for this strategy to be adopted in the smallholder sector, there is need to identify suitable niches in each farming system and farm typology (Woomer 1998). Potential niches for the "best bet" legumes fall into temporal niches which are defined by sequential or simultaneous occurrence of food crops and legumes; and spatial niches, defined by the best place on the farm to plant different legumes; depending on farmers priorities and preferences, soil fertility, farm size, climate etc. In assessing promising organic matter technology systems and species in smallholder cropping systems of Southern Africa, Snapp et al. (1998) identified six temporal niches e.g. grain legumes grown either in rotations or as intercrops; perennial legumes grown as intercrops and in hedgerow systems; perennial legume-Faidherbia albida intercrop; perennial legume-improved fallows; and green manure legumes grown in relay intercrops and rotations (Table 1.4). In terms of their long-term contribution to raising soil fertility, short-term ability to raise crop yields, competition for arable land, labour requirement and ease of adoption by farmers, technologies which are both easy for the farmers to adopt, and can raise crop yields in the short-term often require resources which are in short supply (labour, land and cash). A major challenge for research teams is to be able to identify unexploited or under utilized spatial niches such as farm boundary, hedge rows, contour strips, unproductive fields or degraded lands, conservation bunds etc as entry point for integration of legumes. Such studies should be conducted alongside the evaluation of the trade-offs of the alternative uses of legume species (e.g. soil fertility enhancement, fodder, food, erosion, weed and pest control). A major breakthrough in integrating legumes is the integrated soil fertility management (ISFM) in the farming systems is the identification of four major purposive legume classes: (i) Grain/Food Legumes (cowpeas, soybean, *phaseolus, Cajanus,* chickpea, lentil, fababean, pea, *lablab,* groundnuts etc.); (ii) Fodder/Forage Legumes (*Lablab, Centrosema, Stylosanthes, Desmodium* etc.); (iii) Green Manure/Cover Crops/Herbaceous Legumes (*Mucuna, Crotalaria, Tephrosia, Canavalia* etc.); and (iv) Multipurpose Trees/Tree Legumes (*Leucaena, Sesbania, Calliandra, Ptecarpus, Acacia, Gliricidia, Senna* etc.). Table 1.5 lists the four major legume classes with their respective temporal and/or spatial niches. An evaluation of the "best bet" legumes shows that very few simultaneously satisfy the multipurpose role for their integration e.g. soil fertility, food

Table 1.4 Technologies for enhancing soil organic matter, nitrogen and productivity on small-holder farms in Southern Africa: contributions and constraints

Organic matter technology	Long term contribution to raising soil fertility	Short term ability to raise crop yields	Competition for arable land	Labor requirement	Ease of adoption by farmers
Grain legume: rotations	Low-medium	Medium	Low-medium	Low	Medium
Grain legume: intercrops	Low	Low	Very low	Low-medium	High[a]
Perennial legume: intercrops and hedgerow systems	Medium-high	Low-medium	Medium	High	Low
Perennial legume: Faidherbia albida intercrop	Medium	Very low	Low	Low	Medium
Perennial legume: improved fallows		Low-medium	High	Medium	Low
Green manure: relay intercrops and rotations	High	Medium-high	High	Low-medium	Low-medium
Biomass transfer	Medium	High	High	Medium-high	Low
Animal manure	Medium	High	Nil-low	Medium	High

[a]An important point on high potential for adoption is that these technologies are likely to be adopted for other reasons – provide food, food security, weed control, etc – and the soil fertility effect is primarily a spin-off

Table 1.5 Temporal and spatial niches for major classes of legumes: grain, forage, green manure and tree legumes

	Temporal	Spatial
(i) Grain/Food leg		
Cowpea Soyabean *Phaseolus* *Cajanus* Chickpea Lentil Fababean Garden pea Lablab Groundnuts	**Intercrop (GL)** • Common every year in land scarce and densely populated areas because do not take land away from cereals and scavenge from fertilizer applied on cereals • In low plant densities but aggregate cumulative effect important • Stabilizing effect of inter-crops on food security and • Enhanced efficiency on land use through BNF and recycling of nutrients leached below the cereal rhizosphere (1–3 m) by deep-rooted legumes (Farrell et al. 1996; Hartemink et al. 1996) ▪ Common bean *(Phaseolus vulgaris)* ▪ Cowpea *(Vigna unguiculata)* ▪ Pigeon pea *(Cajanus cajan)* ▪ Others *Vicia faba, Lathyrus sativus, Pisum sativum, Lens enlinaris, Crotalaria ochroleuca* etc. Low BNF capacity and high N harvest index result in little soil N replenishment unless augmented by legume cover crops such as velvet bean, lablab, jackbean, sunhemp (humid/sub-humid) **Rotations (GL)** • Common in maize-based systems sufficiently large holding of land (>1 ha) • Multiple purpose roles include grain and leaves for human food and soil fertility enhancement of nutrients and SOM (if high BNF and low NHI) or legume harvested for green manure or forage) ▪ Soybean *(Glycine max)* ▪ Pigeon pea *(Cajanus cajan)* ▪ Croundnuts *(Arachis hypogaea)* ▪ Dolichos bean *(Dolichos lablab)* ▪ Cowpea *(Vigna unguiculata)* ▪ Bambra nuts *(Voandzeia subterranean)*	Recuperation of Degraded Land (RDL) may be overcome by introducing food legumes that are adapted/tolerant to abiotic and biotic stresses (unproductive plots/fields in a farm) ***Abiotic*** Recent studies by AHI have identified cowpeas varieties that can be incorporated in farming systems/soils with low phosphorus because of their limited response to P; while climbing beans help replenish/recapitalize N-depletion • Under poorly drained water-logged conditions, chick pea integration has been reported to fit in well ***Biotic*** In Southern Africa (e.g. Malawi) high infection of yellow witch weed *(Alectra vogelii)* threatens and limits use of legumes as trap crops (Kabambe and Mloza-Banda 2002) ▪ Least susceptible legumes – pigeon pea, mucuna and sunhemp ▪ Most susceptible – groundnuts and soyabeans In Eastern Africa, crop tolerance of beans to root rots is improved through soil fertility improvement e.g. inorganic inputs as well as organics especially high quality green manure legumes but less effect by sowing on ridges

(continued)

Table 1.5 (continued)

	Temporal	Spatial
(ii) Fodder/Forage/ Pasture leg		
Lablab *Centrosema* *Stylosanthes* *Desmodium* *Vicia*	• In pastoralist/mixed farming systems, forage/fodder legumes are incorporated in alley cropping and contour lines e.g. *Aeschynomena hystrix, Chemaecrista rotundifolia, Stylosanthes calopogonium, S. guianensis, Centrosema pueraria, C. brasilianum, C. pasarrum, C. acutifolium, Neonotomia and Macroptulius atropurpurem* (semi-arid), whereby success is attributed to: ▪ Erosion control ▪ Soil fertility replenishment (as follows) ▪ As cover crops for green manures ▪ As animal feed ▪ Grown with native pasture for fodder • One of the positive elements of legumes is its integration to control weeds and erosion. For example, perennial legumes such as *Desmodium uncinatum* (mid-altitude) and *D. intortum* (high altitude, >2,000 masl) can accumulate 8–10 t DM ha^{-1} within 12 months which makes them fit into the farming system as live mulch covers (besides being fodder and forage), but slow establishment requires more weeding • In higher altitudes (>1,800 masl), *Vicia dasycarpa, V. benhalensis* have been noted to perform well by producing excellent forage and hay but may require inoculation (besides being sensitive to low soil moisture and fertility)	The use of LCCs and shrubby legumes in farm boundaries and hedgerows remain under-explained niches ***Abiotic*** The recuperation of degraded lands in Central Kenya Highlands (e.g. humic acid Andosols with high Al and Mn toxicity) has been found feasible through integration of acid-tolerant fodder trees (*Calliandra calothyrsus; Mimosa scabrella, Chamaesytisus palmensis* and *Alnus acuminata*) • Under poorly drained water logged conditions common with vertisols in East African Highlands, LCCs such as *Sesbania rostrata* have been reported to perform well • In higher altitudes (1900–2400 masl) promising legumes to integrate include vetches *(Vicia sativa, V. dasycarpa)* and lupins *(Lupinns mutabicis)* • In arid and semi-arid areas, drought-tolerant legumes such as *Dolichus lablab (Lablab purpurens)* survives up to 4 months of dry season produces high biomass (3–7 t ha^{-1} in 6 months) palatable (green/dry beans) for both human and animals. Research needs to identify varieties with leaves and seeds less susceptible to pests and diseases • *Vicia amphicarpus, Lathyrus* and *Pisum* are good examples of legumes integrated into ecosystems to mitigate stressful soil/climatic conditions. Research needs include identification of geners for stress tolerance, while simultaneously, utilizing other benefits of legumes (ICARDA)

(continued)

Table 1.5 (continued)

	Temporal	Spatial
*(iii) *GML/LCC/HLS* *Mucuna* *Crotalaria* *Tephrosia* *Canavalia* *Dolichos* *Cajanus* *Stylosanthes* *Sesbania*	**Green manure: relay intercrops and rotations**. Simultaneous occurrence of long duration crops (e.g. forage grasses or cassava) with perennial LCCs. • Silverleaf and greenleaf *desmodium* grown together with napier grass ▪ Improves fodder quality ▪ Stabilizes terrace embarkments (Gatanga, Kenya) • *Arachis pintoi and Desmodium* used as LCCs in Madagascar ▪ Checks erosion ▪ Suppresses weed ▪ Provides fodder N replenishment in upland cereal-based systems of the East African Highlands (cereal monocropping commercially) through relaying LCCs *Mucuna* and/or *Lablab* 120 days after planting the cereal. When the cereal is harvested, the legume continues to grow through the dry season, is incorporated into the soil at the onset of the rains for the benefit of the subsequent cereal crop In the Eastern and Southern African highlands rotation of annual crops with deep-rooted short-term IFS (6 months) has been reported to replenish over 600 kg N ha^{-1}. This is attributed to provision of the above-mentioned N as well as 8–12 kg ha^{-1} of P which is half of the requirement of a maize crop yielding 4 t ha^{-1} of grain	Short and medium-term improved fallows (6–12 months) used for recuperation of degraded land (RDL) e.g. in Kenya (**Niang et al.** 1996), and Rwanda (**Raquet** 1990) besides provision of fuelwood, fodder and staking material. ***Abiotic*** • The RDL may be overcome by incorporating legumes. *Dolichos lablabronga, Macroptilium atropurpurem* and *Canavalia ensiformis* have limited response to P-application hence can be integrated in farming systems/soils characterized by low-fertility P-deficient and P-fixing soils as well as low rainfall • In low fertility ultisols *Aeschynomene hystrix* can be incorporated while *Acacia mangium* may be suitable in low N-soils • Improved fallows for low P acid soils with high Al and Mn levels; ▪ STIFs (6 months) – *Glycine wightii, Calopagonium mucunoides, C. agatiflora* and *C. mucronata* ▪ MTIFs (12 months) – *Tephrosia candida, Desmodium viscose* and *Macroptilium atropurpurem* ▪ S/MTIFs (6–12 months) – *C. mucunoides* and *C. grahamiana* (yielding 6–11 t ha^{-1} of dry matter, 150–300 kg N ha^{-1} and 20–30 kg P ha^{-1}) ***Biotic*** In Eastern and Southern Africa, root knot nematodes and mole rats infested/infected plots/field are often of low productivity. Studies are showing variables results e.g. *Crotalaria grahamania* and *C. ensiformis* if integrated repel mole rats but on the other hand,

(continued)

Table 1.5 (continued)

	Temporal	Spatial
	One of the positive elements of legumes is its integration to control weeds and erosion, but this may be accompanied by disadvantages • One of the disadvantages of legume integration is loss in crop yield in intercropping system (e.g. *Mucuna*) and loss of land for food production during STIFs and MTIFs. The first disadvantage can be overcome by integration of multiple purpose type of legumes e.g. jackbean *(Canavalia ensiformis)*. • In the upland Rice-based farming systems of West Africa the growing of site-specific cover crops and green manures after a rice crop helps to recycle nutrients that have been leached below the rice root zone besides suppressing the noxious weeds. Potential improved fallow legumes in the humid/sub-humid include *Stylosanthes guianeusis, Canavalia ensiformis, Aeschynomena hystrix* while those for savanna include *Pueraria phaseoloides, Mucuna pruriens, Crotalaria juncea* and *Canavalia*	*Tephrosia vogelii* does not fit into the system because of being prone to root knot nematodes in spite of its control of mole rats and occurrence of pesticides in its leaves *Mucuna pruriens and C. ochroleuca and C. juncea* all suppress root knot nematodes (and the latter two also control storage pests) *Lablab purpurens* contains medicinal extracts for treatment of blood pressure, diabetes and wounds Jackbean *(Canavalia ensiformis)* is used by some communities to control moles but its erect nature makes it poor as suppressing weeds *Mucuna* and *Canavalia* species control weeds very well as well as improving soil fertility and moisture retention *(Sesbania* performs poorly)
(iv) Tree leg/MPTS		
Leucaena *Sesbania* *Calliandra* *Ptecarpus* Acacia *Gliricidia* *Senna* *Grevillea?* *Chamae-cytisus* *Cordia* *Erythrina* *Prosopsis* *Milletia* *Pterocarpus* *Markhamia* *Acanthus*	A major attribute of the integration of tree legumes in the ecosystem include their potential to recycle or " pump-up" nutrients, BNF and minimize nutrient losses via leaching, erosion etc. The major niches where they fit include being dispersed in annual crop fields, providing shade and fodder in agroecosystem. **Perennial legume: inter-crops and hedgerow** • Ideal only if it involves use of superior tree ideotypes that product most leaf biomass under a pruning regime and where residue quality is high and negative effects (competition with crops for moisture, nutrients and light) are minimized (Ong 1994).	• Contour grass strips (e.g. guatemala grass) often ineffective in controlling erosion made more effective with a line of legume trees *(sesbania, grevillea)*, fodder/forage legumes *(Desmodium)* attributed to; ▪ Microcontour line efficacy in erosion control ▪ Provides fodder ▪ Provides firewood • Similarly, ineffective terrace bunds/established with napier grass *(Penisetum purpureum)* is improved by incorporating LCCs *(Macroptilium atropurpurens and Desmodium sp.)* attributed to; ▪ Improved erosion control ▪ Improved fodder quality

(continued)

Table 1.5 (continued)

Temporal	Spatial
▪ *Gliricidia sepium* ▪ *Leuceana leucocaphala* ▪ *Senna spectabilis* • Competition problems of intercrops of perennial legumes, because of high management requirements have limited adoption except with *Faidherbia albida* which require a long-term view (7–10 years) (Weil and Mughogho 1993) or where they are combined with grass species e.g. hedgerow managed legumes or fruit trees with physical conservation measures (potentially high erosion areas) **Perennial legume: improved fallows** • Ideal in areas widespread with shorter fallow periods of bush and grass of 1–5 years or where there is continuous cropping without fallows, and where past growing leguminous trees can be integrated as part of improved fallow system ▪ *Sesbania sesban* ▪ *Acacia augustissima* • *Leucaena leucocephala, Cordia alliodora, Erythrina poeppigiana, Markhamia platycalyx, Prosopsis chilensis* are leguminous shade trees that not only provide shade and act as windbreak on plantation crops such as coffee and cocoa, but also produce wood, fix N and help create conducive microclimate • Protein-rich "aerial pastures" of woody shrubs and trees are the most nutritious forage in the dry season in the fodder tree-based system in the semi-arid areas of SSA association of trees and shrubs with natural rangelands or improved pastures. This may be provided by Acacia species *(Acacia mangium)* • *Grevillia robusta, Markhamia platycalyx* etc. are some of the trees often found dispersed in annual crop fields providing poles, fuelwood, fodder, edible fruits, nuts and leaves, medicines and a range of other products without unduly competing with associated annual crops	• Legume species for biomass transfer of live-fence leguminous shrubs have been identified (e.g. studies in Ethiopian highlands). ▪ Acacia decurrens and Sesbania sesban on black cotton soils ▪ Acacia decurrens, Sesbania sesban and Chamaecytisus proliferus on red soils • Legume species identified for use as shade trees in coffee plantations e.g. ▪ *Indigofera constrida* • **Abiotic stress** e.g. tolerance to low pH soils where it grows fast in the humid and sub-humid tropics ▪ *Calliandra calothyrsus* tree legume

(continued)

Table 1.5 (continued)

	Temporal	Spatial
	• In the Smallholder Farming Systems with Plantation Crops (rubber, coconut, cocoa, vanilla, coffee, tea, oil-palm etc.), legumes which are integrated either as alleys along contours or as part of the rotation include *Leucaena leucocephala, Cordia alliodora, Erythrina poeppigiana* etc. Alley cropping in the humid lowlands of West Africa integrates *Milletia thonningii* and *Pterocarpus santalinoides*	

*GML-Green Manure
LCC-4 Cover Crops
HLS-Herbaceous Leg
MPTS-Multiple Purpose Trees

and fodder e.g. *C. ochroleuca* and *Lablab purpurens* as the rest contain toxic constituents. This means that research on identifying legumes that fit in their indispensable niches and at the same time satisfy multiple purpose role is very crucial. Potential legume niches include: cereal-legume intercrops, relay cropping, rotational systems, short/medium term improved fallows, legume cover crops in steep slopes, cover crops under tree crops, fodder systems, on bench terrace, under bananas etc.

The temporal niches for food legumes include their intercropping with cereals and rotation in maize-based systems. Research has also identified cultivars that help overcome environmental constraints and abiotic stresses (soil acidity, nutrient deficiency, salinity, Al and Mn toxicity) and biotic stresses (weeds, pests and diseases) as well as drought. Also identified are forage, fodder and pasture legumes suitable for integration in the pastoral and mixed farming systems, mainly as hedgerow or alley cropping and along contour lines; as well as identification of their abiotic spatial niches. Many green manure species for relay intercropping and rotations (improved fallows) including those that tolerate abiotic and biotic stresses have been identified. Perennial legumes for intercropping, hedgerow and as improved fallows are being evaluated in many regions. This subject is dealt with in detail in subsequent chapters.

1.5 Conclusion

Although there has been many years of research on the role of legumes in enhancing soil fertility e.g. above ground biomass and add over 75 kg of N in the soil as a result narrow C:N ratio and hence rapid mineralization after incorporation in high rainfall agricultural lands, however below ground biomass has insignificant effect on crop yields. Surface mulching is a better practice in low rainfall agricultural lands as it drastically reduces run-off and soil loss and conserves moisture as a result of good ground cover. But research is required to assess the role of legumes green manure in reinforcing physical soil erosion control methods and their long-term effect on soil physical properties. With respect to incorporated materials, there is need to determine appropriate time and method of incorporation and the residues effects after incorporation or mulching. The compatibility of legume species with food crops needs to be determined under different temporal niche arrangement. To ascertain their nutritional contribution, there is need to quantify the relationship between the legume resource quality and their N release patterns and recovery efficiency in terms of N fertilizer equivalency of below and above ground biomass. For high quality legume biomass there is need to identify strategies that help to slow down N release to synchronize with crop uptake. This will help on the development of decision tree/guideliness for use by farmers. Another role of legumes is their potential to suppress obnoxious weeds (e.g. striga) lesions and root knot nematodes, as important biotic stresses. There is also need for research on legume inter and intra-species evaluation under on-farm conditions to mitigate abiotic stresses of toxicity (acidity, Al and Mn), deficiency (especially P) in key agro-ecological zones and farm typologies; and to identify potential for highland legumes. Further research is required on quantification of N fixed by "best bet" green manure legumes as well as determination of their inoculant needs.

In spite of a lot of research on legumes in SSA, only a few species emerge as "best-bet" candidates because of their ability to mitigate abiotic and biotic stresses, N-mining by balancing medium N-harvest index with high BNC capacity, and balance in their offer for alternative uses. For ISFM, there is need to determine their fertilizer equivalence to identify supplantation/supplementation value against mineral fertilizers. Recent studies have shown *Calliandra calothyrsus, Senna spectabilis, Tephrosia, Sesbania* and Pigeon pea to have fertilizer equivalencies of 72%, 68%, 93%, 36% and 33% resulting in maize yield increases of between 43% and 71%. This suggests that on-farm research to target "best-bet" legumes to their appropriate temporal and spatial niches is high priority as well as evaluation of their trade-offs alternative uses. Legumes that demonstrate immediate benefits such as substantial increase in human food and animal feed, wood products and cash to farmers while occupying unquestionable temporal and spatial niches are most likely to be widely adopted. Priority setting for research should be based on this rationale. Priority areas include their role in mixed crop rotations, and carry over effects and under conservation tillage; reasons for lack of use of "best-bet" legumes including technologies for optimization of their uses; understanding the socio-economic factors limiting adoption of the legumes in different agroecosystems; and policies for narrowing the adoption gap as well as yield gap. The value of legumes in improving rural livelihoods is crucial and therefore research needs to be focused on managing factors that determine the integration of legumes into agricultural systems in order to fully realize their benefits. Subsequent Chaps. (2–8) will deal with these issues more specifically.

References

AHI (African Highland Initiative), (1997) Phase II Work Plan and Budget 1998–2000. International centre for research in agroforestry (ICRAF), Nairobi, Kenya

Badeyo MA, Mitsch WJ (ed.) (1998) Agro-ecological restoration of savanna ecosystems. In: Ecological engineering at Eco. Summit 1996. Proceedings of a conference, Copenhagen, Denmark, 19–23 Aug 1996. Ecological Engineering. 1998, 10

Beets WC (1990) Raising and sustaining productivity of smallholder farming systems in the tropics. Ag Be Publishing, Alkmaar

Denich M, Veilhaur K, Hedden-Dunkhors B (2002) New technologies to replace slash and burn in the Eastern Amazon. ZEF NEWS No. 9, Feb 2002

Eilitta M, Carsky R (2002) Increasing mucuna's food and feed uses. In: Target 30. Soil Fert Net. Newsletters, pp 4–5

FAO (1982) Tropical forest resources. FAO forestry paper No. 30, Rome, Italy

Farrell RE, Sandercock PJ, Pennock DJ, van Kessel CDJ (1996) Landscape-scale variations in leached nitrate: relationship to densification and natural NIJ abundance. Soil Sci Soc Am J 60:1410–1415

Gachene CKK, Palm CA, Mureithi JG (1999) Legume cover crop for soil fertility improvement in the East Africa region. Report of an AHI workshop TSBF Nairobi, 18–19 Feb 1999, 25pp

Hartemink AE, Buresh RJ, Jama B, Janssen BH (1996) Soil nitrate and water dynamic in sesbania fallows, weed fallows and maize. Soil Sci Soc Am J 60:568–574

Hilhorst T, Toulmin C (2000) Integrated soil fertility management. information department, ministry of foreign affairs, The Hague, Netherlands

Howard W, Killick T, Kayizzi-Mugerwa S, Savane MA (2001) African poverty at the millennium: causes, complexities and challenges. In: Strategic partnership with Africa. The World Bank, Washington, DC

IPPF/IUCN/Sida/UNFPA/WWF (1998) People and planet: sustaining the Soil Newsletter, Vol 7 (1), 31pp

Kabambe VH, Mloza-Banda HR (2002) Yellow witchweed in Malawi – a threat to the use of legumes as trap crops against striga or for soil fertility enhancement. In: Target, Soil Fert Net Newsletter, 30 pp

KARI (2000) Legume research network project newsletter Issue No. 3, June 2002. Kenya agricultural research institute (KARI), Nairobi

Lanly JP (1985) Defining and measuring shifting cultivation. Unasylva 37(147): 17–21

Niang A, Gathumbi S, Amadalo B (1996) The Potential of short- duration improved fallow for crop productivity enhancement in the highlands of Western Kenya. In: Mugah JO (ed.) People and institutional participation in agroforestry for sustainable developments. Kenya Forestry Research Institute, Nairobi, pp 218–230

NRSP (2001) Poverty reduction through partnerships in natural resources research. In: Natural resources systems programme research highlights 1999–2000. DFID, UK, p 29

Okigbo B (1979) Farming systems and crops of the humid tropics in relation to soil utilization. In: Greenland DH (ed.) Characterization of soils in relation to their classification and management: examples from some areas of the humid tropics. Oxford University Press, London

Ong C (1994) Alley cropping – ecological pie in the sky? Agroforest Syst 22:43–48

Pessarakli M (1999) Handbook of plant and crop stress. Marcel Dokker, Inc., New York

Raquet K (1990) Green manuring with fast-growing shrub fallow in the tropical highland of Rwanda. In: Kotschi (ed.) Ecofarming practices for tropical smallholding. Tropical Agroecology, 5: pp 50–55

Scherr SJ (1997) People and environment: what is the relationship between exploitation of natural resources and population growth in the south? Forum Dev Stud 1:33–58

Scherr SJ (1999) Past and present effects of soil degradation. In: Scherr SJ (ed.) Soil degradation – a treat to developing-country food security by 2020. International Food Policy Research Institute 27, Washington, DC, pp 13–30

Snapp SS, Mafongoya PL, Waddington S (1998) Organic matter technologies for integrated nutrient management in smallholder cropping systems of Southern Africa. Agric Ecosyst Environ 71:185–200

Stocking M (1984) Crisis for Agriculture's Cinderella – shifting cultivation. International Agriculture Development 4 (2):8–9

Waddington SR, Snapp S (1999) Workshop on "Best Bet" soil fertility technologies revisited for Malawi and Zimbabwe. In: Waddington SR (ed.) TARGET, The newsletter of the soil fertility research network for maize-based cropping systems in Malawi and Zimbabwe

Weil RR, Mughogho SK (1993) Nutrient cycling Acacia albida (syn. Faidherbia albida) in agroforesty systems. Technologies for sustainable agriculture in the tropics. ASA Special Publication No. 56 American Society of Agronomy, Madison, pp 97–108

Woomer PL (1998) Legume niches in East African highlands smallhold agriculture: an approach to legume characterization. Department of Soil Science, University of Nairobi, Nairobi

World Bank (1986) *Poverty and Hunger: Issues and Options for Food Security in the Developing Countries*, Washington, DC

World Bank (1996) Natural resource degradation in Sub-Saharan Africa: restoration of soil fertility, Africa region. World Bank, Washington, DC

World Bank (1997) Soil fertility initiative for Africa. The World Bank, Washington, DC

Chapter 2
Socio-Economic Contribution of Legumes to Livelihoods in Sub-Saharan Africa

M. Odendo, A. Bationo, and S. Kimani

Abstract Legumes play important roles in provision of food security, generation of income, and maintenance of environment in most smallholder farming systems in sub-Saharan Africa (SSA). Grain legumes are particularly important human food as they are rich in protein and are sometimes sold for cash income, whilst herbaceous and tree legumes are important livestock feeds. Legumes have the ability to biologically fix atmospheric nitrogen and are usually intercropped in cereals. Cowpea (*Vigna uinguiculata* (L) Walp), groundnut (*Arachis hypogaea*), soybean (*Glysine max L.* Merril) and common bean (*Phaseolus vulgaris L.*) are the most important grain legumes in SSA. Maize-bean intercrop is predominant in eastern Africa, whilst in southern Africa maize is intercropped with cowpeas, groundnuts and bambara nuts. In SSA, annual quantities of groundnut production are highest, followed by cowpea, beans, soybean and peas. Promotion of grain legumes in SSA offers important opportunities for improving smallholder livelihoods and sustainable land productivity. However, the contribution of legumes to food security and poverty alleviation can be substantial when bio-physical and socio-economic constraints are simultaneously addressed. This involves development of innovative strategies for improving access to markets as well as incorporation of consumer preferences and appreciation in the research and development processes.

M. Odendo (✉)
Kenya Agricultural Research Institute (KARI), Socio-economics Division,
P.O. Box 169-50100, Kakamega, Kenya
e-mail: Odendos@yahoo.com

A Bationo
Alliance for a Green Revolution in Africa (AGRA), Soil Health Program,
PMB KIA 114, Airport-Accra, Ghana
e-mail: abationo@agra-alliance.org

S. Kimani
Kenya Agricultural Research Institute, Nairobi, Kenya
e-mail: 2skimani@gmail.com

A. Bationo et al. (eds.), *Fighting Poverty in Sub-Saharan Africa: The Multiple Roles of Legumes in Integrated Soil Fertility Management*,
DOI 10.1007/978-94-007-1536-3_2, © Springer Science+Business Media B.V. 2011

2.1 Introduction

Legumes play a wide role in contributing to food security, income generation, and maintenance of environment for millions of small-scale farmers in sub-Saharan Africa (Tarawali et al 2002). In most parts of sub-Saharan Africa, legumes are usually intercropped with cereals and improve land productivity through soil amelioration. Grain legumes are important human food as they are rich in protein and are sometimes sold for cash income, whilst herbaceous and tree legumes are important livestock feeds (Mureithi et al. 1998; Peoples and Craswell 1992; Morris and Garrity 1993; Muller-Samann and Kotschi 1994; Tarawali et al.2002). However, the ability of legumes to play these roles depends on their ability to biologically fix atmospheric nitrogen (Skerman and Riveros 1988). Grain legumes are more acceptable to farmers than herbaceous legumes, which as cover crops, occupy land meant for food production (Peoples and Craswell 1992; Mafongoya et al. 2003), indicating some tradeoffs that need to be accommodated. Oyewole et al. (2000), for example, reported that farmers preferred cowpea-maize to *Mucuna*-maize double cropping to keep grain producing cowpea in the system although the benefits of cowpea was less than that of *Mucuna.*

Cowpea (*Vigna uinguiculata* (L) Walp), groundnut (*Arachis hypogaea*), soybean (*Glysine max L.* Merril) and common bean (*Phaseolus vulgaris L.*) are the most important grain legumes in sub-Saharan Africa. However, the potential benefits to the soil and subsequent crops from the grain legumes are less (Shulz et al. 2001). Maize-bean intercrop is predominant in eastern Africa, whilst in southern Africa maize is intercropped with cowpeas, groundnuts and bambara nuts, to a lesser extent.

2.2 Trends in Grain Legumes and Cereals Production in Sub-Saharan Africa

Figure 2.1 shows trends in cereals production in sub-Saharan Africa in the past decade. In terms of quantities, cereal production exceeds grain legume production. Production of maize is the highest, followed by sorghum and millet.

Against popular belief, groundnut production is highest, followed by cowpea, beans, soybean and peas (Fig. 2.2). The groundnut production figures reported by Food and Agricultural Organization (FAO) were adjusted at the shelling percentage of 50% and still remains the highest production.

2.2.1 Cowpea

Cowpea is widely cultivated in semi-arid areas of the tropics and subtropics for human as well as animal consumption (Singh and Emden 1979). The world cowpea production was estimated at 3 319 375 Metric Tonnes (MT), of which 75% is from

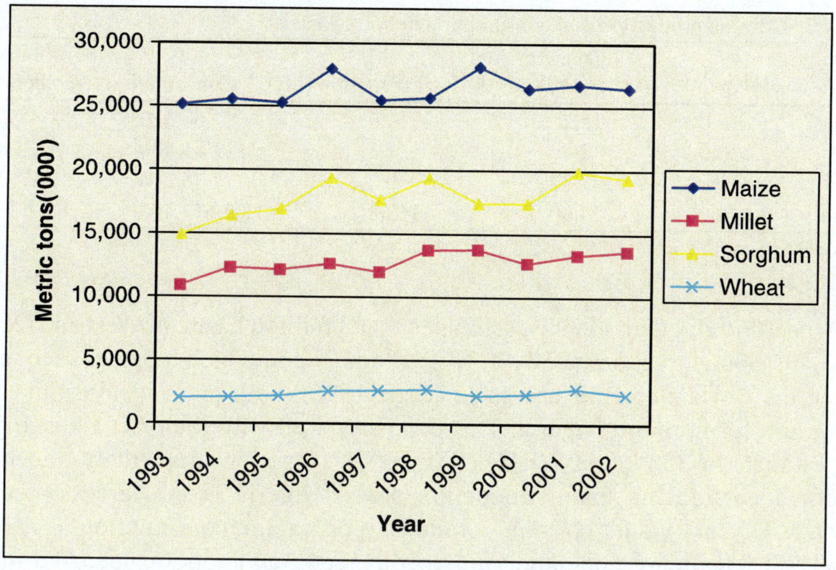

Fig. 2.1 Cereal production in Sub-Saharan Africa (Source: FAOSTAT 2003)

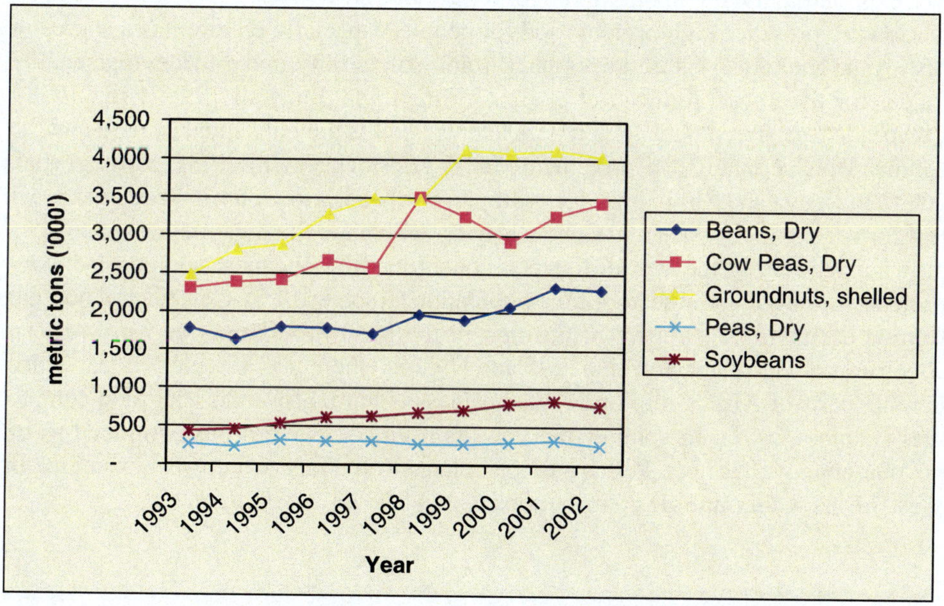

Fig. 2.2 Grain legume production in Sub-Saharan Africa (Source: FAOSTAT 2003)

Africa (FAOSTAT 2000). West Africa is the key cowpea producing zone, mainly in the dry savannas and semiarid agro-ecological zones. The principal cowpea producing countries are Nigeria, Niger, Senegal, Ghana, Mali, and Burkina Faso (Coulibaly and Lowenberg-DeBoer 2001). Out of the 12.5 million hectares (ha) cultivated

Table 2.1 Cowpea production and consumption in West Africa

Countries	Production (t)	Consumption/ capita/year (kg)	Population (MI)	Consumption (t)	Surplus/ deficit (t)
Burkina	10,000	5.2	11.6	60,320	–50,320
Mali	110,060	7.4	11	81,400	28,660
Niger	641,024	7.82	10	78,200	562,824
Nigeria	2,099,000	23.00	113.8	2,617,400	–518,400

Source: Computed from FAOSTAT (2000)

to cowpea, Singh et. al. (1997), estimated that 8 million ha are in West and Central Africa, distributed predominantly in Nigeria and Niger, which together account for 87% of the world's cowpea harvested (Ortiz 1998), with grain yield estimates of 240 kg ha^{-1} for northern Nigeria. This compares well with the FAO average grain yield estimate of 358 kg ha^{-1} (FAOSTAT 2000). However, it is widely traded and consumed outside the main production areas. Nigeria is the largest producer (2,099 MT year^{-1}) and consumer (annual per capita consumption is 23 kg) (Table 2.1). The dense population in Nigeria creates an enormous demand for the crop (FAOSTAT 2000; Coulibaly and Lowenberg-DeBoer 2001). Niger, on the other hand, is the main cowpea exporter in West Africa with an estimated 215,000 MT exported annually, mainly to Nigeria (Lambot 2001).

Cowpea provides high protein food for people, especially children; is a source of cash (grain and fodder), it improves and sustains soil fertility, and provides high quality fodder for livestock (Tarawali et al. 1997; Singh and Tarawali 1997; Ortiz 1998). Cowpea leaves, green pods, green peas and dry grains are consumed as food and the haulms, which contain 20% protein, are fed to livestock (Rachie 1985; Karungi et al. 2000). In Kenya and Zimbabwe, for example, cowpea is grown for its seed and tender leaf for vegetable. In West and Central Africa, which are the most important cowpea growing regions in the world, cowpea constitutes the cheapest source of dietary proteins for low-income sector of the population (Rachie 1985). Cowpea is a popular legume occupying an important position in the local food habits in West African countries such as Niger, Burkina Faso and Nigeria where it is a staple food, and also in other parts of Africa, reinforcing its business potential for industrial food companies (Lambot 2001). In spite of the low grain yields, cowpea has continued to be popular amongst farmers. Perhaps this is related to the fact that cowpea is unique in West Africa, with multiple contributions to households' welfare.

2.2.2 Common Bean

Common bean (*Phaseolus vulgaris* L.) is a major staple in the diet of many, especially the poor in sub-Saharan Africa. Bean is traditionally regarded as a woman's crop because in around 90% of the cases, women are involved in all aspects of its production, storage and commercialisation. Most bean traders in rural and small town settings are women. Eastern, central and southern Africa are the main bean growing regions in Africa. Eastern Africa produces 115,700 tonnes, central Africa 571,000 tonnes and southern Africa 156, 000 tonnes (FAOSTAT 2000). However, bean yield in sub-Saharan Africa is extremely low (<1 t ha^{-1}) compared to South America where yields are over 2 t ha^{-1}. Importance of bean as a food and cash crop is closely associated with human population density. Bean production is diverse, being grown in many different crop associations, primarily by small-scale farmers in eastern and southern Africa. Maize-bean intercropping systems are particularly common in many parts of eastern and southern Africa. Intensity of bean production is highest in areas of high population density, although a significant proportion of the production is in areas of moderately low population density (Wortmann et al. 1998). It is a major staple in eastern and southern Africa where it is recognized as the second most important source of calories (Pachico 1993) and as a source of dietary proteins to people of all income categories, but more so to the poor (Wortmann et al. 1998), hence commonly regarded as the meat of the poor. It is also a valuable source of vitamin-B complex, iron, zinc and other essential minerals. An estimated 3,741,000 ha of beans are sown annually in Africa. Consumption of beans in eastern Africa exceeds 50 kg per capita per year, reaching 66 kg per person per year in parts of Kisii district, one of the most populous districts in Kenya (Jaetzold and Schmidt 1983). Thus beans cover more than 22% and 11% of daily proteins and energy needs, respectively in sub-Saharan Africa (Trutmann et al. 1996; David et. al. 2000).

2.2.3 *Groundnut*

Groundnut (*Arachis hypogaea L.*) is the most important oilseed in the developing world and has been described as the most widespread legume in the world (Norden et al. 1982). Groundnut is grown primarily for its fresh and dry seeds. In Uganda, for instance, it is the most widely grown grain legume after common bean. It is a source of protein and high grade fat, a food source that does not even need processing. The crop contributes significantly to household food security, and since many smallholder farmers in sub-Saharan Africa are women, it has an important bearing on gender issue (Hildebrand and Subrahmanyam 1994). Nearly 19 million hectares are sown to groundnut worldwide and 21 million MT of dry pods harvested. Asia produces 72% of the world's groundnut output followed by Africa with 19% of the production.

2.2.4 *Soybean*

Soybean (*Glysine max L.* Merril) is another important legume in sub-Saharan Africa. It is a relatively new crop in Africa whose importance in the farming systems of sub-Saharan Africa has been increasing rapidly, especially in Nigeria in the last 10–15 years. Between 1988 and 1998, production in Nigeria increased by 170%

from about 150,000 t in 1988 to 405,000 t in 1998 (FAO 2001). Over the same period, average yields doubled from about 340 kg ha^{-1} to 740 kg ha^{-1}, while area under cultivation increased by just 24%. Soybean is now grown in several parts of sub-Saharan Africa including Malawi, Nigeria, Uganda, Kenya, Zambia, and Zimbabwe where it is making significant contributions to rural livelihoods. Due to limited inoculant production capacity in Africa, promiscuous soybean varieties that effectively nodulate with indigenous rhizobia have been successfully grown without inoculants and have demonstrated potential as a vehicle for conveying the benefits of Biological Nitrogen Fixation (BNF) to poor and marginalized communities (Mpepereki et al. 2000). Three main reasons related to the multiple benefits of soybean, account for the increasing importance of soybean in Africa. These are:

(i) High-income generation potential of soybean especially for rural dwellers, especially women and those engaged in processing and marketing of soybean and its products.
(ii) Its positive role in sustainable cropping systems by improving soil fertility and lately serving as trap crop for the parasitic weed *Striga hermonthica* which causes crop damage estimated to the tune of $7 billion annually
(iii) Its health and nutritional benefits. Soybean has about 40% protein, 20% oil, and is rich in essential minerals and vitamins. These make it a very valuable food for human and animals

Soybean farming in Africa, however, faces some key constraints. Though efforts have been made to introduce soybean in African food habits and farming activities, little success has been achieved. Even in Nigeria, where national production is about 326,000 MT it is still considered an industrial crop, difficult to cook, and with undesirable taste (FAOSTAT 2002).

2.3 Grain Legumes in the Context of Agriculture Sector in Sub Saharan Africa

The task facing African Agriculture is how to cope with the needs of a rapidly growing population. It must achieve sufficient growth in food crops not merely to maintain output per person, but also to reduce food calorie deficits and to lower food imports. In the process, agriculture must be a major employer of Africa's growing labour force and compete on the world markets to earn foreign exchange that Africa needs to fuel its economic growth. This should be done while reversing the degradation of the natural resources that threatens long-term production (World Bank 1993). Transforming agriculture and expanding its productive capacity is a prerequisite for improving living standards in sub-Saharan Africa.

Food security has deteriorated since independence in sub-Saharan Africa, and severe food shortages, exceptional in 1960, are now widespread. Food security at the household level is directly influenced by the agricultural performance (World Bank 1993). In many countries, malnutrition is seasonal and increases before the

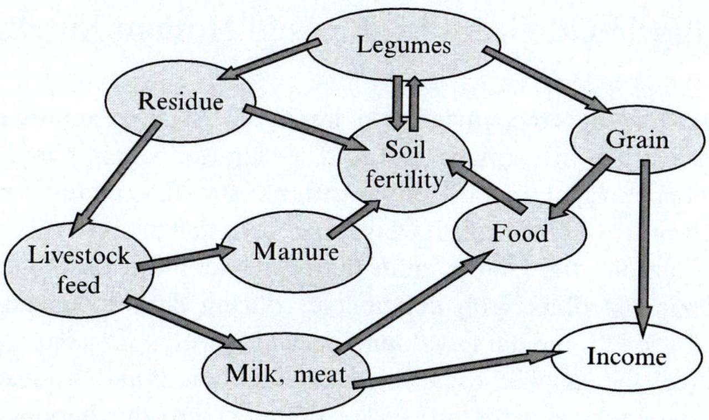

Fig. 2.3 Contributions of legumes to livelihoods in Sub-Saharan Africa. (Source: Bationo 1998)

harvest, when food supplies have dwindled. The gap in food intake widens further in years of drought. In terms of energy value food consumption, in sub-Saharan Africa between 1965 and 1986, it averaged 2,100 calories per person per day, or about 85% of recommended requirements. It is estimated that one-quarter of sub-Saharan Africa population, comprising more than 100 million people obtain on average over good and bad crop years, less than 80% of the daily calorie supply recommended by FAO and World Health Organization (WHO). In drought and other years, the numbers could even be higher. To achieve, food security, food production should grow at about 4%. Integration of cereals, legumes and livestock is key to the realization of food security in sub-Saharan Africa.

To put the contribution of legumes to livelihoods context in sub-Saharan Africa, this chapter discusses the role of legumes in crop-livestock integration. The present and potential contributions of legumes in the farming systems, especially the role of legumes in human nutrition, income generation, soil fertility improvement, and livestock productivity are discussed. A schematic representation of the key contributions of legumes in the farming systems is shown in Fig. 2.3.

Legume residues can be used for soil fertility improvement and for feeding livestock. When animals are fed on leguminous fodder, which inherently contains higher levels of nitrogen (N) compared with cereal-based fodder, the manure produced by such animals contains higher levels of N. Working in central Kenya, Delve et al. (1999) showed that feeding cattle on a legume fodder supplement increased N content by 80%. The higher quality manure when used as fertiliser has been shown to produce higher crop yields (Lekasi et al. 2001). The manure from the cattle can be used to fertilize soils for higher crop yields. The livestock once fed can provide protein through milk and meat (Fig. 2.3). The milk and meat can be consumed as food or sold to the market. Legumes can act as 'green manures' when planted for biological N-fixation, thus increasing nutrient supply in the soil, especially N and improve crop yield (Peoples and Craswell 1992; Muller-Samann and Kotschi 1994). They can also provide good ground cover, which is likely to minimise soil erosion through reduction of raindrop impacts and run off, thus improving soil physical properties (Fig. 2.3).

2.4 Contribution of Grain Legumes in Human Nutrition

Nutritional improvement is recognized as an important aspect of improved livelihoods. Good nutrition is a key investment for future generation's health and productivity. An adult in better physical, mental and nutritional state has a higher probability of begetting an offspring of higher birth weight and better chances to survive (Oniang'o 2001). Leading among the factors contributing to malnutrition is inadequate protein, iron and zinc in diets. Iron deficiencies during childhood and adolescence impair physical growth, mental development, and learning capacity. In adults, iron deficiency reduces the capacity to perform physical labour and is a leading cause of death among women during childbirth. Zinc deficiency results in poor child growth, delayed maturation, poor appetite and impaired immune function. Iron helps to fight anaemia while zinc helps in the body immune system.

Human-Immuno Virus/Acquired Immune Deficiency Syndrome (HIV/AIDS) that attacks the human immune system is a modern-day plague, which has far eluded modern medical sciences' search for a cure and left countless numbers of infected and dying adults and children. It has become increasingly evident that nutritional support plays a vital role in HIV-infected and AIDS patients (Williams and Anderson 1993). In most rural areas in sub-Saharan Africa, HIV/AIDS pandemic has contributed seriously to increasing and sustaining and expanding poverty. Over 7.5 million of the 12 million HIV/AIDS patients in the world live in sub-Saharan Africa (Williams and Anderson 1993). Nutrition is a vital component of care for involuntary weight-loss and wasting of body tissues caused by the disease effects on metabolism, reflected in the severe state of protein-energy malnutrition. Again nutrition is fundamental in all conditions associated with the body's basic immune system and is an intimate and integral component of care through the specific roles of key nutrients in maintaining the immuno-competence of the body. For many AIDS associated diseases, individuals' nutritional status influences their impact on morbidity and mortality, irrespective of disease process. However, it is important to note that though nutrition can play a great role in HIV/AIDS management, the disease is wrought with complications, including inadequate food intake and malabsorption of nutrients, which need to be simultaneously addressed.

Nutritional advantages of legumes, and their potential to increase household incomes if their production is increased, makes them to have a central role in combating poor health and helplessness caused by HIV/AIDS. Legumes, especially, soybean and groundnut have comparatively much higher protein values than the cereals and are particularly important in areas where animal protein is costly and rarely available. Legumes are often referred to as 'meat' of the poor for the ability to supply cheap proteins. Thus, legumes have a strategic role in food security policies of the countries of the developing world and there is little likelihood of decrease in demand. Grain legumes are known to be rich in protein and have an advantage over the cereals (Table 2.2). Some recently produced legumes have been reported to have higher concentrations of iron and zinc. Thus, malnutrition can be alleviated by consumption of grain legumes, which are high in protein, iron and zinc, and relatively cheap compared to animal products. Madamba (2001) analysed nutritive

Table 2.2 The nutritive value of major legumes as compared to other common crops in Sub-Saharan Africa

Product	Moisture	Crude protein	Ether extract	Crude fibre	Carbohydrates
Cowpea dry grain	10	24	1.0	5.7	56
Maize dry grain	11	12	4.1	2.0	70
Groundnut dry grain	7	28	46.3	4.1	13
Soya bean dry grain	7	38	23.6	3.9	23
Beans dry grain	11	25	1.1	4.8	56
Cooked cowpea dry grain	61	10	0.3	1.6	24
Cowpea fresh green seed	59	12	0.3	2.2	25
Cooked cowpea fresh grain seed	62	11	0.6	3.1	20
Fresh tender cowpea leaf	88	5	0.3	1.8	4
Boiled fresh cowpea leaf	88	4	0.6	1.5	5
Cooked maize dry grain	67	4	1.6	0.5	25
Cooked ground nut dry grain	44	17	26.5	3.5	5

Source: Madamba (2001)

value of grain legumes and cereals in Zimbabwe and found out that among the crops and products listed, soya bean dry grain had the highest crude protein (38%), whilst maize had the highest carbohydrates (Table 2.2). Cowpea dry grain had a protein content of 24% and a carbohydrate content of 56%. These nutritive values compare well with 25% protein and 64% carbohydrate reported by Bressani (1985).

Groundnut and soybean dry grain had a higher protein but lower carbohydrates contents than cowpea dry grain. The crude proteins in cowpea dry grain were higher than that in field beans, soybean and groundnuts. The cooked cowpea dry bean had 10% protein, which was lower than that of cooked groundnuts grain with 18%. Conversely, cooked cowpea grain had higher carbohydrates than cooked groundnuts. The high levels of both protein and carbohydrates in cooked dry and fresh grain of cowpea mean that they can provide a balanced diet if eaten on their own. They are better than cooked maize grain or groundnuts. The carbohydrate content in cooked cowpea was comparable with that of maize dry grain, and hence can be used as alternative sources of carbohydrates. Cooked dry grain and fresh seed of cowpea had almost similar chemical compositions, suggesting that eating these legumes as fresh or dry grain has the same nutritional effect. The calorie value of cooked and fresh seed of groundnut and cowpea was comparable to that of cooked maize grain, suggesting that they can be used as alternative sources of carbohydrates if maize is not available. However, like all other legumes, cowpea and groundnut are deficient in essential amino acids, cystine and methionine, but these sulphur containing amino acids are in adequate amounts in cereals. Therefore, legumes are complementary to

Table 2.3 Nutritional status of children in soyabean-producing households and in no-soyabean-producing households in Benue State, Nigeria (January 1997)

Nutritional index	Mean Z scores[a]		t value	Significance level
	Soybean-using households	No-soybean-producing/ using households		
Weight-for-age	–0.083	–0.144	3.89	0.002
Height-for-age	–0.045	–0.668	2.56	0.011
Weight-for-height	0.006	0.004	0.10	0.273

Source: Sanginga et al. (1999)

International standards:[a] Z>–1.00: normal, –1.00>Z>–2.00: mild malnutrition, –2.00>Z>–3.00: moderate malnutrition, Z<–3.00: severe malnutrition

cereals rather than substitutes. Soybean protein is particularly valuable because its amino acid composition compliments that of cereals. Like other legumes, soybean protein is low in sulphur-containing amino acids, cystine and methionine, but high in lysine. Combining soybean with cereals in a diet provides a complete protein diet which contains all the essential amino acids. Other legumes such as peas and pigeon peas are nutritionally similar.

In Africa, soybean received mixed reaction: 'soybean takes too long to cook, taste bad.' However, the real problem may be lack of familiarity with methods to use the soybean effectively. Advances in food technology have made it possible to include soybeans as part of the diet in Africa within a short time. In western Nigeria, for example, soybean-based foods are fed to the sick and malnourished children. First soybean milk is fed to those who are unable to eat solid foods and then supplement the traditional foods with soybean paste (Weingartner 1987). Sanginga et al. (1999), in a study conducted in the northern zone of Benue State, Nigeria, which has the longest history of soybean production in Nigeria, estimated that the State accounts for over 70% of soybean production. Using anthropometrical measurements of 353 children (186 boys and 167 girls) aged between 0 and 12 years, Sanginga et al. (1999) found out that the mean Z score values for weight-for-age and height- for- age show that the nutritional status of children in soybean producing/using households appeared to be significantly higher than those of children in households that did not produce/use soybean (Table 2.3). Communities that produced and consumed soybean had a significantly higher percentage of nutritionally normal and a lower percentage of severely malnourished children than communities that did not produce soybean.

Women control of soybean production was significantly related to the long-term nutritional status and household food security of children (weight-for-age) at the 1% significance level, and to children's height-for-age at the 5% significance level (Sanginga et al. 1999). This implies that that increasing women's production of soybean had robust effects on the long-term food security and better nutritional status for children. Women's control of soybean production was associated with household consumption of soybean, which, in turn, improved the nutritional status of children.

Cash income earned from soybean sales was positively and significantly related to weight-for-height measurements. The finding of this study provide empirical evidence confirming that household production and consumption of soybean contribute significantly to improving the nutritional status of children and provides a strong case for promotion of soybean as a cheap solution for malnutrition and a means of poverty alleviation. All farmers reported the nutritional benefits of soybean. In farmers' opinion "soybean adds blood" and "soybean makes children look fresh".

2.5 Contribution of Legumes to Sustainable Soil Productivity

The need for application of inorganic fertilizers as a requisite for high yields in sub-Saharan Africa was recognized long ago, however, due to lack of capital, inefficient distribution systems, poor enabling policies and other socio-economic factors, there is limited use of chemical fertilizers. Cheaper means of improving soil fertility and productivity is therefore necessary. Cereal legume rotation effects on cereal yields have been reported to be very beneficial on the succeeding cereal crop. In semi-arid eastern Kenya for instance, legume contributions to sloe crop rotation systems were found to range between 8 kg N ha^{-1} for tepary beans and 44 kg N ha^{-1} for *Phaseolus* beans. In the intercrop rotation system, pigeon peas were the highest contributors at 15 kg N ha^{-1} (Nadar and Faught 1984).

Legumes are key to crop-livestock integration because they provide fodder to livestock, and bring nitrogen into the farming system through nitrogen fixation. It is widely recognized that BNF by the legume-r*hizobium* symbiosis is an important component of productivity in tropical agriculture, especially in farmland which is marginal either in terms of distance from the markets, small farm size or the poverty of the farmers. In such resource-poor, smallholder systems, application of large quantities of inorganic fertilizers such as urea is not economically feasible. The use of management techniques that increase the contribution of N to the system through the legume-*rhizobium* symbiosis would improve crop-livestock production levels and their stability.

Part of the nitrogen fixed will remain in the soil in the roots and thereby contribute to the soil fertility for subsequent crops. Some of the fixed nitrogen will eventually return to the soil as manure after residues are fed to livestock. Table 2.4 gives the estimates of N_2 fixed by various legumes under field conditions. It can be noticed that cowpea can fix between 73 and 80 kg N ha^{-1} whereas groundnut can fix up to 240 kg N ha^{-1} and common bean 50–80 kg N ha^{-1}.

Table 2.4 Estimated nitrogen (N_2) fixation by various grain legumes under field conditions

Legume	Estimated N_2 fixed (ka ha^{-1})
Soybean	20–260
Common bean	50–80
Cowpea	73–80
Mung bean	61
Groundnut	72–240
Pigeon pea	69
Chickpea	1–141

In reality, actual N_2 fixation in bean cultivars is generally low when compared with many other legumes. The weak nodulation and N_2 fixation rate of beans as compared to other legumes is partly due to high sensitivity of its N_2 fixing Rhizobia strains to nutritional and environmental stresses. In good environment, beans can derive up to 70% of its nitrogen from the air. This is due to an intrinsic inability of beans to nodulate. Although poor nodulation is frequently observed, soils in most bean growing areas contain large numbers of compatible and effective *Rhizobia*. Selection of adapted *Rhizobium* strains for beans sown directly in pots of soils containing large populations of indigenous, compatible *Rhizobia* has resulted in yield increases when these strains were tested in the field.

In most farming systems in sub-Saharan Africa, maize is traditionally intercropped with legumes such as cowpea, beans, groundnuts, pigeon peas and green grams. The importance of intercrops is widely recognized by farmers. Intercropping results in better use of soil nutrients, especially by intercropping crops that do not compete for the same nutrients, since legumes are able to release some fixed N, they improve soil productivity, and reduce soil erosion due to good ground cover. If calamity occurs, such as drought, diseases and pest invasion, one crop in the intercropping system may survive and provide some food to the household, thus stabilizing effect of crops on food security (Upton 1987; Kanyanjua et al. 2000).

It should be noted, however, that as soil fertility continues to decline, intercropping will become more and more problematic, with regard to increasing crop yields, due to competition for nutrients and water. This will be more critical in marginal semi-arid zones.

2.6 The Role of Grain Legumes in Income Generation

Legumes play an important role in food export market. Kormawa (1996) found out that soybean was more profitable to farmers than other competing and companion crops in the farming system. Both gross margin and benefit cost analyses showed higher returns to soybean production as a sole crop or when intercropped with sorghum. The gross margin from sole soybean and sole groundnuts showed that it was twice as profitable for farm families to invest their labour in growing soybean rather than in growing groundnuts under the then prices in Nigeria. In eastern Africa the price of beans is often double the price of maize.

Kormawa et al. (2001) reported that the price of brown cowpea was higher than that of other crops, except soybean. Amongst the cereals, only rice costs more than cowpea. Maize-cowpea price ratio indicates that the price of maize is 44% lower than that of cowpea (Table 2.5). However some farmers in the past preferred to cultivate more maize than cowpea because the two are complementary as cowpea is often used as a sauce for maize dishes. Most production programs focused on maize. For example there has been a subsidy on fertilizers for maize, while pesticides, the most important input for cowpea production, have not been subsidized.

Table 2.5 Value of different food crops in Nigeria

Commodity	Price (Naira kg^{-1})	Price (ratio[a])	Energy value (ratio±[b])	Protein value (ratio)
Cowpea grain (white)	38	1.00	1.00	1.00
Cowpea grain (brown)	42	1.10	1.00	1.00
Soybean	46	1.21	1.28	1.74
Groundnut	40	1.06	1.86	1.08
Maize	17	0.44	1.11	0.45
Sorghum	16	0.42	1.08	0.41
Millet	18	0.48	1.14	0.24
Local rice	50	1.32	1.08	0.33

Source: Adapted from Kormawa et al. (2001)
[a]Market survey 1999/2000 (average price)
[b]Energy value is in Kcal of edible portion. All ratios are in relation to white cowpea

Table 2.6 Percent respondents who have adopted/dis-adopted improved beans in western Kenya

	Adopted		Dis-adopted	
Bean variety	Kakamega	Vihiga	Kakamega	Vihiga
Bush beans:KK15	42	80	33	6.3
KK22	69	69	7	8.0
KK14	4	5	40	8.4
KK20	2	5	3	4.5
KK8	34	35	24	13.5
Climbers				
Umubano	8	18	75	18.0
Vunikingi	0	0	–	–
Ngwenurare	0	1	0	0.0
Gisenyi	3	2	0	60
Flora	0	2	0	–
Puebla	0	0	0	–

Source: Odendo et al. (2002)

Odendo, et al. (2002) estimated adoption and impacts of root rot resistant bean varieties (KK8, KK14, KK15, KK20 and KK22) and climbing bean varieties (Umubano, Gisenyi, Flora, Puebla and Ngwenurare) introduced in western Kenya in the early 1990s. The study found out that 35–80% of the farmers had adopted new bush bean varieties – KK22, KK15 and KK8. On the whole, a relatively small proportion of farmers (8–18%) adopted climbing bean varieties, especially Umubano and Gisenyi (Table 2.6). Results of this study demonstrated that a simple technology such as improved bean seed has potential to reduce poverty amongst the resource poor farmers because its adoption requires less financial resources and can be tailored to the limited resources available to the poor farmers. Even when farmers do not have cash, they can easily obtain initial seed from other farmers as a gift, a common practice in most rural areas of Kenya.

The bean varieties introduced to western Kenya brought several impacts to farming households. However, quantifying the impacts in the absence of reliable base-

Table 2.7 Partial budget for adoption of an improved bean variety versus a local variety

Variable	Sole maize	Maize-local beans (Alulu)	Maize-improved Beans (KK22)
Average bean yield (kg ha^{-1})	0	484	790
Average maize yield (kg ha^{-1})	1,500	1,350	1,350
Gross field benefits beans (Ksh ha^{-1})	0	8,930	14,576
Gross field benefits maize (Ksh ha^{-1})	8,250	7,425	7,425
Gross benefits (Ksh)	8,250	16,355	22,001
Cost of bean seed (Ksh)	0	625	625
Cost of labour for weeding (Ksh ha^{-1})	2,000	2,500	2,500
Cost of labour for planting (Ksh ha^{-1})	1,000	1,600	1,600
Total costs that vary (Ksh ha^{-1})	3,000	4,725	4,725
Net benefits (Ksh ha^{-1})	5,250	11,630	17,276
MRR%		370	697

Source: Odendo et al (2002)
An average price of maize was Ksh. 5.50 per kg and beans were Ksh.18.45 per kg. (1US$ = Ksh.75); Bean yields are adjusted downward by 30% due to error in extrapolation

line information that depicts the situation prior to the root rot disease is problematic. In such a situation, farmers' perceptions, though subjective, provided the best measure of change (Odendo et al. 2002). The main impacts of the varieties were on food security, household income, bean commercialization and varietal diversity. The new bean varieties, especially bush beans, improved food security and welfare for 98% and 99% of the adopters in Kakamega and Vihiga, respectively. For the vast majority of the adopters, the new bean varieties improved welfare by freeing up money for purchasing other foods. The superior yields of the new varieties increased bean self-sufficiency on average by approximately one month for the long rain harvest, and 2 weeks for the short rains harvest. All wealth groups enjoyed improved bean self-sufficiency, but generally wealthy adopters reported eating beans for a longer period. Improved beans formed nearly 62% of the total quantity of beans sold in the surveyed local markets. Of the new varieties, the most widely sold were KK15 and KK22, by 67% 71.2%) of the traders respectively. Income gains from bean sales of new bean varieties were used for both short-term consumption and productive investments, especially on food, household items, and school related expenses. Education related expenses were the second most important aspect on which bean earnings were spent, after food expenses.

A partial budget analysis for introduction of improved bean varieties in a maize-bean intercropping system is shown in Table 2.7. Changing maize-local bean variety intercrop (the common farmers' practice) to the widely adopted improved bean variety, KK 22 resulted in marginal rate of return (MRR) of 697%. Thus, for each shilling invested in maize-bean intercrop the farmers recovered the shilling plus extra 3.70 and 6.97 shillings for local and improved bean variety, respectively. The findings confirm results of a similar analysis (Nkonya et al. 1997), which shows that growing maize in association with beans has an overwhelming advantage over planting sole maize.

Another related survey focusing on impact of two modern bush bean varieties, K132 and K131, on income, food security, consumption patterns and gender relations through a longitudinal study in a rural maize based farming systems of eastern

Table 2.8 Contribution of soybean to household income (Naira per year, 1996 harvest) and intrahousehold gender redistribution of soybean income in Benue State, Nigeria

	Gboko (High adoption)		Gwer (Low production area)		Totals		
Variable	Men (n=60)	Women (n=43)	Men (n=69)	Women (n=31)	Men (n=129)	Women (n=74)	Total (n=203)
Mean soybean income	23,351	13,374	5,965	3,306	14,051	9,156	12,223
Proportion of soybeans income relative to total farm	58.5	58.3	20.1	22.5	37.9	43.3	39.8
Mean soyabean income redistributed[a]	2,316	1,253	1,148	1,012	1,792	1,165	1,555
Percent of soyabean income redistributed	21.1	18.8	27.8	23.0	22.6	20.7	21.3
Proportion of farmers (%) who redistributed soyabean income	62.7	43.9	54.2	41.5	43.8	37.5	41.7

Source: Sanginga et al. (1999)
n number of respondents
[a]Amount of money in Naira that the farmer gives to his wife or her husband from the sale of soyabean

Uganda (David et. al. 2000) shows that in just 4 years, the new varieties accounted for 74% of the bean area among a sample of 100 adopters. Though adopters did not show significant income gains in the first season of 1998, per capita bean consumption increased significantly over figures recorded before the introduction of the varieties. Most of the increased consumption was attributed to higher productivity of both varieties. Whilst, impact was wealth and gender neutral, the greatest benefits went to households of average wealth who have necessary resources to take advantage of yield increases. The study concludes that modern bean varieties can contribute importantly to poverty alleviation, although gains in quantitative measures of welfare such as income and per capita food consumption may be modest. Overall, the new varieties bettered women's lives by improving household welfare, increasing both household and personal income

A study conducted in Nigeria's southern Guinea Savanna (Sanginga, et al. 1999), using farmers' ordinal ranking of the relative importance of major crops in Nigeria revealed that soybean was ranked first as the most important source of cash income by 42% of men and 47% of the women (Table 2.8). The mean soybean cash income was Naira (N) 14,051 for men compared to N 9,156 for women, and men had significantly higher income than men. Income from soybean often came in one lump-sum payment as the crop was usually sold at once after threshing. The study showed that women earned substantial proportion of their income from soybean, a traditionally male crop.

Analysis of the effects of soybean production on the welfare of farmers applying material style of life (MSL) and human capital investment (HCI) indices shows that majority of the farmers reported substantial increases in MSL items (Sanginga et al. 1999). Soybean income contributed significantly to acquisition of MSL items such as radio (48.5%), mattresses (71.9%), bicycles (27.7%), livestock (58%), and metal roofed houses (14%) for men. Women reported similar investments, although notable differences existed. The majority of the women seemed to spend more of their soybean income on household consumer items (HCI) for themselves and children, paying school fees (89.7%), medical bills (45.9%), buying high value food and other household items (67%).

Cowpea has a high potential to increase farmers' and traders' incomes, thereby contributing to poverty reduction and food security (Bressani 1985). In many regions, cowpea fodder is particularly valued as supplement feed in periods leading up to Muslim festivals when sheep are traditionally slaughtered. Some farmers sell cowpea fodder during dry season when feed shortage is critical, and there have been suggestions that incomes from fodder sales make a substantial contribution to the annual income in such cases (ICRISAT 1991).

2.7 Contribution to Livestock Nutrition and Productivity

In sub-Saharan Africa crop-livestock integration is a common feature of the farming systems. Recognition of the importance of fodder from cowpea led to initiation of joint IITA-ILRI research in 1990 where fodder quantity and quality parameters were included in breeding and selection program. These efforts resulted in identification of promising dual purpose cowpea varieties suitable for dry savannas (Singh and Tarawali 1997). The fodder and husks from cowpea form an important source of protein, fibre, and energy for the livestock. Cowpea residue is an important fodder resource for ruminant livestock (Tarawali et al. 1997). In northern Nigeria, for example, during the first season, farmers usually release their small ruminants into the fields once the grain harvest is completed to enable them to graze the remaining crop residues and weeds. Once these resources are used up, usually by the middle of the dry season, the animals are tethered within the homestead and fed with stored crop residues (Tarawali et al. 1997).

Winrock (1992) estimates that 56% of the goats and 64% of the sheep in sub-Saharan Africa are in dry savannas. Farmers in the dry savannas deliberately grow varieties and use management practices that ensure some cowpea fodder is available for harvest at the end of the growing season, even at the expense of grain production. Harvesting at the end of wet season, before dry season becomes severe, gives the best quality, and this is preserved throughout the storage period. If fodder is harvested late, when the dry season is already underway, quality is poor (Tarawali et al. 1997).

Cowpea fodder as a feed supplement increases animal live weight gain during dry season. A livestock feeding study carried out in northern Nigeria (Binchi Local Government), using the fodder harvested in 1998 to feed small ruminants during the 1998/1999 dry seasons, the average weight gains per animal over the 16-week feeding period were 1.75 kg for best- bet (BB+) option, comprised of minimum inputs in form of fertilizer, with N applied to only to sorghum rows, and an insecticide spray only applied to the cowpea. The other best bet (BB) portion, had no inputs and yielded 0.28 kg, whilst traditional field of sorghum and cowpea (L) yielded 0.03 kg, representing gains of 8%, 1.3%, and 0.1%, respectively(Tarawali et al. 1997). These values are within ranges reported by Tarawali et al. (2001). Economic analysis of the treatments is shown in Table 2.9.

Table 2.9 Summary of partial economic analyses for the three treatments

	BB+	BB	L
Total cost			
Total revenue	32,069	35,181	19,872
Materials	8,746	10,796	4,767
Labor	12,581	16,675	10,644
Total costs	21,327	27,471	15,411
Gross margin	10,742	7,710	4,461
Benefit: cost ratio	1.50	1.28	1.29
Actual cost			
Total revenue	32,069	35,181	19,872
Materials	8,746	10,796	4,767
Labor	3,355	3,617	3,004
Total cost	12,101	14,413	7,771
Gross margin	19,968	20,768	12,101
Benefit: cost ratio	2.65	2.44	2.56

Source: Tarawali et al. (1997)
Values are all in Naira ha^{-1} (at the time of writing, N 100=US$ 1.00); Total costs include the value of family labor, which is not accounted for in the values for actual costs

Schlecht et al. (1995) in a study carried out in Sahelian zone of West Africa, reported that when Zebu cattle were supplemented with one kg of cowpea hay at night and 0.5 kg fresh rice feed meal in the morning per day per animal during second half of the dry season, the supplemented animals gained 95 kg compared to 62 kg for unsupplemented group in 15 months, nearly 67 g/animal per day. In addition to direct benefits of improved livestock production, and health that result from feeding cowpea fodder, the quantity and quality of manure from such better fed animals were improved. Therefore when returned to the soil, the manure can contribute towards maintenance of soil fertility. From one ha of improved cowpea, a farmer could benefit by extra 50 kg meat per annum from better-nourished animals, over 300 kg more cereal grain as a result of improved soil fertility directly from cowpea as well as more quantity and better quality manure from animals.

2.8 Challenges and Opportunities

Promotion of grain-legumes in the African farming systems offers important opportunities for improving smallholder livelihoods, national and regional food security and sustainable land productivity. The contribution of legumes to food security and poverty alleviation can be substantial in sub- Saharan Africa if both biophysical and socioeconomic constraints are addressed. Much research on legumes has been devoted to bio-physical aspects of legume production, especially development of improved varieties and crop management technologies. However, to enhance diffusion, adoption and impacts of the technologies in Sub-Saharan Africa, there is need to simultaneously address socio-economic issues. This entails developing innovative strategies for improving access to legume input and output markets as well as incorporation of consumer preferences and appreciation in the research process. Knowledge of consumer preferences is particularly essential in developing legume markets and for legume breeders to incorporate the key characteristics consumers demand in the improved varieties. Participatory research methods are needed to improve understanding of the inter-linkages amongst the various components of the farming systems.

Poor access to markets is particularly an important constraint bedevilling grain legume production. There are a large number of poor smallholder farmers growing grain legumes and are able to benefit from the opportunity offered by regional markets because of insignificant productions and weak marketing opportunities that have led to high transaction costs and market risks that reduce producer incentives. The challenge is how to organize farmers into groups that are economically feasible marketing units and promoting industrial utilization of legumes so that farmers are encouraged to produce more.

Most likely, the production of grain legumes in sub-Saharan Africa will be in form of intercrops. In most intercrop systems, the application of soil fertility amendments are targeted at the cereal component of the intercrop, whilst the legume is assumed to provide some surplus. There may therefore be need to emphasize more on the soil fertility decline associated with grain legume cereal mixtures, and the associated overall yield decline in such systems.

References

Bationo A (1998) Integrated soil fertility management options for agricultural intensification in Sudano-Sahelian zone of West Africa. Academy of Science Publishers and TSBF of CIAT, Nairobi

Bressani R (1985) Nutritive value of cowpea. In: Singh SR, Rachier KO (eds.) Cowpea: research, production and utilization. Wiley, New York, pp 353–360

Coulibaly, Lowenberg-DeBoer (2001) The economics cowpea in West Africa. In: Challenges and opportunities for enhancing sustainable cowpea production. In: Proceedings of the world cowpea conference III held at the International Institute of Tropical Agriculture (IITA), Ibadan, Nigeria, 4–8 Sep 2000

FAO (Food and Agriculture Organization) (2001) Food and agriculture statistical database. http://apps.fao.org. Accessed 24 June 2008

FAOSTAT (2000) Food and agriculture statistical database. http://apps.fao.org/page/collections?Subset=agriculture

FAOSTAT (Food and Agriculture Organization) (2002) http://apps.fao.org. Accessed 24 June 2008

FAOSTAT (2003) Food and agriculture statistical database http://apps.fao.org/collections. Accessed 24 June 2008

David S, Kirkby R, Kasozi S (2000) Assessing the impact of bush bean varieties on poverty reduction in Sub-Saharan Africa. Evidence from Uganda. Occasional Publication Series No. 131. CIAT, Kampala

Delve RJ, Tanner JC, Kimani SK, Giller KE, Cadisch G, Thorne PJ (1999) Effects of dietary nitrogen source fed to steers on faecal nitrogen, its mineralization, and the vegetative responses in maize to the incorporation of the faeces in soil. In: CIMMYT and EARO 1999. Maize production Technologies for the future: Challenges and Opportunities. Proceedings of the 6th Regional maize conference for Eastern and Southern Africa, Addis Ababa, Ethiopia, 21–25 Sept 1998, pp 282–285

Hildebrand GL, Subrahmanyam P (1994) Genetic enhancement of groundnut: it's role in sustainable agriculture. In: sustainable groundnut production in southern and Eastern Africa. In: Proceedings of a workshop held at Mbabane, Swaziland, 5–7 July 1994. ICRISAT, Patancheru, pp 9–13

ICRISAT (1991) ICRISAT West African Programs Annual Report 1990. International Crops Research Institute for the Semi-Arid Tropics, Niamey, Niger

Jaetzold R, Schmidt H (1983) Natural conditions and farm management information, vol 2, Part A, West Kenya Farm management handbook, Ministry of Agriculture (MOA), Kenya MOA and GTZ

Kanyanjua SM, Mureithi JG, Gachene CKK, Saha HM (2000) Soil fertility management handbook for extension staff and farmers in Kenya. KARI Technical Note series no. 6. KARI and LRNP, Nairobi

Karungi J, Adipala E, Nampala P, Ogenga-Latigo MW, Kyamnywa S (2000) Pest management in cowpea. Part 2. Integrating time and plant density on cowpea field pests' infestation in eastern Uganda. Crop Prot 19:231–236

Kormawa P (1996) Economic potential for production and marketing of soybeans in West Africa. Studien zur Ländlichen Entwicklung 53. Rural Development in Africa, Asia and Latin America. Lit Verlag Münster-Hamburg, Germany

Kormawa PM, Chianu HN, Manyong VM (2001) Cowpea demand and supply patterns in West Nigeria: the case of Nigeria. In: Singh BB, Mohan Raj DR, Dashiell KE, Jakai LEN (eds.) Advances in cowpea research. IITA, Japan International Center for Research in Agricultural Sciences (JIRCAS), Ibadan, pp 376–386

Lambot C (2001) Industrial potential of cowpea. In: Challenges and opportunities for enhancing sustainable cowpea production. In: Proceedings of the world cowpea conference III held at the International Institute of Tropical Agriculture (IITA), Ibadan, Nigeria, 4–8 Sep 2000

Lekasi JK, Tanner JC, Kimani SK, Harris PJC (2001) Managing manure to sustain smallholder livelihoods in the East African Highlands. HDRA Emerson Press, Kenilworth

Madamba R (2001) The nutritive value of indigenous grain legumes and their food role at the household level. In: Adipala E, Tenywa JS, Openga-Latigo MW (eds.) African Crop Science Proceedings (1997), vol 3, pp 1255–1258

Mafongoya PL, Mugendi DN, Jama B, Waswa BL (2003) Maize-based cropping systems in sub humid zone of East and Southern Africa. In: Gichuru MP, Bationo A, Bekunda MA, Goma HC, Mafongoya PL, Mugendi DN, Murwira HM, Nandwa SM, Nyathi P, Swift MJ (eds.) Soil fertility management in Africa: a regional perspective. TSBF, Nairobi

Morris RA, Garrity DP (1993) Resource capture and utilization in intercropping: water. Field Crop Res 34:303–307

Mpepereki S, Javaheri F, Davis P, Giller KE (2000) Soybeans and sustainable agriculture: promiscuous soybean in Southern Africa. Field Crop Res 65:137–149

Muller-Samann KM, Kotschi J (1994) Sustaining growth soil fertility management in tropical smallholdings. Weikerrsheim Verlag Margraf, Germany, pp 197–217

Mureithi JG, Maobe SN, Dyck E, Gachene CKK, Gitari N, Kirungu B, Muli BM, Ojiem J, Saha HM, Tana PO (1998) Screening of legume germplasm in Kenya: effect of rhizobia inoculation on performance of Best-Bet legumes. In: Proceeding of the 16th conference held at Mkonge Hotel, Tanga, Tanzania, 13–19 Dec 1998

Nadar HM, Faught WA (1984) Effects of legumes on the yield of associated and subsequent maize in intercropping and rotation systems without nitrogen fertilizer. East Afr Agric Forest J 44(Special issue):127–136

Nkonya E, Schroeder T, Norman D (1997) Factors affecting adoption of improved maize seed and fertilizer in Northern Tanzania. J Agric Econ 48(1):1–12

Norden AJ, Smith OD, Gorbet DW (1982) Breeding of cultivated pea. In: Peasant science and technology. American Peasant Research and Education Society, Yoakum, pp 95–122

Odendo M, David S and Otsyula R (2002) Impact of improved bean varieties in Western Kenya. Research report. KARI and CIAT, Kakamega, p 41
Oniang'o RK (2001) The role of Africa's agriculture in poverty alleviation, food security and industrial development. Afr Crop Sci Conf Proc 3:1303–1311
Ortiz R (1998) Cowpeas from Nigeria. Outlook Agric 27(2):125–128
Oyewole B, Carsky RJ, Schulz S (2000) On-farm testing of Mucuna and cowpea double cropping with maize in the Guinea savanna of Nigeria. In: Carsky RJ, Keatinge JDH, Manyong VM, Eteka AC (eds.) Cover crops for integrated natural resource management in West Africa, IITA, Ibadan, Nigeria, pp 137–147
Pachico D (1993) The demand for bean technology. In: Henry G (ed.) Trends in CIT commodities 1993. CIAT, Cali, pp 60–63
Peoples BM, Craswell ET (1992) Biological nitrogen fixation: investments, expectations and actual contributions to agriculture. Plant and Soil 141:13–39
Rachie KO (1985) Introduction. In: Singh SR, Rachier KO (eds.) Cowpea research, production and utilization. Wiley, London
Sanginga PC, Adesina AA, Manyong VM, Otite O, Dashielle KE (1999) Social impact of soybean in Nigeria's Southern Guinea Savanna. IITA, Ibadan, Nigeria
Shulz S, Carsky RJ, Tarawali S (2001) Herbaceous legumes: the panacea for West African soil fertility problem? In: Etian G (ed.) Soil fertility maintenance in West Africa. Soil Science Society of America and American Society of Agronomy, Madison, USA
Singh SR, van Emden HF (1979) Insect pests of grain legumes. Annu Rev Entomol 24:255–278
Singh BB, Chambliss OL, Sharma B (1997) Recent advances in cowpea breeding. In: Singh BB, Mohan Raj DR, Dashiell KE, Jakai LEN (eds.) Advances in cowpea research. IITA, Japan international center for Research in agricultural Sciences (JIRCAS), Ibadan, pp 30–49
Singh BB, Tarawali SA (1997) Cowpea and its improvement: Key to sustainable mixed crop/livestock farming systems in West Africa. In: Renard C (ed) Crop residues in sustainable mixed crop/livestock farming system. ICRISAT/ILRI/ CABI, Wallingford, UK. pp 79–100
Skerman PJ, Riveros F (1988) Tropical Forage Legumes, vol 2, 2nd edn, FAO Production and Protection Series. FAO, Rome, Italy, pp 103–119
Tarawali SA, Singh BB, Peters SM, Blade SF (1997) Cowpea haulms as fodder. In: Singh BB, Mohan Raj DR, Dashiell KE, Jakai LEN (eds.) Advances in cowpea research. IITA, Japan International Center for Research in Agricultural Sciences (JIRCAS), Ibadan, pp 313–325
Tarawali SA, Larbi A, Fernández-Rivera S, Bationo A (2001) The role of livestock in the maintenance and improvement of soil fertility. In: Sustaining soil fertility in West Africa. SSSA Special Publication No. 58. Soil Science Society of America and American Society of Agronomy, Madison, USA, pp 281–304
Tarawali, SA, Singh, BB, Gupta SC, Tabo R, Harris, Nokoe F, Fernandez-Rivera S, Manyong VM, Makinde K, Odion EC (2002) Cowpea as a key factor for a new approach to integrate crop-livestock systems research in dry savannas of West Africa. In: Challenges and opportunities for enhancing sustainable cowpea production. Proceedings of the world cowpea conference III held at the International Institute of Tropical Agriculture (IITA), Ibadan, Nigeria, 4–8 September, 2000
Trutmann P, Voss J, Fairhead J (1996) Local knowledge and farmers' perceptions of bean diseases in Central African Highlands. Agric Hum Values 13:160–179
Upton M (1987) African farm management. Cambridge University Press, New York
Weingartner KE (1987) Processing, nutrition and utilization. In: Soybeans for tropics: research, production and utilization. Wiley, Chichester, pp 149–178
Williams SR, Anderson SL (1993) Nutrition and AIDS. In: Nutrition and diet therapy, vol 7. Mosby, St. Louis, pp 701–707
Winrock (1992) Assessment of animal agriculture in sub-Saharan Africa. Winrock institute for agricultural development, Morrilton, Arkansas, USA
World Bank (1993) Sub-Saharan Africa from crisis to sustainable growth: a long term perspective study. World Bank, Washington, DC
Wortmann CS, Kirkby RA, Eledu CA, Allen DJ (1998) Atlas of common bean (Phaseolus vulgaris, L.) production in Africa. CIAT and Pan-African Bean Research Alliance (PABRA). CIAT Publication no. 297, African Bean program, communications Unit

Chapter 3
Inter and Intra-Specific Variation of Legumes and Mechanisms to Access and Adapt to Less Available Soil Phosphorus and Rock Phosphate

S.M. Nandwa, A. Bationo, S.N. Obanyi, I.M. Rao, N. Sanginga, and B. Vanlauwe

Abstract Legume integration in African agricultural production systems is among the options with demonstrated potential for sustainable land management. Best bet legumes for promotion in cropping systems should play multiple roles in soil fertility, food and fodder provision. The main constraint to growth and productivity of legumes i.e., phosphorus deficiency, is variably addressed through use of inorganic P fertilizers and slow reactive phosphate rock, genotype adaptation to low P soils and role of phosphate solubilizing micro-organisms (PSMs). Research findings using grain legumes show greater yields for rotated cereal and higher land equivalent ratios for intercropping than cereal monocropping systems across different environments. The current research priority is to enhance recovery of soil P by legumes to increase productivity of cropping systems in low-nutrient environments. A number of research priorities are outlined among them being the integration of physiological research with genetic improvement.

S.M. Nandwa
Kenya Agricultural Research Institute – National Agricultural Research Laboratories (KARI-NARL), Nairobi, Kenya (Deceased)

A. Bationo (✉)
Alliance for a Green Revolution in Africa (AGRA),
Soil Health Program, PMB KIA 114, Airport-Accra, Ghana
e-mail: abationo@agra-alliance.org

S.N. Obanyi
Kenya Agricultural Research Institute – National Agricultural Research Laboratories (KARI-NARL), Nairobi, Kenya

I.M. Rao
International Center for Tropical Agriculture (CIAT), Cali, Colombia

N. Sanginga • B. Vanlauwe
Tropical Soil Biology and Fertility (TSBF) Institute of CIAT, Nairobi, Kenya

A. Bationo et al. (eds.), *Fighting Poverty in Sub-Saharan Africa: The Multiple Roles of Legumes in Integrated Soil Fertility Management*,
DOI 10.1007/978-94-007-1536-3_3, © Springer Science+Business Media B.V. 2011

3.1 Introduction

In Sub-Saharan Africa (SSA), the per capita food production continues to decline against a background of a rapidly growing population. The failure to match food supply to demand is being attributed to land degradation associated with the intensification of land use for agricultural production without adoption of proper land management practices and external inputs (Henao and Baanante 1999). Attainment of food security through increasing per capita agricultural production, is therefore closely linked with reversing agricultural stagnation, safeguarding the natural resource base e.g. vegetation, water, biodiversity and soil fertility and reducing population growth rate (Cleaver and Schreiber 1994; Deckers 1993). Declining soil fertility is associated with widespread disappearance of soil fertility restoration practices such as fallowing as well as inadequate and/or inappropriate nutrient adding and saving practices, resulting in non-sustainable land management (Hartemink and Wood 1998; Greenland 1994). Amongst options likely to result in sustainable land management is legume integration in the agricultural production systems (Nandwa 2001). There are three main ways in which legumes can be integrated in agricultural production systems.

3.1.1 Dual Purpose Use of Legumes

By definition, dual-purpose grain legumes- cereal rotations using improved legume germplasm, serves two purposes; provision of net N benefits to the cropping systems; and provision of grains. This type of practice can enhance cereal yield and supply the farmer with immediate products that can either be consumed or sold to earn income. The importance of legumes as components of cropping systems in Sub-Saharan Africa has been severally acknowledged in agro-ecosystems even where N fertilizer is widely used. The importance of legumes in crop rotations has been and is still well recognized, implying that factors other than N_2 fixation are equally important.

The multiple roles of legumes in SSA as sources of food, feed, soil fertility replenishment and income generation underpins their importance from both the biophysical and socio-economic point of view. It is increasingly being recognized, that crop production practices that minimize the use of expensive imported fertilizers while maintaining and/or increasing yields are likely to enhance sustainable agricultural production systems.

Food and grain legumes include *Vigna, Phaseolus, Cajanus, Rachis, Glycine, Cicer* etc. In SSA, many legumes are cultivated primarily because of their importance as a source of protein and caloric contribution to the human diet. Low harvest index legumes have been found suitable to contribute to increased crop production in low nutrient environments because they can be used to produce a seed crop, generate high N-content plant residues and fix N_2. Nevertheless, there is substantial need for P in the N_2 fixation process.

In Benin (Mono Province), higher adoption of legumes use has been reported and attributed to mucunas ability to reduce noxious spear grass weed (*Imperata cylindrica*) to less than 10% of its initial density in farmers fields. Although the satisfaction of the multiple purpose role of legumes is crucial, an evaluation of the "best bet" legumes has shown that very few of them can simultaneously satisfy the multipurpose roles of soil fertility, food and fodder. This is partly because many of them contain toxic constituents e.g. L-Dopa (Carsky et al. 2001) and anti-nutritional factors. This is part of the reason for poor adoption of *Mucuna pruriens* as a green manure or cover crop, inspite of extensive research and promotion. Mucuna has been reported to control nut grass, *Cyperus rotundus*, besides reducing nematode population (Carsky et al. 1999).

Nitrogen contribution by N-fixing legumes and other N-fixing plants has been widely exploited in enhancing productivity of agro-ecosystems, in terms of species for intercropping, relay cropping, agroforestry, and for use in improved fallow technology. Past experiments on legume intercropping have shown that many legumes hardly fix any nitrogen under low soil available P conditions (AHI 1997). Moreover, competition for water, light and other nutrients between annual crops and perennial legumes, make the intercropping of the latter an inappropriate technology, for soil fertility restoration, except for soil conservation purposes (Place et al. 1994). Similarly, competition has been observed in systems where food legumes have been intercropped with maize as a risk minimization strategy, in case of drought. Nevertheless, deep-rooted legume species planted as improved fallows contribute to soil fertility restoration through biological nitrogen fixation (BNF), as well as from nutrients capture from the subsoil. In a study in Western Kenya, some species like *Dolichos lablabbronga, Macroptilium atropurpurem* and *Canavalia ensiformis*, responded negatively to P application, an indication that they can do well in P deficient soil conditions. This indicates the inter- and intra-specific variation of legumes to access less available soil phosphorus. Thus, an understanding of the mechanisms of legume adaptation to low phosphorus soil is crucial. In systems with limited per capita land, legume species likely to be adopted are those that are grown for food (vegetable and grain) or forage purposes, but also yield high above ground (leaf biomass) and below ground biomass e.g. *Arachis* and *Cajanus* or leguminous cover crops such as *Crotalaria, Dolichos, Mucuna*, etc.

3.1.2 Addressing phosphorus through legume selection

Acute deficiency in phosphorus (P) has been observed in 65% of soils from 42 countries in the tropics, whereby 45.7% were considered as those with low available P (Koala et al. 1988; Tandon 1987). Due to P deficiency, it's relative immobility in the soil, any extension of the rhizosphere and/or increase in solubilization within the rhizosphere, is likely to favor P acquisition. This problem of P deficiency is further exacerbated by P-fixation by some soils, especially those with high iron, aluminium and calcium oxides, where P is strongly bound and made less available for crop uptake. Some studies have indicated that total soil P is often 100- fold more than the fraction of soil P normally available for uptake by crop plants (Al-Abbas and Barber 1964). Yield losses of about 29–45% of crops due to P deficiency have been commonly reported.

One of the challenges in soil fertility, especially of inherently low P-soils is to sustain sufficient, soluble and exchangeable P in the soil. Knowledge from research indicates that physiological traits of some legume plants exist with the potential to increase the transformation of P from its fixed form (inorganic and organic) to forms that can be taken up by plant roots. Mineral fertilizers may be used to supply required P for crop uptake and growth. However its use in SSA is variable e.g. from 234 kg NPK ha^{-1} $year^{-1}$ in Egypt to 99 kg in Swaziland, 46 kg in Kenya and is less than 10 kg ha^{-1} $year^{-1}$ for most countries in the region (Henao and Baanante 1999). Even where fertilizers are used, they are mostly applied to cash crops and plantation crops (coffee, tea, cocoa, cotton, tobacco, sugar cane, oil palm etc.), because of the high profitability. For most food crops, there is unfavorable crop/fertilizer price ratio. Therefore, most food crops are continuously grown with little or no replenishment of nutrients removed through crop harvests, or other losses, thus resulting in the depletion of natural capital stocks below critical levels. To maintain current average levels of crop production without soil nutrients depletion, Africa will require to use approximately 11.68 million metric tonnes of NPK fertilizers annually.

Phosphorus deficiency is the primary limitation to legume production in the tropics (Sanchez et al. 1997). Hitherto, it is desirable to identify plant species or genotypes with ability to solubilize nutrients from sparingly soluble sources and thus introduce available forms of nutrients into the nutrient cycle. Direct application of phosphate rocks (PR) are possible (Rajan et al. 1996), but they are also often sparingly soluble, hence there is need to search for crop genotypes with improved ability to mobilize P from the sparingly soluble soil P-sources.

A number of strategies and options are being recommended and their implementation explored, on how to meet such targets. Options for nitrogen replenishment in SSA are receiving a lot of attention and include use of leguminous tree fallows, improved fallows, agroforestry trees and herbaceous leguminous cover crops grown in situ. Other options include green manure that plays a major role in both N capture and internal nutrient cycling in ways that are compatible with most farmer constraints. Examples of such complementary approach is on-going work in Western Kenya, in South Western Uganda and Eastern Zambia. In Western Kenya, options used include integrated use of inorganic P (rock phosphate or triple super phosphate) and organic nutrient sources either through short duration fallows of *Crotalaria* or *Tephrosia* or biomass transfer from *Tithonia* green manure. In South- Western Uganda, the options include stabilization of terrace risers with napier grass/*Calliandra* contour hedgerows, integrated with use of inorganic P on terraces and tree fallows with shrub or cover crops on terraces. There is also use of plants that are known to employ diversity of strategies and novel plant systems to improve mineral nutrition. This is often attributed to morphological and eco-physiological traits or characteristics.

Genotypes of common bean (*Phaseolus vulgaris*), pigeon pea (*Cajanus cajan*) and cowpea (*Vigna* sp) have been reported to differ (inter- and intraspecific variations) in their ability to grow under limited supply of P, through mobilising of P from sparingly soluble P fractions. Another complementary option entails refraining from application of higher doses of fertilizers to less responsive indigenous crop varieties or cropping systems (liable to be leached or volatilized).

The key to successful implementation of these options requires that the approaches rely primarily on locally available resources and improve upon them with fertilizers e.g fortification of composts with rock phosphate or lime; improve manure quality with livestock mineral supplement; use of improved legume seed inoculant etc. In conclusion, it can be said that integrated nutrient management (INM) practices that optimize nutrient inputs while concurrently minimizing nutrient losses are most likely to contribute towards increased efficiency of the applied nutrient, manifested in high crop response.

3.1.3 Inorganic P Fertilizers and Slow Reactive Phosphate Rock as Strategies to Overcome P Deficiency

The two primary challenges in soil fertility management research are: (i) to improve and maintain or sustain crop productivity to meet demand for food, fibre, fuel and materials for agro-based industries, and (ii) to enhance the quality of land, water and other natural resources. The major forms of soil fertility research in Africa are therefore to replenish mined nutrients on one hand, and identifying ways to solve the problem of imbalanced plant nutrition on the other. Currently, the latter is regarded to be a serious degradation problem in high input agro-ecosystems. Judicious management and balanced application of plant nutrients is a strategy that should overcome the hazards of nutrient depletion through mining as well as environmental pollution e.g. deterioration in biodiversity and ecosystem function (Swift and Anderson 1993).

Two approaches may be used to maximize P use under conditions of low P supply. Firstly, through increasing biomass or grain yield for a given unit of P applied (P use efficiency), and secondly, through increasing the root systems ability to acquire P. To overcome the prohibitive factor of cost of imported and processed fertilizers, unprocessed rock phosphate can be employed to supply P, which is often effective in acid soils (Khasawneh and Doll 1978; van Straaten 2002; Bationo and Kumar 2002). Direct application of indigenous phosphate rocks can be an economic alternative to the use of processed imported P fertilizers (Rajan et al. 1996). The choice for use of either processed P fertilizer or indigenous rock phosphates is dependent on both relative agronomic efficiency (RAE) and the cost of phosphates used (TSBF-CIAT 2001). Phosphorus recovery and use efficiency is low in many production systems because of the inability of many crops to access less available P. However, some legumes have been found to access deficient or sparingly soluble soil phosphorous in several ways: (i) improving soil physical properties through degradation of soil clods or by penetration of their tap root system into the soil profile; (ii) improving soil tilth and water infiltration and thus benefiting the growth of succeeding crops; (iii) scavenging for occluded and unavailable P. The use of phosphate rock can be augmented with the integration of P- efficient use crops (legumes) to enhance chances of sustaining agricultural productivity (Breman and van Reuler 2002).

One strategy of increasing the supply and availability of P from slow release rock phosphate (PR) through interactions with the rhizosphere of herbaceous and grain legumes has been demonstrated. This strategy was demonstrated in field trials in Nigeria and Benin using an improved cowpea variety with the ability to enhance the availability of P from PR in a cowpea- maize rotation (Lyasse et al. 2002). Results obtained indicated that the particular cowpea used may enhance the availability of P from PR without necessarily taking advantage of this source of P. This is because in the first season, the cowpea responded to the application of TSP, but the addition of PR did not influence its agronomic performance, though the maize following the cowpea seemed to have benefited from the PR applied to the preceding cowpea. Many studies have shown that addition of PR leads to significant site and species dependant changes. For example, four cowpea varieties tested to assess their genotype specific potential to utilize PR as the source of P in P deficient soils in Nigeria, showed significant genotype variation in terms of growth performance. One cowpea variety (IT90K-59) stood out as reacting positively to the application of PR and was able to deplete the stable P fraction (non – Olsen P). The P uptake and grain yield was higher with PR than without, while the reverse was true with the remaining three varieties.

These studies, also confirmed that plant residues or green manures with high P content incorporated into soil, could largely increase total soil P amount and the available P (Nziguheba et al. 2000). The occurrence of numerous deposits of PR in SSA provides a potential outlet and a way to reduce the dependence on processed and imported super – phosphates. The main sources of P supply to crops are 'Agricultural P' (the available P) and the 'capital P' that is sorbed relatively strongly by the soil. Addition of PR can increase both 'Capital and Agricultural P' in the soil. This is enhanced by the addition of inorganic and organic sources of P. The soil recapitalization concept is based on a similar understanding as that of capital stocks and service flows in economics (Izac 1993) which is comparable to that of nutrient pools and fluxes in soil science (Sanchez et al. 1997), and hence nutrient fluxes can be analogous to service flows. Since the fluxes subtract from the nutrient capital, they are thus analogous to the concept of capital depreciation. Similarly, Agricultural P (immediately available to the crop) is analogous to the liquid capital in the bank.

From the foregoing, PR adds to the Agriculture P as well as the Capital P since PR continues to supply P to the crop beyond the year of application. Thus enrichment of capital P ensures steady replenishment over a longer period. On the other hand, addition of soluble fertilizers, increases the Agricultural P; but with time, however, the organic compounds in the soil and the clay fraction react with the soluble P and transform it into the relatively unavailable forms. It is estimated that at least 80% of the soils of tropical Africa have low capital stocks of P, hence the need to use PR to improve the capital stock of P in these soils, should be considered as an improvement in natural resource capital (concept of soil fertility recapitalization) (Buresh et al. 1997). Application of slow release PR should not preclude the addition and use of the mineral fertilizers, not withstanding the fact that highly reactive PR can supply much of the P needed by the crop in the first year of application.

In conceptual terms, a one-time application of a highly soluble P source dramatically improves 'Agricultural' P and Capital P. However the conversion of P from Capital P to Agricultural P gradually declines. Because of the long- term residue effectiveness of PR, it is expected that a one time application would initially increase the Agricultural P but to a much lesser extent than the mineral phosphate fertilizer. However, with time, the contribution of the PR to the Capital P pool would increase. From studies on the residual effectiveness of Togo PR as a percentage of the residual effectiveness of single super phosphate, a one- time application of PR improved with time as compared to SSP. One of the merits for combined use of PR and mineral P fertilizers in the recapitalization strategy of P in the soil, is that, initially the contribution to the 'capital P' pool is from mineral fertilizer and less from the PR. Subsequently, however, PR becomes the main contributor to the 'capital P' pool thus reducing the mineral P fertilizer required for optimum performance of the crop. For example, in Zimbabwe, there is evidence indicating an increase in the solubility of Dorowa PR by inclusion of double super phosphate (DAP). This combination showed the potential for reducing the amounts of DAP which is usually the more expensive purchased soluble mineral fertilizer (Muza et al. 1997). The mechanisms of how the combination of Dorowa PR and DAP increases the solubility of the PR are not well understood.

3.2 Screening Legumes for of Phosphorus Use Efficiency

The efficiency of crop selection or screening for nutrient-stressed environments can be enhanced if adaptive morphological, physiological and developmental plant traits can be identified. A large number of green house and field experiments have been conducted to investigate this option. Using a ^{32}P isotope to compare the uptake, translocation and distribution of P, Adu-Gyamfi et al. (2002) conducted experiments to screen pigeon pea genotypes (hybrids and improved, local, drought-tolerant and breeding lines). The hybrid pigeon pea and germplasm accessions from arid and high altitude areas were the most adapted to both low P and sufficient P soil conditions, where they produced the highest biomass and grain yield. Physiologically, this has been attributed to the re-translocation of P from old leaves to the growing parts before senescence. This is supported by studies by Fujita et al. (2003) who reported that P limitation resulted in the partitioning of more photo-assimilates to roots and less to leaves and stems of pigeon pea (*Cajanus cajan*) and mash bean (*Vigna aconitifolia*).

Another study was conducted to analyze P acquisition mechanisms of pigeon pea. These authors reported results showing that root exudate organic acids responsible for P-mobilization were citric and piscidic acids. However, this could not fully explain the genotypic variation in low P availability of pigeon pea. In Sierra Leone,

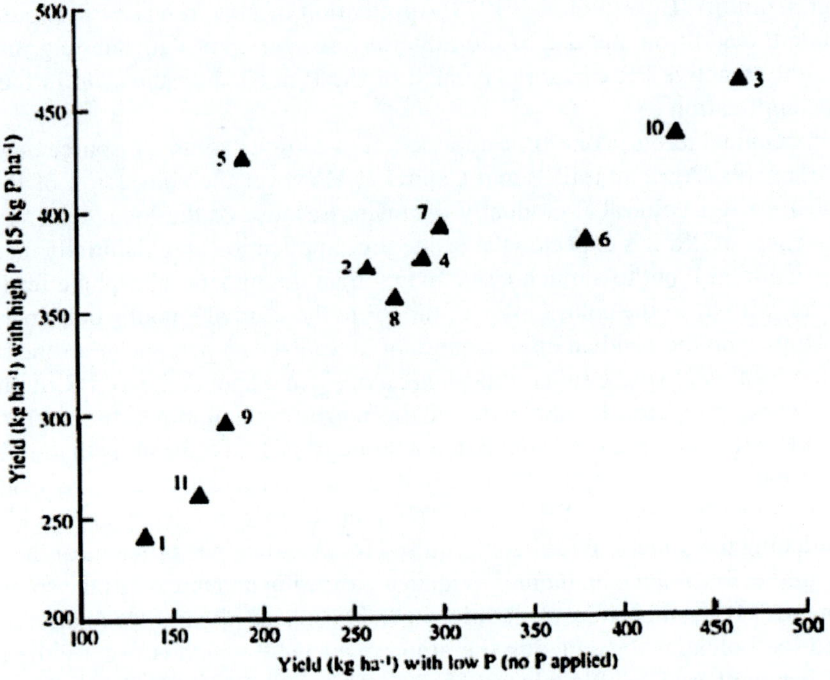

Fig. 3.1 Grain yield of cowpea at low and high phosphorus levels, Sierra Leone, 1991/92. (Source Ahloowalia et al. 1995)

11 genotypes of cowpea *(Vigna)* were screened under field conditions for phosphorus use efficiency in a soil that was low in available P (4 mg kg^{-1} available P) and the results found (Figs. 3.1 and 3.2) indicated that three genotypes were superior in grain yield at both low and high P levels (Ahloowalia et al. 1995). The study showed that high use efficiency in terms of above- ground matter per unit of P did not always translate into high grain yield.

In Nigeria 20 genotypes of cowpea were screened at several P-deficient soils under low P (no P applied) and high P (60 kg P ha^{-1}) conditions. Results showed that there were threefold differences in grain yield at low P (Ahloowalia et al. 1995). The P application had little effect on increasing the yield of 60% of the genotypes that produce high yields at low P levels, but showed positive effects on those that produce low yields at low P levels. Little effect on application of P was attributed to the soil's high capacity for P fixation. In such circumstances, the root appears to play a major role in enabling plants to grow well in P deficient soils: high yielding genotypes have a tendency to produce bigger root systems than those of low yielders.

In screening 364 genotypes of common bean, Sinclair and Vadez (2002) reported there was desirable genetic potential in legumes to increase P extraction under low P conditions. Piscidic acid excreted in pigeon pea is a strong chelator of ferric iron but not Ca. This explains the higher efficiency of pigeon pea in P acquisition from Alfisols compared to that from Vertisols (Ae et al. 1990). These results suggest that the potential is high for genetic manipulation of plants to enhance their productivity under P-deficient conditions. This may be achieved through selection of morphological characteristics associated with an increased photosynthetic surface area and

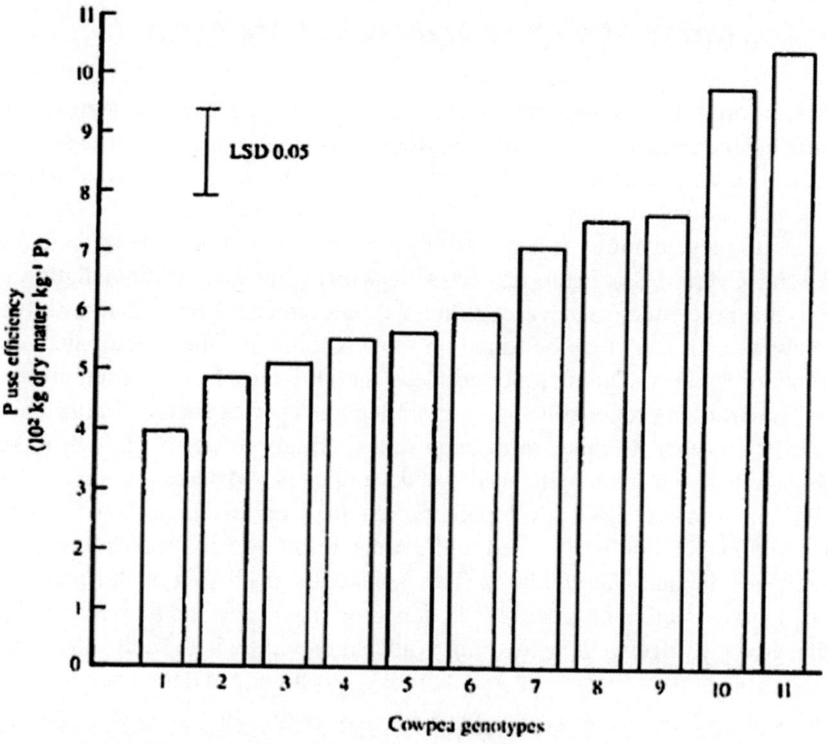

Fig. 3.2 Genotypic differences in phosphate use efficiency of cowpea, Sierra Leone, 1989/90. (Source Ahloowalia et al. 1995)

deep rooting to overcome reduced availability of P. Further, it is possible to combine these approaches to genetically manipulate plants such that they have not only a lower internal demand for P but also a greater ability to scavenge P from sparingly available sources.

3.3 Mechanisms of Legume Adaptation to Low P Soils

Legume species or genotypes have been studied extensively with a view to identify those that can best solubilize P derived from sparingly soluble fertilizers as well as slow-release P sources such as rock phosphates. These studies have shown that in some cases, plants grow more roots or re-translocate nutrients from older leaves to younger ones and/or respond to P stress by depleting vacuole pools. This stress can be due to P-starvation or low P-nutrient environments, which trigger increased P synthesis. There are three broad categories of mechanisms by which plants can increase their access to native or applied soil nutrients including: increasing their absorbing surface; favourably modifying the absorption mechanisms to increase uptake from low ambient concentrations and rhizosphere modification to increase nutrient availability. Such studies have identified how these mechanisms and their attributes are affected by environmental, physiological and biochemical factors (Bar – Yosef 1991).

3.3.1 *Increasing Absorptive Area: Root Morphology*

Increases in root biomass and soil contact, is one of the mechanisms legumes use to overcome the problem of low P mobility. Root proliferation or root hair development, or mycorrhizal associations that allow capture of nutrients from beyond the rhizosphere. Deep-rooting legumes, agroforestry trees and herbaceous species use this mechanism to recycle nutrients acquired from below the rooting zone of annual crops (Irizarry and Rivera 1983). In this respect, roots with higher specific root length, smaller diameters are more effective in P acquisition through capillary action. In more advanced cases, massive root hairs may be found on rootlets following the development of proteoid roots or rootlets. This is reported to be the attribute of crop adaptation to low nutrient and drought prone soils e.g. in grain legume species such as Lupin (Gardner et al. 1981). Another mechanism of acquiring sparingly low P is through increased root: shoot ratio as a result of enhanced root development and decreased shoot development. It is believed that shoot development is readily decreased under low P conditions so that additional carbon is allocated to the roots, resulting in increases in root carbohydrates (Sinclair and Vadez 2002). This apparently results from the premise that through natural selection or breeder selection, crop plants develop nutrient uptake systems that are very effective at recovering available P (and N) in the soil. Consequently, if P is available at the root surface and there is a physiological deficiency in the plant, roots have the capacity for uptake rates that are many times greater than uptakes observed under adequate P levels. Yan et al. (1995) reported that common bean genotypes that performed best on low P available soils were those that produced the greatest root mass, especially branched roots and the presence of root hairs.

3.3.2 *Mycorrhizal Associations*

Another approach to increase the P-absorbing surfaces is through non-functional symbiosis formed between roots of plant species and mycorrhiza (Smith and Gianinazzi-Pearson 1988). The stimulation of plant growth with mycorrhizal infection is often attributed to enhanced P uptake, withstanding other limiting nutrients. This has led to the phenomenon of mycorrhizal dependency; namely the degree to which a host relies on the mycorrhizal condition to produce maximum growth at a given level of soil fertility (Gerdemann 1975); such is exhibited by many cultivated legumes (Thompson 1991). In most cases P deficiency stimulates growth of mycorrhizae, resulting in increased exudation of reducing sugars and amino acids (Graham et al. 1981).

3.3.3 *Modification of the Absorption Mechanisms*

Plant species and genotypes have been known to have different abilities to extract nutrients from low ambient concentrations, namely their threshold levels (Barber 1979).

Table 3.1 Response of grain yield and P uptake to phosphorus fertilizer application in an Alfisol and a Vertisol fields, ICRISAT Asia Center, rainy season, 1987

Crop	Soil	P fertilizer application (kg ha^{-1}) 0	9	18	Standard error
		P uptake at the flowering stage (kg ha^{-1})			
Sorghum	Alfisol	2.00	3.43	7.38	±1.19
	Vertisols	6.21	6.61	9.35	±2.23
Pigeonpea	Alfisol	3.18	3.73	6.91	±1.28
	Vertisols	2.46	2.50	4.04	±0.69
		Seed yield at harvest (kg ha^{-1})			
Sorghum	Alfisol	87	673	2101	±725
	Vertisols	3043	3364	3853	±570
Pigeonpea	Alfisol	929	727	1113	±393
	Vertisols	248	457	674	±158

Source: Ae et al. (1991a)

3.3.3.1 Root Exudates: Organic Acids and Phosphatases

Unavailable P occurs in both in-organic and organic forms. Availability of in-organic P is enhanced under acidic conditions in the soil (acidification) and response to P starvation. One of the positive physiological traits of legumes is the ability of roots to release both organic acids and hydrogen ions to acidify the rhizosphere so as to increase P availability. The ability to solubilize Ca – P complexes has been attributed to the ability of the phosphorus solubilizing microbes (PSMs) to reduce the pH of their surroundings, either by the release of protons or organic acids (which directly dissolve the mineral phosphate by anion exchange of PO_4^{2-} or by chelating Fe and Al ions). Many microorganisms can reduce the pH of the medium and solubilize P from rock phosphate (Nahas 1996; Gyaneshwar et al. 1998). In certain cases P solubilization has been found to be induced by P starvation (Gyaneshwar et al. 1999).

In India, at ICRISAT Asia Center, field and pot screening experiments were conducted using vertisols (dominated by Fe and Ca-bound P) and Alfisols (Fe-bound P) to investigate the potential of pigeon pea, soybean and chick pea, besides sorghum, millet and maize (Ae et al. 1991a, b) to access sparingly soluble P. Results on grain yield, P uptake of pigeon pea and sorghum (Table 3.1), shoot phosphorous contents (mg pot $^{-1}$) of both legumes and cereals (Table 3.2) showed that in the absence of P application, pigeon pea exhibited much greater dry matter production and P uptake in the Alfisol than Vertisol while the reverse was true for sorghum. Results obtained showed that pigeon pea could take up 2.5–7.0 times more P from Fe PO_4 than other crops, when 80 mg kg^{-1} of P was applied, indicating the unique ability of pigeon pea to solubilize Fe PO_4.

The release of organic P is ascribed to the catalytic action of phosphate enzymes directly from plants or from microbes, stimulated by plant exudates. Crops such as pigeon pea and groundnuts are capable of accessing poorly available P in the soil through several mechanisms. One of the mechanisms is root exudation in which the

Table 3.2 Shoot phosphorus contents (mg po^{-1}) of several crop species at the grain filling stages after growth in potted Alfisols and Vertisols without phosphorus fertilizer

Soil	Pigeonpea	Sorghum	Soybean	Pearl Millet	Maize
Alfisol	5.72	0.59	1.40	0.64	0.51
Vertisols	2.34	3.91	6.53	5.38	6.13
Standard Error	±0.82	±0.39	±0.20	±0.34	±0.25

Source: Ae et al. (1991b)

exudates (organic acids and phenols) are regarded as mediators of mineral acquisition. This results from the ability of organic acids from root exudates to modify the rhizosphere for increased nutrient uptake or from the establishment of symbiotic interactions that enhance nutrient acquisition. The organic acids are able to solubilize non-available soil calcium, iron and aluminum phosphates. Similarly, through chelating, extra cellular enzymes release P from organic compounds and there-by increase nutrient availability. Tolerance and growth in low soil P levels by chickpea has been attributed to the ability of exuded citric acid to solubilize unavailable forms of P which is calcium bound. In lupin and pigeon pea, tolerance to low soil P levels has been attributed to their ability to solubilize occluded P by exuding citric acid, primarily as a chelating substance that solubilizes iron bound P (highly insoluble or sparingly soluble). In many instances, P-deficiency induced enhancement of organic acid excretion, which is confined to certain root zones (e.g. sub apical root zones in rape, proteiod root zones in annual legumes as white lupin, and the zone around the cluster root in some tree species such as Bankra). Temporal and localized rhizosphere acidification by excretion of organic acids into the rhizosphere of soil-grown plants might be more important and abundant in higher plants than so far reported from past studies.

3.3.3.2 Growth Regulators and Extension of Growth Period

Rhizosphere microorganism produce growth regulators such as indoleacetic acid (IAA) and gibberellins (GAs) that, promote root growth thereby increasing the capture of P and other nutrients. Another approach to increase overall uptake is to extend the duration of the growth so that the crop has a longer time frame to accumulate P. This is the major mechanism in climbing beans, which commonly have a longer season than bush types.

3.3.3.3 Influencing Rhizosphere pH and Redox Potential

Microorganisms enhance nutrient availability though excretion of OH^- and HCO_3-, evolution of carbon dioxide by respiration and exudation of organic acids thus influencing the rhizosphere pH and redox potential. This facilitates the growth of plants in low pH (3–5) soil. This is the main mechanism by which chickpea and soybean lowers rhizosphere pH (better than maize), as dictated by cation/anion uptake ratios in alkaline soils. In alkaline calcareous vertisols nutrient availability is achieved primarily by acidification of the soils. Emerging evidence from many studies indicate that there may be genotypic differences within species (intraspecific relatives) for acidifying the soil and extracting P (Sinclair and Vadez 2002). Substantial variation in yield obtained when 364 genotypes of the common bean were grown under P-deficient conditions indicated that there existed considerable genetic potential among legumes to increase P extraction in low availability conditions.

3.3.3.4 Active Loading of Ureides

The P deficiency tolerance mechanism developed by certain bean cultivars and *Acacia mangium* (Ribet and Drevon 1995) has been explained by an increased nodule ureides concentration and activity. These are compounds transported from nodules into the xylem for transport to the plant shoot.

3.4 Role of Phosphorus Efficient Grain Legumes in Cropping Systems

Integration of grain legumes (improved) into the cropping systems plays two major roles: of providing grain and nutrients. Efficient P use grain legumes may be defined as those resulting in high yield increase per kg fertilizers added, subject to P sources, environmental factors, crop and soil management. This definition can be extended to include those that respond negatively to P application, indicating that they are efficient up-takers, do well in P deficient soil conditions. This includes those that mobilize P from sparingly low P sources. Because of this attribute, legumes form a major component of the rotation or relay systems in SSA. Past research indicates that genotypes of common bean (P. vulgaris), pigeon pea (*C. cajan*) and cowpea (*Vigna* sp) differ in their ability to grow under limited P supply (Sanginga et al. 2000).

3.4.1 Cereal-Legume Rotations

Studies investigating the effect of cereal-legume rotation on cereal yields have consistently shown the benefits of cereal-legume rotations over continuous cereal cultivation (Bationo and Ntare 2000; Bagayoko et al. 2000). Table 3.3 shows maize yields under monocropping and following intercropping with a legume (beans) in Western Kenya. In smallholder cropping systems in Southern Africa, Snapp et al. (1998) reported on contributions of grain legumes (e.g. cowpea, beans, groundnuts, etc. grain legume rotations or intercrops) in raising soil fertility and crop yields but

Table 3.3 Long term maize production under monocropping and intercropping with beans in Western Kenya

Area	Season	Maize mono crop	Maize intercrop with beans	% change	Number of seasons
Yala swamp	1	3,457	3,128	10	4
	2	3,480	2,558	26	5
Siaya	1	2,208	2,155	2	5
	2	1,784	1,597	10	2
Ukwala	1	1,645	1,839	–12	4
	2	1,184	1,093	8	3
Rodi kopani	1	1,441	1,100	24	5
	2	1,971	1,510	23	5
Rongo	1	2,004	2,010	0.3	3
	2	1,300	1,281	1	3
Mukuyu	1	2,306	1,925	17	4
korondo	2	1,595	1,925	–21	4
Oxugis	1	2,324	1,383	40	5
	2	1,509	1,087	28	6
Mumias	1	1,092	1,706	–56	6
	2	390	1,084	–178	5
Kakamega	1	4,524	5,221	–15	6
	2	1,807	1,573	13	4
Vihiga	1	4,191	5,035	–20	4
	2	2,021	2,760	–37	4
Mwililia	1	3,681	4,516	–23	5
	2	1,110	1,213	–9	4
Bukiri	1	1,400	2,184	–56	5
	2	892	1,044	–17	4
Alupe	1	2,645	3,466	–31	6
	2	1,671	2,169	–30	4
Kamakoiwa	1	3,290	4,514	–37	4
Tongaren	1	1,495	269	49	5

also reported on their constraints including competition for arable land, labour requirements and farmers inability to adopt them. An important point on high potential for adoption of grain legume technologies is that they are likely to be adopted for other reasons (providing food, food security, weed control, etc.), and that the soil fertility effects is primarily a spin-off. The rotation of cereals and cowpea and soil amendment with crop residue has been reported to drastically increase crop yields, besides phosphorus use efficiency (PUE) in West African semi-arid tropics (Bationo and Kumar 2002). In investigating the effect of P and cropping systems on pearl millet grain, total dry matter and cowpea fodder yields in Niger, they reported a linear positive response to application of P. The rotation of pearl millet with cowpea significantly increased both the pearl millet (from 149 to 252 kg ha^{-1} PUE in moving from continuous cultivation to rotation) and cowpea fodder (from 40 kg ha^{-1} PUE in continuous cultivation to 65 kg ha^{-1} in rotation). In all cases, pearl millet rotated with cowpea resulted in higher yields than continuous millet.

3.4.2 Legume Intercrops

Legumes are an important component of cropping systems in SSA in terms of cropping sequence. In spatial and temporal context, they may be included in one extreme system as intercrops (multiple or poly culture) while on the other extreme they are used in rotations (sole cropping). The benefits and limitations of pure stand cropping or monocropping as opposed to crop mixtures (intercropping, relays etc.) has been demonstrated in many recent studies (FURP 1994). The FURP results showed that in 46% of the cases mono cropped maize yielded higher than intercropped maize in Western Kenya. However in terms of land equivalence ratio, the intercrop was more beneficial than sole cropping because of the added yield from the bean crop.

3.4.3 Legume Relay Cropping

In tropical West African soils, continuous cropping with a maize/cotton relay cropping system is common (Breman and Van Reuler 2002). Using this system as a baseline scenario, crop and weed residues returned to the soil has been reported to amount to 8 tonnes DM ha^{-1}. Studies have shown that in the system, organic inputs may be enhanced by alternative intensive systems of legume relays e.g. maize relayed with cowpea, *cajanus,* mucuna and *senna*. The role of the legumes would be to increase the level of organic inputs to 12 tonnes DM ha^{-1} $year^{-1}$ and hence their equilibrium SOM level will be 50% higher. Due to the predominant drier conditions (compared to the humid forest zone), the increase in SOM after 20 years has been found to be by 0.33% C only in the top 15 cm of the soil. This was in the order of maize/mucuna ≥ maize/Cajanus > maize/cotton with Senna mulch > maize/cotton (base line scenario) > maize/cowpea.

3.4.4 Legume Cover Crop Fallows

Besides rotation and intercropping, legumes may also be introduced into the cropping systems to replace natural fallows. From studies conducted in West Africa, Tian et al. (2000) reported higher maize grain yields (1.3 t ha^{-1}) as a result of legume cover crop. This was attributed to a contribution of 50–60 kg ha^{-1} of fertilizer for maize from Pueraria fallow. In other studies, increased productivity has been attributed to positive contribution to soil physical properties. For example, from a study in Southern Nigeria, Salako et al. (2000) found that Pueraria residue reduced more soil loss due to erosion than cowpea residue. This was partly attributed to increased water infiltration in the Pueraria fallow. Other benefits of the fallow may include improved soil microbial properties and suppression of weeds and pests. However, these benefits may be constrained by lack of arable land for fallowing, or competition with food crops that are relay cropped, intercropped or rotated.

3.4.5 *Intensive Production of Leguminous Fodder or Fodder Banks*

Legumes play an important role in the integrated crop-livestock systems. It is generally known that the highest stimulation of fodder intake by leguminous supplementation is obtained if the N: DOM (digestible organic matter) ratio of roughage is less than 0.017. A particularity of ruminant feeding is that low DOM content of fodder cannot be compensated for by eating more; on the contrary, the lower the content, the lower the intake. This is especially true in the savannah than in the Sahel where legumes are likely to stimulate fodder intake and animal production in the dry season, when the amount of fodder is insufficient. In places where there is abundant space, integration of fodder banks may be a better option for pastoral animal production.

In intensive production systems, it may be more recommendable to integrate high N-content and relatively low production legumes e.g. special variety of cowpeas. These systems may be more affordable than grass-based fodders (sweet sorghum, maize etc), which only do well with application of N, P and K fertilizers. This conclusion is based on the argument that, the cost of 1 kg of fertilizer P is considerably less than (8 times) the cost of 1 kg fertilizer N (based on the fact that the legumes fix on average an extra 4 kg N for each kg of fertilizer P in the legume fodder system). This supposes a 50% recovery of fertilizer N producing in fodder grasses. In this case, legumes become a more favourable option for livestock intensification.

3.5 Future Research Priorities

This review has elaborated on the need to integrate legumes in agroecosystems to enhance sustainable cropping systems in SSA through improvement of acquisition and utilization of P from sparingly available sources. It is generally agreed that to promote nutrient uptake and increase yields in low nutrient environments, a multi-pronged approach is required. Such an approach should include selection and breeding for useful nutrition biological traits; genetic manipulation of cellular processes (uptake and assimilation); of enhancing and optimizing growth and productivity. Research should investigate factors affecting P uptake, especially through exploration of a large soil volume (proliferation), factors for changes in pH, the release of chelating agents and phosphate enzymes. Equally of priority is research on intra-specific strategies involving use of land races, farmer varieties and wild relatives in the primary gene pool.

Overall in the research priorities is the integration of physiological research with genetic improvement. Also to better understand the role of the root cell wall in plant nutrition, it is necessary to test more crops to see if other crops have similar strategies to acquire P from low P soils. A major shortcoming of present knowledge is inadequate quantification of the various adaptive mechanisms for acquiring P, so that it can enable calculation of by how much a proposal manipulation will enhance P acquisition and estimate returns on research investments.

Recent advances in the understanding of genetic control of nutrient acquisition mechanisms has opened increased unrealized possibilities for favorable genetic manipulation of the mechanisms through standard plant breeding procedures or through new opportunities in the field of molecular biology. Genetic manipulation of root exudates to enhance P uptake is a promising priority research agenda since many candidate exudates have been identified and their metabolic pathways and genetic control understood. Besides there are large differences among and within plant species (inter and intraspecific variation) in the type and amount of different compounds exuded (Hale et al. 1971). This enhances the possibility of manipulating root exudates to form rhizosphere microorganisms beneficial to P mobilization.

Other unexploited areas include adaptation in low P environments through acquisition of large thin leaves, long main stems and a low leaf to stem weight ratio (Nakano et al. 2002). A major focus of research for low nutrient environments will be on the economy of P availability and accumulation by legumes. Therefore, approaches to enhance recovery of soil P by legumes will likely be a high priority research to increase productivity of whole cropping systems in low-nutrient environments.

Hash et al. (2002) have shown that there are prospects for using conventional techniques and molecular biological tools to enhance performance of "orphan" crop plants on soils low in available phosphorus. Use of tools for marker-facilitated mapping and manipulation of major genes as components of integrated soil fertility management systems can contribute to ability of legume crops like chick pea (*Cicer arietinum*,) groundnuts (*Arachis hypogea L.*); pigeon pea (*Cajanus cajan* L.); and cereals like sorghum [(*Sorghum bicolor* L.) and pearl millet (*Pennisetum glanium* L.) R. Br] to acquire P from sources with limited availability.

Further research focus in the microbial solubilization of Fe and Al phosphates, as well as mobilization of the phosphate reserves in the soil will augment the role of phosphate solubilizing micro-organisms (PSMs) in agriculture. Such focus should include the detailed understanding of the biochemical and molecular mechanisms of solubilization. The understanding of the genetic basis of P solubilization could help in transforming more rhizosphere competent bacteria into PSMs.

References

Adu-Gyamfi JJ, Ishikawa S, Nakamura T, Nakano H (2002) Genotypic variability and physiological characteristics of crop plants adapted low-nutrient environment. In Food Security in Nutrient Stressed Environments: Exploiting Plants' Genetic Capabilities. Adu-Gyamfi JJ (ed.) pp 42–43. Summary and Recommendations of an International Workshop, September 1999, ICRISAT, Patancheru, India

Ae N, Arihara J, Okada K (1991a) Phosphorus uptake mechanisms for pigeonpea grown in Alfisol and Vertisols. In: Johansen C, Lee KK, Sahrawat KL (eds.) Phosphorus nutrition of grain legumes in the semi-arid tropics. International Crops Research Institute for Semi Arid Tropics (ICRISAT), Patancheru, AP 502 324, India, PP 91–98

Ae N, Arihara J, Okada K (1991b) Phosphorus response of chickpea and evaluation of phosphorus availability in Indian Alfisols and Vertisols. In: Johansen C, Lee KK, Sahrawat KL (eds.)

Phosphorus nutrition of grain legumes in the semi-arid tropics. International Crops Research Institute for Semi Arid Tropics (ICRISAT), Patancheru, AP 502 324, India, pp 33–41

Ae N, Arihara J, Okada K, Yoshihara T, Johansen C (1990) Phosphorus uptake by pigeon pea and its role in cropping systems of the Indian subcontinent. Science 248:477–480

AHI (1997) Africa highland initiative, phase 1 report ICRAF, Nairobi

Ahloowalia BS, Kumarasinghe KS, Sigurbjoernsson B, Maluszynski M (1995) Genotype selection for improved phosphorus utilization in crop plants. In: Johansen C, Lee KK, Sharma KK, Subbarao GV, Kueneman EA (eds.). Genetic manipulation of crop plants to enhance integrated nutrient management in cropping systems. Phosphorous: Proceeding of an FAO-ICRISAT expert consultancy workshop, 15–18 March 1994, ICRISAT Asia Center Patancheru, India, pp 49–74

Al-Abbas AH, Barber SA (1964) A soil test for phosphorus based upon fractionation of soil phosphorus: I. Correlation of soil phosphorus fractions with plant -available phosphorus. Soil Soc Am Proc 28:218–221

Bagayoko M, Buerkert A, Lung G, Bationo A, Romheld V (2000) Cereal/legume rotation effects on cereal growth in Sudano-Sahelian West Africa: soil mineral nitrogen, mycorrhizae and nematodes. Plant Soil 218:103–116

Barber SA (1979) Growth requirement for nutrients in relation to demand at the root interface. In: Harley JL, Scott-Russell R (eds.) The soil root interface. Academic, London, pp 5–20

Bar-Yosef B (1991) Root excretions and their environmental effects: influence on availability of phosphorous. In: Waisel Y, Eshal A, Kafkafi U (eds.) Plant roots, the hidden half. Marcel Dekker, New York, pp 529–558

Bationo A, Ntare BR (2000) Rotation and nitrogen fertilizer effects on pearl millet, cowpea and groundnut yield and soil chemical properties in a sandy soil in the semi-arid tropics, West Africa. J Agric Sci 134:277–284

Bationo A, Kumar KA (2002) Phosphorus use efficiency as related to sources of P fertilizers, rainfall, soil, crop management, and genotypes in the West African semi-arid tropics. In: Adu-Gyamfi JJ (ed.) Food security in nutrient-stressed environments: exploiting plants genetic capabilities. Kluwer Academic, Dorstrech/Boston/London, pp 145–154

Breman H, van Reuler H (2002) Legumes: when and where an option? No panacea for poor tropical West African soils and expensive fertilizers. In: Vanlauwe B, Diels J, Sanginga N and Merckx R (eds.) Integrated Plant Nutrient Management in Sub-Saharan Africa, Wallingford, UK: CABI

Buresh RJ, Smithson PC, Hellums DT (1997) Building phosphorus capital in Africa. In: Buresh RJ, Sanchez PA, Calhoun F (eds.). Replenishing soil fertility in Africa. SSSA Special Publication 51, Madison, pp 111–151, SSSA and ASA, Madison, WI

Carsky RJ, Oyawole B, Tian G (1999) Integrated soil management for the savanna zone of West Africa: legume rotation and fertilizer N. Nutr Cycl Agroecosyst 55:95–105

Carsky RJ, Becker M, Hauser S (2001) Mucuna cover crop fallow systems: potential and limitations. In: Tian et al. (ed.). Sustaining soil fertility in West Africa. SSSA special publication 58, Madison

TSBF – CIAT (2001) The Annual Report of the soil water and nutrient management programme, TSBF – CIAT, Nairobi, Oct 2001, 125pp

Cleaver KM, Schreiber GA (1994) Reversing the spiral, the population, agriculture and environment nexus in Sub-Saharan Africa. World Bank, Washington, DC

Deckers J (1993) Soil fertility and environmental problems in different ecological zone of the development countries in Sub-Saharan Africa. In: Van Reuler H, Prins WH (eds.) The role of plant nutrient and sustainable food production in Sub-Saharan Africa. Vereninging van Kunstmest Producenten, Leidschedam, pp 37–52

Fujita K, Okada M, Lei K, Ito J, Ohkura K, Adu-Gyamfi JJ, Mohapatra PK (2003) Effect of P-deficiency on photoassimilate partitioning and rhythmic changes in fruit and stem diameter of tomato (Lycopersicon esculentum) during fruit growth. J Exp Bot 54:2519–2528

FURP (1994) Fertiliser use recommendation, 1-2-3. KARI, FURP, Nairobi

Gerdemann JW (1975) Vascular– arbuscular mycorrhizae. In: Torrey JG, Clarkson DT (eds.) The development and function of roots. Academic, New York, pp 575–591

Gardner W, Parbery D, Barber D (1981) Proteoid root morphology and function in Lupinus albus. Plant Soil 60: 143–147

Graham JH, Leonard RT, Menge JA (1981) Membrane mediated decrease in root exudation responsible for phosphorus inhibition of vesicular arbuscular mycorrhizae formation. Plant Physiol 68:548–552

Greenland DJ (1994) Soil science and sustainable land management. In: Syers JK, Rmmers DL (eds.) Soil science and sustainable land management in the tropics. CAB international, Wallingford, pp 1–15

Gyaneshwar P, Naresh KG, Parekh LJ (1998) Effect of buffering on the rock phosphate-solubilizing ability of microorganisms. World J Microbiol Biotechnol 14:669–673

Gyaneshwar P, Paresh LJ, Archana G, Pook PS, Hutson MD (1999) Involvement of a phosphorus starvation indivisible glucose dehydrogenese in soil phosphere solubilization by Enterobacter asburiae. FEMS Microbiol Lett 17:223–229

Hale N, Foy C, Shay F (1971) Factors affecting root exudation. Advan Agron 23:89–10

Hartemink AE, Wood AW (1998) Sustainable land management in the tropics. The case of sugarcane plantations 16th world congress of soil science. ISSS, Montpellier, p 7. (Symposium no 14 papers no 1539)

Hash CT, Schaffert RE, Peacock JM (2002) Prospects for using conventional techniques and molecular biological tools to enhance performance of "orphan" crop plants on soils low in available phosphorus. In: Adu-Gyamf JJ (ed.) Food security in nutrient stressed environments: exploiting plants genetic capabilities. Academic, Dordrech/Boston/London, pp 25–36

Henao J, Baanante C (1999) Estimating rates of nutrient depletion in soils of agricultural lands of Africa. International Fertilizer Development Center (IFDC), Technical Bulletin, 48, pp 0–76

Irizarry H, Rivera E (1983) Nutrient uptake and dry matter accumulation by intensively managed pigeon peas grown on a corozal clay, an ultisol. J Agric Sci (P R) 67:188–196

Izac AMN (1993) Ecological economics of investing in natural resource capital in Africa. In: Buresh RJ, Sanchez PA, Calhoun F (eds.). Replenishing soil fertility in Africa. SSSA, Special Publication 51, Madison, pp 237–251

Khasawneh FE, Doll EC (1978) The use of phosphorous rock for direct application of soils. Adv Agron 30:159–206

Koala S, Sims JR, El-Altar H, El-Halfaw M (1988) Phosphorus deficiency in the semi-arid tropics and implications for grain legume production. In: Summerfield RJ (ed.) World crops: cool season legumes. Kluwer, London, pp 205–216

Lyasse O, Tossah, BK, Vanlanme B, Dials J, Sanginga N, Merckx R (2002) Option for increasing availability for low reactive phosphate rock. In: Vanlauwe B, Diels J., N. Sanginga, Merckx R (eds.) Integrated plant nutrient management in Sub-Saharan Africa: from concept to practice, pp 225–237

Muza L, Mugwira L, Muzezewe J, Nyalanda L (1997) Soil fertility bibliography for maize based cropping systems in Zimbabwe. Department of Research and Specialist Services, Ministry of Agriculture and Crop Science Department, University of Zimbabwe, Harare

Nahas E (1996) Factors determining rock phosphere solubilization by micro-organisms. World J Microbiol Biotechnol 112:567–572

Nakano H, Adu-Gyamfi JJ, Nakamura T (2002) Genetic adaptation of crop plants to low-nutrient environments: morphological and ecophysiological characteristics of adaptation. In: Jr Adu-Gyamfied JJ (ed.) Food security in nutrient-stressed environments: exploiting plants genetic capabilities. Kluwer Academic, Dordrecht/Boston/London, pp 59–66

Nandwa SM (2001) Soil organic carbon (SOC) management for sustainable productivity of cropping and agro-forestry systems in Eastern and Southern Africa. Nutr Cycl Agroecosyst 61: 143–158

Nziguheba G, Merckx R, Palm C, Rao M (2000) Organic residues affect phosphorous availability and maize yields in a Nitisol of Western Kenya. Biol Fertile Soils 32:328–339, pp 117–128

Place F, Mwanze S, Kwesiga F (1994) A cost benefit analysis of improved fallows in Eastern province of Zambia, ICRAF, Nairobi. Kenya. Proteoid root morphology and function in Lupinus albus. Plant Soil 60:143–147

Rajan SSS, Watkinson JH, Sinelair AG (1996) Phosphate rocks for direct applications in soils. Adv Agron 57:77–159

Ribet J, Drevon JJ (1995) Increase in permeability to oxygen diffusion and in oxygen uptake of soybean nodules under limiting phosphorus nutrition. Physiol Plant 94:298–304

Salako FK, Kirchohofi G, Tian G (2000) Conservation of tropical Alfisol using legume cover crop intercropping and residue mulching. In: Proceeding of tillage at the threshold of the of the 21st century: looking ahead, fifth conference of the international soil tillage research organization (ISTRO), Fortworth, TX, 27 July 2000. Science 248:477–480

Sanchez PA, Shepherd KD, Soule MJ, Place FM, Buresh RJ, Anne-Marie N, Izac A, Mokwunye U, Kwesiga FR, Ndiritu CG, Woomer PL (1997) Soil fertility replenishment in Africa: An investment in natural resource capital. In: Buresh RJ, Sanchez PA, Calhoun F (eds.) Replenishing soil fertility in Africa. SSSA Special Publication No. 51. Madison, Wisconsin: Soil Science Society of America, pp 1–46

Sanginga N, Lyasse O, Singh BB (2000) Phosphorus use efficiency and nitrogen balance of cowpea breeding lines in low P soil of the derived savanna zone in West Africa. Plant Soil 220: 119–128

Sinclair TR, Vadez V (2002) Physiological traits for crop yield improvement in low N and P environments. In: AduGyamfi JJ (ed.) Food security in nutrient stressed environments: exploiting plants genetic capabilities. Kluwer Academic, Dordrecht/Boston/London, pp 9–23

Smith SE, Gianinazzi-Pearson V (1988) Physiological interactions between symbiots in vesicular-arbuscular mycorrhizal plants. Annu Rev Plant Physiol Plant Mol 39:221–244

Snapp SS, Mafongoya PL, Waddington S (1998) Organic matter technologies for integrated nutrient management in smallholder cropping systems of Southern Africa. Agric Ecosys Environ 71:185–200

Swift MJ, Anderson JM (1993) Biodiversity and ecosystem function in agricultural systems. Pages 15–41 In: Schulze ED and Mooney HA (eds.) Biodiversity and ecosystem function. Springer-Verlag, Berlin, Germany

Tandon HLS (1987) Phosphorus research and agricultural production in India. Fertilizer, development organization, New Delhi, India, 160 pp

Thompson JP (1991) Improving the mycorrizal addition of the soil through cultural practices and effects on growth and phosphorus uptake by plants. In: Johansen C, Lee KK, Sarawal KL (eds.) Phosphorus nutrition of grain legumes in the semi-arid tropics pantancheru, A. P. 502/324. International Crops Research Institute for the Semi Arid Tropics, India, pp 117–137

Tian G, Kolawole GO, Ishida F, Carsky RJ (2000) On-farm study of puerania cores crop fallow system in the desired savanna of West Africa. In: RJ Casky et al. (eds.) cover crop for the natural resource management in West Africa. Proceedings of a workshop organized by IITA and CIEPA, Cotonou republic of Benin, 26–29 Oct 1999, pp 185–193

Tian G, Hauser Kautica LS, Ishida F, Chianu JN (2001) Purenia cover crop fallow systems: benefits and applicability. In: Tian G et al (eds.) Sustaining soil fertility in West Africa. SSSA Special Publication, Madison

Van Straaten P (2002) Rocks for crops: agrominerals of Sub-Saharan Africa. ICRAF, Nairobi, p 338

Yan Y, Lynch JP, Beebe SE (1995) Genetic variation for phosphorus efficiency of common bean in contrasting soil types: I. Vegetative response. Crop Sci 35:108–1093

Chapter 4
Legume Based Cropping and Soil Biodiversity Dynamics

N.K. Karanja, J.M. Kimenju, A.O. Esilaba, J. Jefwa, and F. Ayuke

Abstract The soil is home to an enormous diversity of organisms, many of which are beneficial, while a small and equally diverse proportion contains harmful organisms. Legume cropping and husbandry practices impact negatively or positively on diversity. These changes in biodiversity may be quantitative and/or qualitative. Sustainable management of diversity in soils involves an ecosystem approach which encourages the establishment of equilibria between the organisms in the soil. It is evident that the diversity of beneficial organisms in the soil can be harnessed to improve and sustain crop productivity with minimal external inputs. Legumes will continue to play a central role in the supply of plant nutrients and in the establishment of a self-regulating ecosystem which is based on a diverse range of organisms.

N.K. Karanja (✉) • F. Ayuke
Department of Land Resource Management and Agricultural Technology, University of Nairobi, P.O. Box 30197-00100, Nairobi, Kenya
e-mail: nancy.karanja@cgiar.org; fredrick.ayuke@yahoo.com

J.M. Kimenju
Department of Plant Science and Crop Protection, University of Nairobi, P.O. Box 30197-00100, Nairobi, Kenya

A.O. Esilaba
Kenya Agricultural Research Institute (KARI), Headquarters, P.O. Box 57811-00200, City Square, Nairobi, Kenya
e-mail: aesilaba@gmail.com

J. Jefwa
CIAT-Tropical Soil Biology and Fertility, P.O Box 30677-00100, Nairobi, Kenya
e-mail: j.jefwa@cgiar.org

A. Bationo et al. (eds.), *Fighting Poverty in Sub-Saharan Africa: The Multiple Roles of Legumes in Integrated Soil Fertility Management*,
DOI 10.1007/978-94-007-1536-3_4, © Springer Science+Business Media B.V. 2011

4.1 Introduction

Soil biodiversity refers to the variety of different living organisms and their behaviour patterns (plants, animals' micro-organisms and their genes, their interactions) within the soil environment (UNEP 2001). It may be defined as the variety and variability of living organisms and the ecosystems in which they occur (Campbell and Puri 2002). This diversity may further be expressed in terms of differences in taxonomic, functional, ecological and genetic differences. The diversity in soils is several orders of magnitude higher than that above ground (Heywood 1995) and plays a critical role in maintaining soil health and quality. It has an essential role in ensuring that soils can be managed for pollution abatement, carbon sequestration, flood control, land remediation, soil restoration and waste disposal.

Land management strategies such as cropping systems change soil habitats and thus the soil. For example improved soil cover as a result of a specific cropping pattern would affect soil biodiversity dynamics by increased presence of soil organisms. Leguminous crops are widely cultivated under varying ecological and cropping systems in Africa. Cropping systems are dominated by cereal-legume mixtures with legumes playing a pivotal role, particularly in the form of nitrogen supply to cereals and grasses. Legume-based cropping systems support enormous diversity of soil microbial communities that may impact on the sustainability of agricultural ecosystems (Kloepper et al. 1992; Kandji et al. 2001). A legume based cropping system influences soil biodiversity through its effect on nutrient cycling, crop diseases and pest damage. The paper aims to enhance scientific understanding of the implications of legume integration on the diversity of soil organisms and make suggestions with regards to sustainable soil management practices. Details of effect of selected legume systems on the dynamics and diversity of soil biota are presented in a series of sections in this chapter.

4.2 Biological Nitrogen Fixation

Biological nitrogen fixation is the ability of living organisms to convert the inert dinitrogen gas of the atmosphere (N_2) into nitrogen-containing organic compounds, such as ammonia (NH_3) and nitrate (NO_3^-), through asymbiotic, associative or symbiotic processes. Nitrogen thus fixed becomes available to all forms of life through the nitrogen cycle. It provides about two thirds of the nitrogen, which the biosphere needs. The contribution of legume N fixation to the N-economy of any ecosystem is mediated by: the efficiency of the N_2 fixing system; the contribution of BNF to the soil N pool; and the total amount of N_2 fixed that actually is recycled by human practices and animal manure into the system (UNEP 2001).

Several opportunities to enhance BNF inputs are available across different agroecosystems and socio-economic conditions, *inter alia*: through altering the number of effective symbiotic or associated organisms in the system (inoculation); screening and selection of the appropriate legume crop (selecting high BNF species well adapted to environmental conditions); and management practices that enhance N_2

fixation and recycling of net N_2 inputs into the cropping system (rotation, green manure application, no-tillage) and strategic use of legumes.

4.3 Potential of Symbiotic Legume N_2 Fixing System

Legumes through BNF have the potential to contribute to the improvement and sustainability of soil biodiversity. BNF is known to occur in legumes and has been a significant subject of international research. This followed the stimulus given when it became possible to study nitrogenase in cell-free preparations during the early 1960s. Later, the genes controlling the expression of nitrogen fixation became available for laboratory study and a new burst of research activity was initiated. Since the 1960s, improvements in agricultural technology became evident and perhaps the main practical impact of BNF has been the provision of effective *Rhizobium* cultures for crop and pasture legumes, whereby the two forms of symbiotic associations, have resulted in the development of structures called root nodules where the dinitrogen fixation process takes place.

4.4 Diversity of Legume Nodulating Rhizobia in Soil

Initial classification of rhizobia into 'species' was made purely on phenotypic grounds and largely on the ability of the legumes to nodulate particular legumes, which gave rise to the concept of "cross-inoculation" groups. A cross-inoculation group can be defined as a "group of legume host species nodulated specifically by one set of rhizobial species, and not by any rhizobial species that could induce nodules on legumes not belonging to that cross-inoculation group. A useful concept, when considering the nodulation range of both rhizobia and their legume hosts, is that of host specificity or promiscuity.

Recently, phylogenetic analysis which uses the 16S rRNA gene, has become the standard for the classification of bacteria, while sufficient sequences had become available in the past 7 years to allow this method to be usefully applied to the rhizobia. This new classification is independent of phenotypic traits and has confirmed a number of taxonomic divisions. The four main genera, namely *Azorhizobium*, *Bradyrhizobium*, *Mesorhizobium* and *Rhizobium*, are clearly distinguished, as are many of the species that were previously recognised amongst the fast-growing rhizobia. Using PCR-RFLP of the 16S rRNA gene and sequence analysis of a 230 bp fragment of this gene, five bacterial genera namely *Agrobacterium, Bradyrhizobium, Mesorhizobium, Rhizobium* and *Sinorhizobium*, have been identified among root-nodule isolates from *Acacia* spp., *Sesbania sesban, Prosopis juliflora, Macroptilium atropurpureum, Phaseoulus vulgaris* and *Vigna unguiculata* which are grown in diverse soils in Kenya (Odee et al. 2002).

A recent study on the types of rhizobia that nodulate 17 *Crotalaria* spp. in Senegal has led to the identification of a new genus – *Methylobacterium* (Sy et al. 2001).

Table 4.1 Correlation coefficients (r^2) of abiotic factors and population sizes of indigenous *Rhizobium* and *Bradyrhizobium spp*

	Abiotic factors[a]				
Rhizobial species	CEC	Ca	Mg	K	Sum
R. loti	0.58*	0.43	0.61**	0.81**	0.51*
R. leg. Bv. Viceae	0.99**	0.74**	0.84**	0.85**	0.79**
R. leg bv. Trifolii	0.38	0.77**	0.80***	0.70**	0.79***
R. leg. Bv. Phaseoli	−0.07	−0.14	−0.15	−0.29	0.27
Bradyrhizobium sp.	−0.22	−0.17	−0.15	−0.19	−0.15

$*p<0.05$; $**p<0.01$; $***+p=0.001$
[a]Probability level of relationship

Samba et al. (1999) reported a total of 117 strains of both slow and fast growing rhizobia from root nodules of nine *Crotalaria* species in Senegal. Some *Crotalaria* spp. were effectively nodulated by fast growing strains, while the slow growing strains nodulated *Crotalaria*, as well as a host of other legumes namely *Fanherbia albida* and *Indingofera microcarpa*. Conversely, none of the two strain types nodulated either *Sesbana rostrata* or *Acacia raddiana*. These two species are normally nodulated by fast-growing rhizobia, which belong to the genus *Azorhizobuium, Mesorhizobium* and *Sinorhizobium*. The introduction of exotic species sometimes requires an introduction of its associated microbes. For example, Galiana et al. (1998) in Senegal, found positive effects of inoculation on *Acacia mangium* tree growth, which were inoculated with rhizobium strains that were isolated from its natural distribution in Australia. Nodulation also varied across sites with test species frequently showing higher nodulation ability in native soils. *Sesbania sesban* was the most prolific nodulating species, while *Acacia tortilis* was erratic in nodulation.

A similar study was carried out to characterize the rhizobia isolates, which are associated with woody legumes. Using over 480 rhizobial isolates, phenotypic characterization grouped the isolates into *Rhizobium* and *Bradyrhizobium* species (Table 4.1). Earlier reports had described rhizobia which are associated with woody legumes as either of the slow growing type or the cowpea miscellany. Later, it was shown that the tropic soil constitute a large reservoir of *bona fide* and potentially new species within the genera *Azorhizobium, Bradyrhizobium, Rhizobium* and *Sinorhizobium* (Odee et al. 1997).

Another study, which used the 16S RNA gene sequencing analysis (Bala 1999), showed that the rhizobia nodulating *C. calothyrsus, G. sepium, L. leucocephala* and *S. sesban*, belong to the genera *Agrobacterium, Mesorhizobium, Rhizobium* and *Sinorhizobium*. The specificity or promiscuity of rhizobia or bradyrhizobia has been tested for different rhizobia strains. In Egypt, phenotypic diversity was studied among 13 Rhizobium strains, which were selected from a total of 160 Rhizobium isolates from the root nodules of *Leucaena leucocephala* (Swelim et al. 1997). In that study it was found that Rhizobium strains from *Leucaena* could nodulate some *Lucaena* sp. and not others. Some of the *Leucaena* strains were found to nodulate the common bean *Phaseolus vulgaris*, while other *Leucaena* strains were found to nodulate *Glycine max* and *Vigna sinensis*. Likewise, *Sesbania* strains nodulated

Leucaena. This broad host characteristic of rhizobia enables tree and crop legumes to share common diversity of microsymbionts.

The key factors, which are responsible for the variations in population sizes, are extant naturalised legumes, season of sampling and soil fertility management practises, such as the use of rhizobia inoculants. A number of studies, which assess the populations and diversity of legume nodulating rhizobia, have been carried out. For example, Woomer et al. (1997) reported a study where rhizobial populations were determined at 46 sites in 8 African countries. Total population sizes were greatest in highland soils and least in semi–arid lowlands and differed significantly between ecological zones.

4.5 Arbuscular Mycorrhiza Fungi

Arbuscular mycorrhizal fungi (AMF) form a mutually beneficial association with plants and are key to soil processes (Harley and Smith 1983; Hawksworth 1991; Power and Mills 1995; Bull 1996). The AMF are globally widespread and are associated with most plant species (Harley and Smith 1983). Mycorrhizae can effectively bridge the gap across the zone of nutrient depletion that forms near the root, probably through external hyphae or root physiology (Li et al. 1991). They are reported to have the ability to transfer nutrients from one host plant to another (Whittingham and Read 1982; Francis et al. 1986 and Haystead et al. 1988) in a source to sink relationship. The hyphae play an important role in the formation of soil aggregates that are greater than 240 μm diameter (Hamel et al. 1997). AMF may also influence plant biodiversity (van der Heijden et al. 1998), and help to control pests (for example, nematodes) and fungal pathogens (Azcon-Aguilar and Barea 1996), as well as affect the fitness of plants in polluted environments (Hildebrandt et al. 1999). The importance of soil biological processes in soil fertility and their relationship with management practices in the tropics, was emphasized (Sanginga et al. 1992). The benefits of mycorrhizal symbiosis, particularly in legume-based systems, are numerous.

All legumes, N_2-fixing and non N_2-fixing, in farming systems, require adequate P in the soil. The process of nodulation is P mediated, with the P content in nodules of AMF inoculated plants generally being greater than that of non-AMF plants (Kawai and Yamamoto 1986). The uptake of K is increased by AMF (Osonubi et al. 1992; Atayese et al. 1993).

4.6 Role of Agroforestry Systems in Dynamics of AMF

The mycorrhizal association has been evaluated predominantly in agroforestry farming systems in Africa. The inclusion of trees in cropping systems, particularly woody legumes, is an efficient way to recycle nutrients and replenish soil organic matter (Atayese et al. 1993). AMF are known to improve the plant uptake of immobile nutrients, particularly P in soils that are infertile (Stribley et al. 1980; Reddell and Warren 1986) and NH_4^+ (Chambers et al. 1980a, b).

Mycorrhizal association is recognized as vital for cassava production (Sieverding 1991). Cassava is cultivated in a wide variety of infertile soils, usually without the application of fertilizers (Howeler 1985) while adequate P and K are vital for cassava tuber production (Osonubi et al. 1995).

Studies have also been carried out on the distribution and dynamics of mycorrhizae in legume based systems in Nigeria, found prolific arbuscular mycorrhizal colonization in the roots but no nodules (Tomlinson et al. 1995). Shepherd et al. 1996 found forest and grassland soils to have narrower species distribution than most farm soils indicating some degree of ecosystem adaptation.

Leguminous species differ in their effects on AMF species. In the agroforestry experiment with the two *Sesbania* species, agroforestry decreased AMF species diversity. The effects of *S. sesban* and *S. macrantha* were similar, which suggests that mycorrhizal associations by plant species, are a heritable trait. The decrease in species diversity in the *Sesbania* agroforestry systems, compared to maize monocrop systems, was linked to root turnover, with the turnover higher in the annual maize crop than with the two tree leguminous species. The *Sesbania* species were also left longer as improved fallow while the maize monocrop system was left bare after maize harvest. These results were confirmed by observations that were made by Guardarrama and Sanchez (1999) in a tropical rainforest where they linked high spore density and species diversity in the dry season to root decay. The contrasting observations made in *Gliricidia*/maize intercrop and *Sesbania* spp/maize intercrop systems, were linked to the root biomass. The root biomass in the two *Sesbania* was higher than in *Gliricidia* as evidenced by the large quantities of root hairs in the two *Sesbania* spp., compared to *Gliricidia*. Baylis (1970) observed that plants with denser root hairs, associate less with mycorrhizal fungi, while Ocampo et al. (1980) demonstrated less colonization owing to high root biomass and vice versa. The two agroforestry experiments showed that root density, mulch application and root longevity are important characteristics that would influence mycorrhizal associations.

It is evident that the mycorrhizal symbiosis, in combination with legumes, can be harnessed to improve crop productivity, maximize root functions and also reduce fertilizer input. However, there is still a need to establish the distribution and functions of the AMF species in different habitats and different land use systems, in order to facilitate inoculation programs. With the improved methods and technologies in utilization, approaches to studying AMF should be streamlined in order to derive maximum benefits from the association.

4.7 Diversity of Macrofauna

Soil macrofauna, in association with other microorganisms, are important agents in the generation and maintenance of the biological, chemical and physical characteristics. They enhance biodegradation and humification of organic residues in several ways: (1) by breaking down organic residues and increasing the surface area for microbial activity; (2) by producing enzymes, which break down complex bio-molecules into

Table 4.2 The effect of organic materials on abundance of soil macrofauna in soils at Maseno, Western Kenya

Treatment	Earthworm Number m^{-2}	Termite	Other macrofauna
Control	99(9.5) c	299(14.0) b	43(6.3) d
Fertilizer	132(11.0) c	348(16.0) b	90(9.4) c
S. spectabilis	572(22.5) a	737(25.4) a	391(15.5) a
T. diversifolia	339(18.5) b	652(21.4) ab	309(14.0) b
SED	(1.0)	(6.3)	(1.1)

Means followed by the same letter in a column are not significantly different at $p<0.05$. Values in parenthesis are square root transformed

Table 4.3 Effect of tree fallow on mean earthworm and microarthropod populations during a wet season in Nigeria

Treatment	Earthworm Numbers m^{-2}	Other macrofauna
Contour cropping	58b	21900c
Natural bushfallow	194b	32600ab
L. leucocephala	118a	27900bc
A. leptocarpa	100a	45000a
S. siamea	142a	45300a

Source: Tian et al. (2000)
Analysis of variance based on log (X+1) transformed data. Values in a column followed by the same letter are not significantly different at $p=0.05$

simple compounds to form humus, and (3) by improving the soil environment for microbial growth and soil-plant interactions (Lal 1988). They also mediate the synthesis and breakdown of soil organic matter (humus) and influence soil water availability to plants by modification of soil structure and water regimes (Blanchart et al. 1999). Their effect on physical properties may be through the construction of mounds, nests, galleries and surface sheeting (Mando et al. 1996; Asawalam et al. 1999; Sarr et al. 2001). Soil fauna communities show a variety of reactions to changes that are induced by land management. Practices such as mechanized land clearing and burning, continuous tillage, monoculture and the use of agrochemicals, adversely affect invertebrate fauna population and/-or diversity of their communities (Lal 1988).

Restoration of micro-arthropod activities of degraded soils depends on the management strategies that are adopted. Agroforestry systems, such as alley cropping, natural fallows (bush fallows), planted fallows and biomass transfer systems can restore activities of organisms such as earthworms, termites, ants and other microarthropods (Ayuke 2000; Tian et al. 2000). Ayuke (2000) found that organic residues from *Senna spectabilis* and *Tithonia diversifolia* increased the population of earthworms by 400 and 240% over no input control respectively, while termites increased by 150 and 120% when the two different organic materials were added to the soil (Table 4.2). Tian et al. (2000) similarly found higher earthworm and microarthropod population under planted fallows than under continuous cropping systems and attributed this to higher litter fall, lower temperature and higher soil moisture (Table 4.3).

4.8 Role of Legumes on Dynamics of Soil Borne-Pathogens and Pests

Legumes affect dynamics of soil borne diseases through cultural factors, such as residues, organic matter and cover as well variety. Root rot diseases are among the major constraint for the production of pulse crops in Africa (Table 4.4). The diseases are caused by one or a combination of *Pythium* Spp, *Fusarium solani* s.sp. *phaseoli, Rhizoctonia solani, Aphanomyces eutichus, macrophomina phaseolina* and *Sclerotium rolfsii* (Rusuku et al. 1997; Allen et al. 1996). Depending on specific cultural practice, legumes may lead to a build up or reduction in the occurrence of soil-borne pathogens and pests.

4.9 Diversity of Nematodes in Legume Based Cropping Systems

In areas under intensive crop cultivation, build-up of pests and diseases, particularly those that are soil borne, is the norm rather than the exception. Leguminous plants are generally susceptible to root-knot nematodes, which suggest that farm-

Table 4.4 Soil borne diseases associated with major pulse crop in Africa

Crop	Disease	Causal agent	Reference
Common bean *Phaseoli vulgaris*	Root rot	*Fusarium solani f.sp phaseoli,*	Kiprop (2001)
		Rhizoctonia solani	Muriungi (1997)
		Phythium spp.	Nzungize et al. (2011)
	Texas root rot	*Aphanomyces eutichus*	Rusuku et al. (1997)
	Black root rot	*Macrophomina phaseolina*	Muriungi (1997)
	Southern blight	*Screrotium rolfsii*	Punja (1988)
Pigeon peas	Fusarium wilt	*Fusarium udum*	Kiprop (2001)
Cajanus cajan	Phytopthora blight	*Phytopthora drechsleri*	Kannaiyan et al. (1981)
Soyabean	Root rot	*Macrophomina phaseolina*	Abd El-Ghany (1998)
Glycine max	Root rot	*Fusarium solani*	Iqbal et al. (2005)
		Pythium spp.	Frazer (1999)
		Phytopthora spp.	Frazer (1999)
	Southern blight	*Sclerotium rolfsii*	Frazer (1999)
	Rhizoctonia rot	*Rhizoctonia solani*	
Chik peas	Fusarium wilt	*Fusarium oxysporum f. sp. Ciceri*	Haware et al. (1990)
	Dry root not	*Rhizoctonia bataticola*	Haware et al. (1990)
	Black root not	*Fusarium solani*	Haware et al. (1990)
	Damping off	*Pythium spp.*	Haware et al. (1990)
	Collar rot	*Sclerotium rolfsii*	Haware et al. (1990)

ers assume a significant risk when they introduce leguminous fallow crops in their farming systems (Cadet et al. 1996). Kibani and Msabaha (1995) have reported that sunhemp (*Crotolaria* spp) increased the population of *Meloidogyne* spp among other nematodes when grown in rotation with maize. Alley cropping with, *Leucena* sp, which is used throughout the tropics as a multipurpose tree, as well as for soil fertility improvement, showed no effect on the populations of parasitic nematodes (Kang et al. 1999). Vincente et al., (1986) found no galls on the roots of *L. leucocephala* and concluded that it was resistant to *Meloidogyne incognita* and *M. javanica.* Similarly, velvetbean (*Mucuna pruriens*), when tested alongside tomato cv. Rutgers and okra cv. Clemson spineless, had only a few galls of *M. arenaria* and *M javanica* but none from either race of *M. incognita* (McSorley et al. 1994). However, when velvetbean was followed in rotation by *Arachis hypogaea* (peanuts), no effect on nematode populations was reported, except the free-living species and *Trichodorus* spp (Moura et al. 1997). *Tephrosia purpurea* (Roxb.) DC, a common green manure plant for soil fertility improvement, was identified as a host of root-knot nematodes, namely *Meloidogyne hapla, M. incognita* and *M. javanica* (Ponte et al. 1992). The reniform nematode, *Rotylenchus reniform,* also showed host preference for *Tephrosia* spp, rather than cacao, when intercropped (Yuen 1979). *Sesbania sesban* is highly susceptible to root-knot nematodes, while its continued cultivation has led to a buildup of nematodes in the soils of west Kenya (Desaeger and Rao 1999). When *Sesbania* spp were grown as rotation fallow, an increase in the population of *M. javanica* and *M. incognita*, coupled by low yields and of french beans following this fallow were recorded. In another study, a 12-month Sesbania fallow resulted in the build-up of root-knot nematode densities to levels where yield of a subsequent bean crop, was reduced by up to 87% (Desaeger and Rao 1999).

A study to determine the population dynamics of plant parasitic nematodes under different cover crops and continuous maize crops was conducted by Deseager and Rao (2000). Crotalaria fallow and continuous maize cultivation reduced the population of root-knot nematodes (*Meloidogyne* spp.), while the population of lesion nematodes (*Pratylenchus* zeae) increased to levels that could limit growth of the two plant species. The low root-knot nematode (*Meloidogyne* spp.) and high lesion (*Pratylenchus* spp.) nematode populations in plots under *Crotalaria*, were not unexpected (Desaeger and Rao 1999). It is now clear that *Crotalaria* is suppressive on root-knot nematodes but act as suitable host for lesion nematodes (Deseager and Rao 2000). Lesion nematodes (*P. zeae* and *P. brachyurus*) are the most important nematode pests of maize in Kenya (Kimenju et al. 1999), therefore, incorporation of Crotalaria into maize cropping systems would be counter productive. Kandji et al. (2001) have shown that nematode communities in different cropping systems including populations of plant parasitic nematodes, were lower under long term natural fallows when compared with improved fallows or continuous maize/bean cultivation. The free-living nematodes were least and most abundant in areas that are under continuous maize/bean cultivation and long-term natural fallows, respectively. The ratio of free-living plant parasitic nematodes was highest, intermediate and lowest under long-term natural fallows and improved fallows, continuous maize/bean cultivation, respectively. The variability in the nematode community structure in different agroecosystems is an indication that cropping systems and land management practices have a significant impact on nematodes (Kandji et al. 2001). The lower proportion of plant parasitic nematodes in longterm natural fallows explains why nematodes are seldom a major problem in traditional farming systems that are characterized by short cultivation phases, which are separated by long periods of bush regrowth. Changes that take place when soil is continuously under cultivation may tilt the natural balance in favour of plant pathogenic nematodes (Brown 1987; Kandji et al. 2001).

4.10 Soil Borne Insect Pests of Leguminous Plants

The most widely cultivated pulses in Africa include common beans, pigeon pea, cowpea, groundnuts, chickpea and soybean (Allen et al. 1996). As shown in Table 4.5, each pulse crop is a host of one or more soilborne insect pests. The majority of pests, such as the beanfly, cutworms, termites and white grubs have broad host ranges and are cosmopolitan in occurrence (Allen et al. 1996; Wightman et al. 1990; Jackai et al. 1990; Lateef and Reed 1990).

4.11 Dynamics of *Striga* in Legume-based Systems

Striga represents one of the most important biological constraints to food production in sub-Saharan Africa (Sauerborn 1991). The greatest damage is caused by *Striga hermonthica* (Del.) Benth and *Striga asiatica* (L.) Kuntze on cereal crops (sorghum, corn, pearl millet, rice, etc.) and *S. gesnerioides*, which is parasitic mainly harmful to legumes (cowpeas, peanuts, etc.). Although *Striga* is distributed in more than 40% of the arable land in Africa, south of the Sahara, however, precise figures on the area that is infested with *Striga* and the magnitude of losses to food production, are not available for most countries. Estimates suggest that losses to cereal production are as high as 311 million US Dollars annually (Sauerborn 1991). *Striga* is most problematic in the drier areas of Africa, where fertility and organic matter content of the soils are low and where cropping intensity is high (Vogt et al. 1991). Parasitic plants that reduce the productivity or quality of agricultural and forest crops are found in eight families. By their parasitic habits, these weeds damage their crop hosts much more severely than non-parasitic weeds and contribute to the low productivity of staple cereal or pulse crops in smallholder systems within sub-Saharan Africa. The family, Scrophulariaceae, includes eight genera of root hemiparasites that damage crops. The most notable, S*triga* and *Alectra,* are weeds of cereal and pulse crops in the dry tropics and sub-tropics, particularly in Africa where they are among the most serious agricultural pests.

Table 4.5 Insect pest of leguminous plants that have atleast one soilborne stage

Crop	Insect pest (scientific name)	Common name	Reference
Cowpea	Ophiomyia phaseoli	Beanfly	Abul-Nasr and Assem (1968)
Vigna unguiculata Walp	*Acheta assimilis*	Crickets	Singh et al. (1990)
	Agrotis ipsilon	Cutworms	Santos and Quindere (1988)
Common bean	*Ophiomyia spp*	Bean stem maggot (beanfly)	Allen et al. (1996)
Phaseolus vulgaris L.	*Phyllophaga spp*	White grub	Cardona and Karel (1990)
	Agrotis ipsilon	Cutworm	CIAT (1994)
	Cerotoma facialis	Bean bettle	–
	Elasmopalpus Lignosellus	Lesser cornstalk borer	Salinas (1976)
Pigeonpea	*Odontotermes spp*	Termites	Lateef and Reed (1990)
***Cajanus** cajan (L) Mellsp*	Microtermes spp	Termites	
	Lachnosterna spp	White grub	Lateef and Reed (1990)
	Adoretus spp	White grub	Lateef and Reed (1990)
	Sphenoptera indica	Stem borer	
	Rivellia angulata	Nodule-damaging fly	Sithananthamet al. (1981)
Groundnut	*Microtermes spp.*	Termites	Wightman et al. (1990)
Arachis hypogaea L.	*Odontotermes spp.*	Termites	Wightman et al. (1990)
	Hilda patruelis	Groundnut planthoper	Weaving (1980)
	Phenacoccus solani	Mealy bugs	Wightman et al. (1990)
	Gonocephalum spp	Wireworms and false warms	Wightman (1989)
	Phyllophaga spp	White grubs	Wightman et al. (1990)
	Agrotis spp.	Cutworms	Wightman et al. (1990)
Soybean	*Ophiomyia spp*	Beanfly	Talekar (1987)
Glycine max	Macrotermes spp and Odontotermes spp.	Termites	Jackai et al. (1990)
	Rivella quadrifasciata	Nodule fly	Jackai et al. (1985)
Chickpea	Ondontotermes spp and Microtermes spp.	Termites	
	Gonocephalum spp.	Soil beetle	
	Agrotis ipsilon (Hufnagel)	Cutworms	

4.11.1 Striga Gesnerioides

Striga gesnerioides (Willd.) Vatke is common throughout the drier regions of Africa in natural vegetation. It is a serous constraint to cowpea production in the dry Sahelian and savanna zones of West and Central Africa and is reported occasionally as affecting tobacco in southern Africa but rarely sweet potato as was reported most recently, in Ethiopia (Parker 1991). Is spite of the extensive host range of this species on members of a number of families, particularly Convolvulaceae and Leguminosae, individual populations tend to exhibit extreme host specificity (Musselman and Parker 1981), which is expressed after parasite germination. Following infection by *S. gesnerioides*, there is a rapid reduction in cowpea leaf area at a stage when the canopy of uninfested plants continues to expand. Leaves of heavily infested plants develop necrotic patches, while the lamina and petiole become desiccated and the leaf is lost from the plant. While the reduction in host photosynthetic capacity is probably contributory, the dependence of the parasites on the cowpea plant for carbon is the main factor, which leads to reduced grain yield (Gouws et al. 1980; Graves et al. 1992).

4.11.2 *Alectra* Species

Some 30 species of the largely African genus, *Alectra*, are recognized, of which there are three parasitize broadleaf crops. *Alectra vogelii* Benth is of greater significance, since it occurs in Africa on a wide range of pulse crops. The distribution of *A. vogelii* is spread throughout semi-arid areas of tropical Africa and sub-tropical southern Africa. *Alectra vogelii* attacks many leguminous crops; cowpea in southern and West Africa and groundnut in East and West Africa, are particularly important hosts. Bambara, common bean, mung bean and tepary are also attacked in parts of the parasite's distribution. It has occasionally been reported from chickpea, soyabean and runner bean; many legume fodder crops, including *Dolichos lablab,* siratro and velvet bean, can also be parasitized (Riches et al. 1992). Pigeon pea appears to be the only widely grown pulse that is not attacked, while host preference varies between different regions. Populations from Mali, Nigeria and Cameroon attack cowpea and groundnut, while those from Botswana and northern Transvaal attack mung bean, in addition to cowpea and groundnut. Populations that were sampled from Kenya, Malawi and eastern Transvaal parasitize bambara, as well as crops that are susceptible in other regions (Riches et al. 1992).

Infestation by *A. vogelii* or *S. gesneriodes* are is a major cause of low yields of cowpea in much of semi-arid Africa. Production of the crop has traditionally been in multiple cropping systems in which low populations of landrace cowpeas are planted in mixtures with cereals. An increase in the importance of parasitic weeds on cowpea during the past 25 years, has been associated in some areas with a change to sole cropping by the introduction of, potentially higher yielding, but highly susceptible varieties, and elsewhere owing to an increase in the area and frequency of cowpea production.

4.12 Conclusion

The soil is home to an enormous diversity of organisms, many of which are beneficial, while a small and equally diverse proportion contains harmful organisms. Legume cropping and husbandry practices impact negatively or positively on diversity. These changes in biodiversity may be quantitative and/or qualitative. Sustainable management of diversity in soils involves an ecosystem approach which encourages the establishment of equilibria between the organisms in the soil. It is evident that the diversity of beneficial organisms in the soil can be harnessed to improve and sustain crop productivity with minimal external inputs. Legumes will continue to play a central role in the supply of plant nutrients and in the establishment of a self-regulating ecosystem which is based on a diverse range of organisms.

References

Abd El-Ghany (Hiffaa) S (1998) Physiological, pathological and biochemical comparative studies on Macrophomina phaseolina isolates from different hosts. Menofia J Agric Res 23(4): 885–906

Abul-Nasr S, Assem M (1968) Studies on the biological processes of the bean fly, *Melanogromyza phaseoli* (Tryon) (Diptera: Agromyzidae). Bulletin de societe entomologique de c Egypte Ser Econ 11:151–159

Allen DJ, Ampafo JKO, Wortmann CS (1996) Pests, diseases and nutritional disorders of the common bean in Africa. CIAT, Cali, 132 pp

Asawalam DO, Osodeke VE, Kamalu OJ, Ugwa IK (1999) Effects of termites on the physical and chemical properties of the acid sandy soils of Southern Nigeria. Commun Soil Sci Plant Anal 30(11–12):1691–1696

Atayese MO, Awotoye OO, Osonubi O, Mulongoy K (1993) Comparison of the influence of hedgerow woody legumes and cassava at the top and base of a hillslope in alley cropping system. Biol Fertil Soils 16:198–204

Ayuke FO (2000) Diversity, abundance and function of soil invertebrate fauna in relation in relation to quality of organic residues. M. Phil thesis of Moi University, Eldoret, Kenya.

Azcon-Aguillar C, Barea JA (1996) Arbuscular mycorrhizas and biological control of soil-borne plant pathogens-an overview of the mechanisms involved. Mycorrhiza 6:457–464

Bala A (1999) Diversity of rhizobia which nodulates fast growing tree legumes in tropical soils. Ph.D thesis, Department of Biological Sciences, Nye College, University of London

Baylis GTS (1970) Root hairs and phycomycetous mycorrhizas in phosphorus-deficient soil. Plant Soil 33: 713

Blanchart E, Albretch A, Alegre J, Duboisset A, Gilot C, Pashanasi B, Lavelle P, Brussaard L (1999) Effects of earthworms on soil structure and physical properties. In: Lavelle P, Brussaard L, Hendrix P (eds.) Earthworm management in tropical agroecosystems. CAB International, Wallingford, pp 149–172

Brown RH (1987) Control strategies in low-value crops. In: Brown RH, Kerry BR (eds.) Principles and practice of nematode control in crops. Academic, London, pp 351–387

Bull AT (1996) Biotechnology for environmental quality: closing the circles. Biodivers Conserv 5:1–25

Cadet P, Duponoss R, Sengor K (1996) Relation between nematodes and Acacia spp. A synthesis of preliminary information in Senegal. Agrofor Abstracts 12:2–6

Campbell C, Puri G (2002) Soil bio diversity. Information and advisory notes no 51. Scotish natural heritage publication, 6 pp
Cardona C, Karel AK (1990) Key insects and other invertebrate pests of beans. In: Singh SR (ed.) Insect pests of tropical food legumes. Wiley, Chichester
Chambers CA, Smith SE, Smith FA (1980a) Effects of ammonium and nitrate ions on mycorrhizal infection, nodulation and growth of *Trifolium subterraneum*. New Phytol 85:47
Chambers CA, Smith SE, Smith FA, Ramsay MD, Nicholas DJD (1980b) Symbiosis of *Trifolium subterraneum* with mycorrhizal fungi and Rhizobia trifolii as affected by ammonium sulpahte and nitrification inhibitors. Soil Biol Biochem 12:93
CIAT, Centro Internacional De Agricultura Tropical (1994) Field pests of beans in Africa: study guide to be used as a supplement to the audiotutorial unit on the same topic. Scientific content: Kwasi Ampofo production: Carmen Llanos, Alexandra Walter, CIAT, Cali, 48 p
Desaeger J, Rao MR (1999) The root-knot nematode (*Meloidogyne* spp.) problem in sesbania fallows and scope for managing it in western Kenya. Agrofor Sys 47:273–278
Deseager J, Rao MR (2000) Parasitic nematode populations in natural fallows and improved cover crops and their effects on subsequent crops in Kenya. Field Crop Res 65:41–56
Frazer NL (1999) One stop mycology. Mycological Research 103, pp 372–384
Francis R, Finlay RD, Read DJ (1986) Vesicular-arbuscular mycorrhiza in natural vegetation systems. IV. Transfer of nutrients in inter- and -intra-specific combinations of host plants. New Phytol 102:103–111
Galiana A, Grahousa GM, Chaumont J, Lesueur D, Prin Y, Mallet B (1998) Improvement of nitrogen fixation in *Acacia mangium* through inoculation with Rhizobium. Agrofor Syst 40: 297–307
Gouws J, Visset JH, Grobbelaar WP (1980) Some aspects of the bidirectional translocation of ^{14}C-labelled metabolites between *Alectra vogelii* Benth. and *Voandzeia subterranea* (L.) Thou. Zeitschrift fiir Pfianzen-physiologie 99:225–33
Graves JD, Press MC, Smith S, Stewart GR (1992) The carbon canopy economy of the association between cowpea and the parasitic angiosperm *Striga gesnerioides*. Plant Cell Environ 15:283–8
Guardarrama P, Sanchez AFJ (1999) Abundance of arbascular mycorrhizal spores in different environment in a tropical rain forest, Veracruz, Mexico. Mycorrhiza 8:267–270
Hamel C, Dalpe Y, Furlan V (1997) Indigenous populations of arbuscular mycorrhizal fungi and soil aggregate stability are major determinants of leek (*Allium porrum* L) response to inoculation with *Glomus intraradices* Schenck & Smith or *Glomus versiforme* (Karsten) Berch. Mycorrhiza 7(4):187–196
Harley JL, Smith SE (1983) Mycorrhizal symbiosis. Academic, London
Haware MP (1990) Fusarium wilt and other important diseases of chickenpea in the Mediterranean area. OptionsM'editerran'eennes. S'erie S'eminaires 9:63–166
Hawksworth DI (1991) The biodiversity of microorganisms and invertebrates: its role in sustainable agriculture. CAB International, Wallingford
Haystead A, Malajczuk N, Grove TS (1988) Underground transfer of nitrogen between pasture plants infected with vesicular-arbuscular mycorrhizal fungi. New Phytol 108:417–423
Heywood VH (1995) Global biodiversity assessment. Cambridge University Press, Cambridge
Hildebrandt U, Kaldorf M, Bothe H (1999) The zinc violet and its colonization by arbuscular mycorrhizal fungi. J Plant Physiol 154:709–717
Howeler RH (1985) Potassium nutrition of cassava. In: Munson RD (ed.) Potassium in agriculrure. American Society of Agronomy, Madison, pp 819–841
Iqbal MJ, Yaegashi S, Ahsan R, Shopinski KL, Lightfoot DA (2005). Root response to Fusarium solani f. sp. glycines: temporal accumulation of transcripts in partially resistant and susceptible soybean. Theor Appl Genet 110:1429–1438
Jackai LEN, Dashiell KE, Shannon DA, Root WR (1985) Soybean production and utilization in subsaharan Africa. In: Shibles R (ed.) World soybean research conference III. Westview Press, Boulder, pp 1193–1202
Jackai LEN, Panizzi AR, Kundu GG, Srivastava KP (1990) Insect pest of soybean in the tropics. In: Singh SR (ed.) Insect pests of tropical food legumes. Wiley, Chichester, pp 91–156

Kandji ST, Ogol CKPO, Albrecht A (2001) Diversity of plant-parasitic nematodes and their relationships with some soil physico-chemical characteristics in improved fallows in western Kenya. Appl Soil Ecol 18(2):143–157

Kang BT, Caveness FE, Tian G, Kolawale GO (1999) Long term alley cropping with four hedgerow species on an alfisol in S.W. Nigeria. Effect on crop performance, soil chemical properties and nematode populations. Nutr Cycl Agroecosyst 54:145–155

Kannaiyan J, Nene YL, Raju TN, Sheila VK (1981) Screening for resistance to Phytophthora blight of pigeon pea. Plant Disease 65:61–62

Kawai Y, Yamamoto Y (1986) Increase in the formation and nitrogen fixation of soybean nodules by vescicular-arbuscular mycorrhiza. Plant Cell Physiol 27:399–405

Kibani TH, Msabaha RP (1995) Studies on the effect of crop rotation on nematode and Fusarium wilt of cotton. Research and training newsletter, Dar es Salaam 10:1–4, 19–21

Kiprop EK (2001) Characterization of Fusarium udum Butler isolates and pigeonpea (*Cajanus cajan* (L) Millsp.) resistance in Kenya. PhD thesis. University of Nairobi, Nairobi, p 235

Kimenju JW, Karanja NK, Macharia I (1999) Plant parasitic nematodes associated with common bean in Kenya and the effect of Melodogyne infection on bean nodulation. Afr Crop Sci J 7(4):503–510

Kloepper JW, Rodriguez-Kabana R, McInvoy JA, Young RA (1992) Rhizosphere bacteria antagonistic to soybean cyst (Heterodera glycines) and root-knot (Meloidogyne incognita) nematodes: identification by fatty acid analysis and frequency of biological control activity. Plant Soil 139:75–84

Lal R (1988) Effects of macrofauna on soil properties in tropical agroecosystems. Agric Ecosyst Environ 24:101–116

Lateef SS, Reed W (1990) Insect pests on pigeon pea. In: Singh SR (ed.) Insect pests of tropical food legumes. Wiley, Chichester

Li X, George E, Marschner H (1991) Phosphorus depletion and Ph decrease at the root-soil and hyphae-soil interfaces of VA mycorrhizal white clover fertilized with ammonium. New Phytol 119:397–404

Mando A, Stroosnijder L, Brussaard L (1996) Effects of termites on infiltration into crusted soil. Geoderma 74:107–113

McSorley R, Dickson DW, Brito JA (1994) Host status of selected tropical rotation crops to four populations of root-knot nematodes. Nematropica 24:45–53

Moura RM, Moura AM, Macedo MEA, Sylva EG (1997) Influence of three crop combinations in nematode populations associated with sugar cane. Nematropica-Brasileira 21:75–83

Muriungi SJ (1997) Bean root rot complex, its management by microbial agents and plant resistance. MSc thesis, University of Nairobi, Kenya

Musselman LJ, Parker C (1981) Studies on indigo witchweed, the American strain of *Striga gesnerioides* (Scrophulariaceae). Weed Sci 29:594–6

Nzungize J, Gepts P, Buruchara R, Male A, Ragama P, Busogoro JP, Baudoin JP (2011) Introgression of Pythium root rot resistance gene into Rwandan susceptible common bean cultivars. African J Plant Sci vol. 5(3):193–200

Ocampo JA, Martini J, Hayman DS (1980) Influence of plant interactions on vesicular-arbuscular mycorrhizal infection. I. Host and non-host plants grown together. New Phytol 84: 27

Odee DW, Haukka K, McInroy SG, Sprent JI, Sutherland JM, Young JPW (2002) Genetic and symbiotic characterization of rhizobia isolated from tree and herbaceous legumes grown in soils from ecologically diverse sites in Kenya. Soil Biol Biochem 34:801–811

Odee DW, Sutherland JM, Makatiani ET, McInroy SG, Sprent JI (1997) Phenotypic characteristics and composition of rhizobia associated with woody legumes growing in diverse Kenyan conditions. Plant Soil 188:56–75

Osonubi O, Atayese MO, Mulongoy K (1995) The effect of vesicular-arbuscular mycorrhizal inoculation on nutrient uptake and yield of alley-cropped cassava in degraded Alfisol of Southwestern Nigeria. Biol Fertil Soils 20:70–76

Osonubi O, Bakare N, Mulongoy K (1992) Interaction between drought stress and vesicular-arbuscular mycorrhiza on the growth of *Faidherbia albida* (syn. *Acacia albida*) and *Acacia nilotica* in sterile and non-sterile soil. Biol Fertil Soils 14:159–165

Parker C (1991) Protection of crops against parasitic weeds. Crop Prot 10:6–22
Ponte JJ, Franco A, Leal DB (1992) New host plants of root-knot nematodes. Nematropica 5:21–23
Power ME, Mills LS (1995) The keystone cops meet in Hilo. Tree 10:182–184
Punja ZK (1988) Sclerotium (Athelia) rolfsii a pathogen of many plant species. In: Sidhu GS (ed.) Advances in Plant Pathology. Vol. 6, Genetics of Plant Pathogenic Fungi, pp 523–534. Academic Press, London
Reddell P, Warren P (1986) Inoculation of acacias with mycorrhizal fungi: potential benefits. In: Turnbull JW (ed.) Australian acacias in developing countries, ACIAR proceedings No. 16. Brown Prior Anderson Press, Victoria, pp 50–53
Riches CR, Hamilton KA, Parker C (1992) Parasitism of grain legumes by *Alectra* species. Ann Appl Biol 121:361–70
Rusuku G, Buruchara RA, Gatabazi M, Pastor-Corrales MA (1997) Occurrence and distribution in Rwanda of soilborne fungi pathogenic to the common bean. Plant Dis 81:445–449
Salina PJ (1976) Presencia de *Elasmopalpus lignosellus* (Zeller) (Lepidoptera: Pyralidae) en los Andes venezolanos. Agron Trop 26:71–76
Samba RT, Lajudie P, Gillis M, Neyra M, Barelto MMS, Dreyfus B (1999) Diversity of rhizobia nodulating *Crotalaria spp*. from Senegal. Symbiosis 27:259–268
Sanginga N, Mulongoy K, Swift MJ (1992) Contribution of soil organisms to the sustainability and productivity cropping systems in the tropics. Agric Ecosyst Environ 41:135–152
Santos JHR, Quindere MAW (1988) Biological importance and management of Brazillian cowpea pests. In: Watt EE, Araujo JPP (eds.) Cowpea research in Brazil. IITA/EMBRAPA, Brasilia, pp 267–300
Sarr M, Agbogba C, Russell-Smith A, Masse D (2001) Effect of soil faunal activity and woody shrubs on water infiltration rates in a semi-arid fallow of Senegal. Appl Soil Ecol 16: 283–290
Sauerborn J (1991) Parasitic flowering plants: ecology and management. Verlag joseph Margraf, Weikersheim, 127 p
Shepherd KD, Jefwa J, Wilson J, Ndufa J, Ingleby K, Mbuthia KW (1996) Infection potential of farm soils as mycorrhizal inocula for *Leucaena leucocephala*. Biol Fertil Soils 22:16–21
Sieverding E (1991) Vesicular-arbuscular mycorrhiza management in tropical agro systems. German Technical cooperation (GTZ), Eschborn
Singh SR, Jackai LEN, dos Santos JHR, Adalla CB (1990) Insect pests of cowpea. In: Singh SR (ed.) Insect pests of tropical food legumes. Wiley, Chichester, pp 43–90
Sithanantham S, Kumar Rao JVDK, Reed W, Dart PJ (1981) Studies on nodule damage in pigeon pea. In: ICRISAT, proceedings of the international workshop on pigeon peas, vol 2. ICRISAT, Patancheru, 15–19 Dec 1980, pp 329–335
Stribley DP, Tinker PB, Snellgrove RC (1980) Effect of vesicular-arbuscular mycorrhizal fungi on the relations of plant growth, internal phosphorus concentration and soil phosphate analyses. J Soil Sci 31:655–672
Swelim DM, Hashem FM, Kuykendall LD, Hegazi NI, Wahab SMA (1997) Host specificity and phenotypic diversity of Rhizobium strains nodulating *Leuceana, Acacia* and *Sesbania* in Egypt. Biol Fertil Soils 25:224–232
Sy A et al (2001) Methylotrophic *Methylobacterium* nodulate and fix nitrogen in symbiosis with legumes. J Bacteriol 183:214–220
Talekar NA (1987) Insects damaging soybean in Asia for the tropics, research production and utilization. Wiley, Chichester, pp 25–45
Tian G, Olimah JA, Adeoye GO, Kang BT (2000) Regeneration of earthworm populations in a degraded soil by natural and planted fallows under humid tropical conditions. Soil Sci Soc Am J 64:224–228
Tomlison H, Taklehaimanot Z, Traore A, Olapade E (1995) Soil amelioration and root symbioses of *Parkia biglobosa* (Jacq) Benth. in West Africa. Agrofor Syst 19:241–252
UNEP (2001) Soil biodiversity and sustainable agriculture. Paper submitted by the Food and Agriculture Organization of the United Nations to the subsidiary body on scientific, technical and technological advice seventh meeting Montreal, 12–16 Nov 2001

van der Heijden MAG, Klironomos JN, Ursic M, Moufoglis P, Streitwolf-Engel R, Boller T, Wiemken A, Sanders IR (1998) Mycorrhizal fungal diversity determines plant biodiversity, ecosystem variability and productivity. Nature 396:69–72

Vincente NE, Acosta N, Schroder EC (1986) Reaction of *Leucaena leucocephala* to populations of *Meloidogyne incognita* and *M. javanica* from Puerto Rico. J Agric Univ Puerto Rico 70:157–158

Vogt W, Sauerborn J, Honisch M (1991) Striga hermonthica distribution and infestation in Ghana and Togo on grain crops. In: Ransom JK, Musselman LJ, Worsham AD, Parker C (eds.) Proceedings of the 5th international symposium of parasitic weeds. CIMMYT, Nairobi, pp 373–377

Weaving AJS (1980) Observations of *Hilda patruelis* Stal. (Homoptera: Tettigometridae) and its infestation of the groundnut crop in Rhodesia. J Entomol Soc S Afr 43:151–167

Whittingham J, Read DJ (1982) Vesicular-arbuscular mycorrhiza in natural vegetation systems. III Nutrient transfer between plants with mycorrhizal interconnections. New Phytol 90:277–284

Wightman JA (1989) The contribution of insects to low groundnuts yields in Southern Africa. In: ICRISAT, proceedings of the third regional groundnut workshop in Southern Africa, Lilongwe, Malawi, March 1988. ICRISAT, Patancheru

Wightman JA, Dick KM, Ranga Rao GV, Shanower TG, Gold CG (1990) Pests of groundnut in the semi arid tropics. In: Singh SR (ed.) Insect pests of tropical food legumes. Wiley, Chichester, pp 243–322

Woomer PL, Karanja NK, Mekki EI, Mwakolombe B, Tembo H, Nyika M, Silver M, Nkwiine C, Ndakidemi P, Msumali G (1997) Indigenous populations of rhizobia, legume response to inoculation and farmer awareness of inoculants in East and Southern Africa. African crop science conference proceedings, vol 3, pp 297–308

Yuen PM (1979) Nematodes associated with *Theobroma cacao*. Malaysian Agricultural Resource and Development Institute (MARDI). Malaysian Res Bull 7:54–58

Chapter 5
Strategies to Adapt, Disseminate and Scale Out Legume Based Technologies

D.N. Mugendi, B.S. Waswa, M.W. Mucheru-Muna, and J.M. Kimetu

Abstract The full potential of legume based technologies can only be achieved if farmers widely adopt these technologies. Widespread adoption of legume-based technologies calls for the identification of spatial and temporal niches of legumes in the farming systems. The entry point may be immediate provision of food and fodder but also contribution to soil fertility improvement and erosion control among others. Whichever the objective, there is need to demonstrate the immediate benefits to be accrued by the adopters if large scale adoption is to be achieved. There is need to address challenges hindering widescale adoption of legume technologies such as availability of improved germplasm, poor markets and policy. Whereas the national systems play a key role in promoting legumes, the private sector continues to effectively complement government efforts to increase availability and accessibility to quality seed. Flow of knowledge cannot be achieved through a narrow prescriptive approach, but requires development and testing of a battery of possible interventions for soil fertility improvements suited to the specific

D.N. Mugendi (✉)
Department of Environmental Sciences, Kenyatta University, Nairobi, Kenya
e-mail: dmugendi@yahoo.com

B.S. Waswa
Tropical Soil Biology & Fertility (TSBF),
International Centre for Tropical Agriculture (CIAT),
P.O. Box 30677, 00100 Nairobi, Kenya
e-mail: bswaswa@cgiar.org

M.W. Mucheru-Muna
Department of Environmental Sciences, Kenyatta University, Nairobi, Kenya
e-mail: moniquechiku@yahoo.com

J.M. Kimetu
Institute for Sustainable Energy, Environment and Economy (ISEEE),
University of Calgary, Alberta, Canada
e-mail: jmkimetu@ucalgary.ca

A. Bationo et al. (eds.), *Fighting Poverty in Sub-Saharan Africa: The Multiple Roles of Legumes in Integrated Soil Fertility Management*,
DOI 10.1007/978-94-007-1536-3_5, © Springer Science+Business Media B.V. 2011

agroecological environment together with the farmers. Better and innovative mechanisms are required for sharing of knowledge between all those involved in trying to improve the productivity of smallholder agriculture. This should cover the use or both print and audio media as complement to the traditional approaches such as the use of field days and onfarm demonstrations. The focus of this chapter is therefore to look at how the above issues have been addressed in adapting and disseminating legume-based technologies in Africa.

5.1 Introduction

In the past, the problem of declining soil fertility was addressed through agricultural practices like fallowing, crop rotation, application of farmyard manure, compost and to some extent the application of inorganic fertilizer. Nowadays, fallowing is no longer possible because of diminishing farm sizes. Introduction of perennial cash crops has reduced chances to rotate crops with contrasting plant nutritional characteristics (Stahl 1993). The quantities of farmyard manure produced in smallholder farms are small and are not likely to have major impact on soil fertility improvement (Maobe et al. 1998). Compost making is labour intensive and it is mostly prepared for kitchen gardens. Production of inorganic fertilizer depends very much on energy, which is costly and unlikely to become cheaper soon. Soil-improving herbaceous and forage legumes have been shown to have potential to improve soil fertility in various parts of the world (Franzel 1999). Improved tree fallows appear to have the great potential in areas of medium cropping intensification, that is, where population pressure on land is just starting to be perceived, and soil fertility is declining, but farmers still have extra land to fallow (Raintree and Warner 1986). However, recent findings show that fallowing is very common practice in areas of high population density in the tropics (Place and Dewees 1999). For instance, in Western Kenya where the population density ranges from 500 to 1,200 persons square kilometre and farm size average less than 1 ha, researchers found that about 50% farmers fallow their land periodically and that they have strong interest in testing improved fallows (Swinkles et al. 1997). The use of different food, herbaceous (green manure) and forage legumes in cropping systems, either as intercrops or in rotations with other crops, for improving soil fertility is a well known practice in the tropics. Herbaceous legumes that serve the single purpose of improving soil fertility however have not been widely adopted by smallholder farmers because they can not afford to grow them at the expense of food crops on their limited land holdings (Rao and Mathuva 2000). Dual-purpose legumes that produce food and feed [eg. cowpea (*Vigna unguiculata* L. Walp.); groundnut (*Arachis hypogea* L.); pigeon pea (*Cajanus cajan* L Millsp.] and forage legumes (eg. *Calliandra calothyrsus*, *Sesbania sesban*, *Staylosanthes* spp., *Trifolium* spp., *Vicia* spp.) are attractive particularly to smallholder farmers who practice mixed crop/livestock systems. Besides generating cash income to the farmers through sale of grain and/or livestock products (milk, meat, manure), these legumes increase the yields of subsequent cereal crops grown in rotation by improving the soil chemical, physical and biological properties (Haque et al. 1995; Pisanelli et al. 2000), provide fuelwood, reduce striga (*Striga hermonthica* Bench., a common parasitic plant of maize) infestation (Amadalo et al. 1998) and reduced weeds following the fallow.

There is often a direct conflict between the need to assure today's and tomorrow's food supply in our efforts to build sustainable soil fertility. Farmers discount the value of a benefit that will only be achieved several years from when investments are made. The most suitable legumes, from the soil fertility perspective, are often the most difficult for the farmer to adopt. Legumes remain marginal in many of the maize-based systems of the sub-Saharan region, however, for several reasons. Low P levels in the soils inhibit legume growth, and so a minimum amount of P must be added. Much of the work underlying legume-based technologies has been done on research stations. Insufficient account has been taken of the need to tailor these technologies to farming circumstances where labour is in short supply. The fertilizer needed to "kick start" the system may be too costly or unavailable, and there are often difficulties in obtaining legume seeds (Giller et al. 1994). Finally, the family may not be able to release land from staple food crops for the production of legumes.

Although there is little argument about the need for more sustainable soil fertility management, there is less clarity on how it might be achieved. Many legume-based resource management technologies are potentially available, but few are actually easy for farmers to adopt. To understand why this is so, requires understanding of the circumstances under which increasing numbers of smallholders in Africa operate. With this information we can then be able to review the various technologies for enhancing soil fertility and suggest how they might be made more attractive to farmers. It is from this understanding that improvements in technology development and transfer would facilitate smallholders' adoption of better soil fertility management practices.

5.2 Broad Legume Based Technologies

The emerging challenge in the use and transfer of legume based technologies is how to fit these technologies into a particular farming system considering that regional factors may vary widely. This therefore requires that work on legume technology dissemination focus on areas not only with the greatest need, but on areas with the greatest chance of success for transferring the technology to the farm. As such, suitable legume niches in the existing farming systems need to be identified. Potential niches for the best-bet species in smallholder systems fall into temporal niches, as defined by sequential (rotations) or simultaneous (intercropping) occurrence of food crops and legume species; and spatial niches as defined by the best place on the farm to plant the different legume species (Gachene et al. 2000). It is based on this niche demarcation of the legume-based systems that the various legume-based technologies are identified. There are various legume-based technologies commonly found in the tropics as discussed in the previous chapters but these technologies can be divided into two major classes: simultaneous systems, in which the components are grown in mixtures, and sequential systems or rotations. The primary interactions between plant species in simultaneous systems are direct (interspecific competition or facilitation) whereas interactions in sequential systems are principally due to modification of the environment for the succeeding plants often termed as residual effects (Giller 2001). A further type of interaction, allelopathy, which is the negative effect of one species on another due to release of toxic substances, can occur between plants growing together or in sequence (Rizvi et al. 1999). Thus residual effects can either be positive, boosting growth of a subsequent crop, or negative, where growth of a subsequent crop is depressed (Giller 2001). Whether it is best to mix species rather than grow them separately depends on the balance between complementality in growth and competition, as well as the priorities given to yields from the different plants. The system that will work also depends on the soil fertility, farm size, climate and most importantly farmer's priorities and preferences. The broad legume based technologies are discussed below.

5.2.1 Simultaneous Systems

The simultaneous systems include all types of intercropping, alley cropping and relay cropping.

5.2.1.1 Intercropping

This is a system where two or more crops are grown simultaneously. Intercropping cereal crops with grain legumes is the most common cropping system in the East African region, e.g., maize-common bean (*Phaseolus vulgaris*), maize-cowpea (*Vigna unguiculata*), and maize-pigeon pea (*Cajanus cajan*) intercrops. For soil fertility improvement and maintenance such traditional systems have low potential for adding N to the system. N benefit can be improved by replacing the grain legume with legume cover crops such as velvet bean, lablab, sunnhemp and jackbean.

5.2.1.2 Alley Cropping /Hedge Row Intercropping

Alley cropping or hedgerow intercropping is the growing of crops in alleys formed by multipurpose trees (MPTs). By continually retaining fast-growing mostly N_2-fixing MPTs on crop-producing fields, their soil-improving attributes (e.g., recycling nutrients, controlling erosion on sloping land, etc.) will create soil conditions similar to those in the fallow phase of shifting cultivation.

The trees help to enhance crop production in several ways. The leafy branches can be lopped and applied to the soil as a mulch or green manure to enhance the growth of crops in the alley (Wilson et al. 1986). The lopping during the growing period for the crops reduces competition for light, water and nutrients between the tree hedges and crops (as lopping reduces the growth rate of the trees). Maize yields have been reported to increase markedly, when the prunings were incorporated in the soil (Kang et al. 1981; Mafongoya and Nair 1997; Mugendi et al. 1999). Alley cropping has also been reported to assist in recovering some of the leached N below the roots of the annual crops by the action of their deep roots (Hartemink et al. 1996; Kindu et al. 1997; Jama et al. 1998; Mugendi et al. 2000). The alleys also act as windbreaks during dry seasons reducing the desiccating effects of winds.

5.2.1.3 Relay Cropping

Relay cropping involves the growing of trees or herbaceous legumes in continuous cereal systems. The trees or herbaceous legumes are planted when the cereal is almost mature and the trees are left to grow after the cereal has been harvested until the next planting season. The trees or herbaceous legumes are harvested prior to the following cropping season and the green biomass and/or litter is incorporated into the soil during ridging. The relay cropping system was developed to respond to the needs of the highly populated areas of Sub Saharan region where fallowing periods are becoming almost extinct (Place 1995).

5.2.2 Sequential Systems or Rotations

5.2.2.1 Biomass Transfer

This technology involves ex-situ production of biomass away from the cropping land in designated areas or hedges or around or within the farm. The biomass is then harvested and applied as mulch or incorporated into the soil before and/or after the crop is planted. In this technology the interactions between the components is only through the nutrient release from applied mulch and the uptake by the crop. The system eliminates competition but often requires substantial labor inputs during biomass transfer activities. More land is also taken out of production resulting in some opportunity costs and there are additional costs of weeding the separated biomass bank and increased time spent on spreading the biomass.

5.2.2.2 Improved Fallows

Fallow systems are those in which crops are grown with short, intervening fallow periods during which the land is left to regenerate. Fallow systems can be identified as systems with one-third to two-thirds of the land under cultivation at any one time (Giller 2001). In improved fallows, trees are used in the same way as green manure to give a large amount of high-quality organic residues that release nutrients for subsequent crops. Tree fallows have the added advantage of providing fuelwood and can be more effective than alley cropping in preventing soil erosion on sloping lands (Hoang Fagerstrom et al. 2001). Legume cover crops such as *Crotalaria spp., Mucuna spp., Desmodium spp., Canavalia ensiformis, Sesbania sesban* and *Tephrosia vogelli* can be used for short- and medium-term improved fallows. Short term fallows are here defined as less than 6 months and medium-term as 6–12 months (Gachene et al. 1999). In Rwanda and Kenya, legume cover crops and legume shrub fallows have been used to restore soil fertility in nutrient-depleted fields more rapidly than the traditional fallows (Raquet 1990; Niang et al. 1996) and hence, allow shortening of the required fallow period, providing fuelwood, fodder and staking materials.

5.3 Socio-Economic Feasibility and Acceptance of Legume Based Technologies

Smallholder agriculture in the tropics is highly complex and aims at achieving self-sufficiency in food production at the same time averting any possible risks that may lead to reduced food production. Whether a legume-based technology will be adopted or not depends on the perception of the farmer in terms of the balance between the positive and negative aspects. The negative aspects of the various systems are mainly loss in crop yield in intercropping systems and loss of land for food production in short- and medium-term fallows. Positive effects include weed and erosion control, additional products such as food, fodder, and fuel, and increased crop yields. The net effect on labour requirements, overall yields and economic benefit will vary among farming systems and agroecological zones and must be assessed in participation with the farmers.

Cases of adoption of legume-based technologies can be identified across a wide range of agroecological zones in countries of sub Sahara Africa. Grain legumes have the fewest adoption problems and are widely grown by farmers, mainly for home consumption of the seed and sometimes leaves. However, the more productive (high harvest index) grain legumes add relatively little organic matter and N to the soil, since much of the above-ground dry matter and almost all the N is removed from the field in the grain (Giller et al. 1994). Species that combine some grain with high root biomass and shoot-leaf biomass, such as pigeonpeas (*Cajanus cajan*) and dolichos beans (*Dolichos lablab*) offer a useful compromise of promoting farmer adoption and improving soil fertility. Smallholder farmers may rapidly adopt a new legume crop, or varieties of legume crops with which they are already familiar, if it fits a different niche (Giller 2001). For example, cultivation of the climbing varieties of *Phaseolus vulgaris* in the highlands of Rwanda and Burundi led to a rapid uptake by smallholder farmers, from cultivation on small plots by only 5% of farmers in 1986 to over 40% of bean producing households, numbering some 500,000 farmers in 1994 (Sperling and Loevinsohn 1993; Sperling et al. 1994). In Nigeria there has been a rapid adoption of a multi-purpose variety of cowpea from an individual farmer to over 4,000 farmers representing more than 90% of the farmers in two villages and more than 60% in a third village (Inaizumi et al. 1999) in just 4 years. As reported by Kwesiga et al. (1999) the use of *Sesbania sesban* improved fallows in Malawi increased from 200 to more than 3,000 farmers in 3 years pointing to how high farmers' interest can be in adopting a new technology.

Land availability does influence the adoption rates of legume-based technologies. In Eastern Province of Zambia, where the population density is about 25–40 persons km^{-2} and farm size averages 4 ha, the uptake of improved fallows is very high (Kwesiga et al. 1999). However, adoption is less likely among farmers with very small plots of land. Identification of unexploited or underutilized niches, such as conservation bunds, farm hedges, and unproductive fields may provide an entry point for the introduction of the legume technologies. Although there was concern about the small farm sizes in the study area, the average improved tree fallow sizes in 17 study villages increased from 134 to 247 m^2 between 1997 and 1999 (DeWolf et al. 2000). In a study on the adoption of *Mucuna pruriens* fallows in southern Benin, West Africa, the removal of incentives to adopt did not adversely affect the trends in the spread of the technology. However, farmers considered insecure land tenure a constraint to adoption for this non-perennial species (Honlonkou et al. 1999). Land tenure has also been identified as a constraint to the adoption of alley cropping technologies.

Franzel et al. (2000) carried out a survey in Zambia in 1999 of 108 farmers that planted improved fallows in 1994/1995 to assess the farmers' experiences in managing the technology. The survey revealed that tree nursery establishment was not a major problem because of the relative availability of low-lying moist areas. Sesbania had the highest survival rate than tephrosia since the latter is direct seeded. Over time farmers have managed to increase the land area devoted to fallows from an average of 0.04 to 0.07 ha between the first and third plantings. In their work in Western Kenya, Pisanelli et al. (2000) observed that farmers' management of the practice differed considerably from researcher's recommendations, particularly with respect to fallow length and time of planting trees relative to crops.

Gender issues are very important in the adoption of the legume based technologies. In their work in Western Kenya Pisanelli et al. (2000) observed that females rated improved fallows significantly higher than males on improving soils and on reducing weeds. On the other hand, men's ratings on the importance of seed production were higher than women's. Thus, the benefits that are perceived to be accrued by either men or women from any legume based technology will affect the adoption rate. Pisanelli et al. (2000) and Franzel and Scherr (2002) in their work in Western Kenya and Zambia respectively, indicated that females planted smaller areas than males. Despite the small acreage planted, the females rate of increase in area under improved fallows between 1998 and 1999 was much higher (86%) as compared to that of males (4%) (Pisanelli et al. 2000). Female farmers provide most of the labour for African food production and the percentage of households that are female-headed ranges from 18% in central Kenya, to about 30% in Zambia and 50% in western Kenya (Franzel and Scherr 2002). Franzel and Scherr (2002) argue that one would expect that females' use of agroforestry practices would be lower than that

Table 5.1 Profitability of 2 year *Sesbania sesban* improved fallow compared to continuous cropped maize in Eastern Zambia (farmer managed trials, N = 12)

Option	Returns to land: net present value (US$/ha)	Returns to labour: net returns per workday (US$/ha)
Continuous unfertilized maize (5 years)	5	0.42
Continuous fertilized maize (5 years)	160	1.02
Sesbania sesban fallow (2 year fallow, 3 years maize)	203	0.93

Source: Place et al. (2002)

of males, for two reasons. First, female household heads tend to have lower incomes than male household heads. Secondly those choosing participants for experiments and distributing planting materials, usually extension staff, tend to select men. Thus, even if the technology itself is gender neutral, adaptive research and dissemination programmes are often biased towards males (CIMMYT 1993). In contrast however, Pisanelli et al. (2000) in Western Kenya found that 55% of the fallows were cut by women, 35% by men, and 10% by mixed groups, suggesting that improved fallows do not appear to be less acceptable to women for physical or cultural reasons. Data from Zambia indicate that 32% of the males and 23% of the females in the sampled villages were planting improved fallows, and that there were no significant differences between the two proportions (Franzel and Scherr 2002). In fact, one of the villages where womens' groups were particularly strong, the proportion of females using improved fallows was higher than that of males. Moreover, whereas single females are often disadvantaged relative to female heads of household whose husbands live away (Bonnard and Scherr 1994), the Zambia data showed that the same proportions of these two groups were testing the technology. Peterson (1999) however, identified factors such as lack of awareness or germplasm as the major limitations to the trying out of the improved fallows. Farm size or labour were found to be less significant deterrents.

A survey conducted in western Kenya to assess management and innovation in the use of improved fallow and its profitability and acceptability with farmers of different characteristics revealed that farmers had no trouble in establishing their fallows (DeWolf et al. 2000; Pisanelli et al. 2000). Most farmers (70%) incorporated the fallows in an existing crop to save on land preparation and weeding although in 28% of these cases, farmers reported a negative effect on that season's maize crop. Nonetheless, 79% of farmers reported that subsequent crop yields were positively affected by the fallows, through soil fertility improvement and weed reduction (notably striga). Place et al. (2002) reported that in a follow-up of 12 farmers who agreed to compare continuous unfertilized maize, continuous fertilized maize and maize following a 2 year fallow in Zambia revealed that the improved fallows delivered almost twice as much total maize yield as the unfertilized maize, though cropped for two fewer seasons. The net present value from improved fallow system was $160 per hectare as compared to only $5 for continuous cropping without fertilizer (Table 5.1). The fertilized maize system was far superior in total maize production, but the improved fallow system performed better in terms of discounted returns to land.

Wealth can influence adoption in various ways. Poor farmers may be more risk averters, have less access to information, have a higher discount rate and thus a short term planning horizon, and have less capacity to mobilize resources (Hoekstra 1985; CIMMYT 1993). Infact, most adoption studies find strong correlations between wealth and adoption. In Western Kenya, poor farmers increased their areas under improved fallows at much greater rates than did better-off farmers. This strong interest probably reflects their limited access to other soil fertility inputs, such as mineral fertilizer and animal manure (Pisanelli et al. 2000). However, results from Zambia confirm that the higher the wealth status, the greater the use of improved fallows. But, on the other hand, about 20% of the poor and very poor households in four sample villages planted improved fallows during the first 4 years of testing, which suggests that there are no absolute barriers preventing low-income farmers from doing this (Franzel and Scherr 2002).

5.4 Markets and Scaling Out of Legume-Based Technologies

For many decades now legume based technologies extension has often been 'supply driven'; that is, the emphasis has been on increasing production levels of trees and crops, with little attention paid to trends in product demand or price (Scherr and Franzel 2002). Recent studies and experience suggests that product market conditions and market institutions can play a critical role in farmer adoption and the economic contribution from legume based technologies (Leaky and Newton 1994; Dewees and Scherr 1996). Recent developments in Africa highlight an increasing trend toward liberalized domestic markets and an opening up of their economies to the forces of international trade (Jones et al. 2002). As the focus shifts towards outward-oriented policies, export markets seem an important source of economic growth for these countries. These trends provide new opportunities and challenges for poor smallholder farmers in developing countries. However, to take advantage of these opportunities, smallholder farmers must be able to participate in productive activities in which they have a competitive advantage. This implies access to well-organized marketing, distribution and post-harvest systems, effective market information, and technologies that allow them to be price- and quality-competitive (Jones et al. 2002). Unfortunately, smallholder farmers in the tropics face high transaction costs and uncertainty arising from missing or incomplete input and product markets, high access barriers and costs of information, and other market imperfections that restrict market access. As observed by Eicher (1999) policymakers face the challenge of determining and fostering the most productive roles for public, private, and no-governmental organizations in supporting African farmers, traders and agribusinesses. Collective responsibility of the above actors is thus critical to facilitation of access to technology, information, capital and marketing arrangements.

A more balanced approach to agriculture is now being advocated which includes the need not only to promote food crops, but also cash crops to generate income that can be used to purchase food. As such therefore agricultural researchers and extension workers need to think beyond the farm gate and develop strategic partnerships with market players and policymakers to effect real change. In areas where for example soybeans is widely grown such as Mozambique and Malawi, production has been sustained through a strong emphasis being put to training for local community consumption (Giller 2001). This has been a key factor in driving the expansion of soybean as a smallholder crop, but continued production by smallholder farmers on a substantial proportion of their land depends on access to viable markets for sale of the surplus produce. If a secure market is not available, soybean tends to take place alongside other minor food legumes and is cultivated on a small area of the farm.

Traditional marketing channels have high marketing and distribution costs, as most cereal and in fact legume crops, pass through several intermediaries with little value being added before the produce reaches the end-users. The rural assemblers bulk the crop and sell it to middlemen/traders in local market centers. The crop is then taken to larger produce markets where it is sold to transporters who deliver it to processors and exporters. In this marketing channel, farmers receive the lowest share of final consumer prices while urban processors receive the highest. As observed by Jones et al. (2002), it is difficult to determine whether these margins reflect traders' profits or point to inefficiencies in the marketing system without complete knowledge of marketing costs (transport, storage, processing), other transaction costs, and marketing risks. If this trend were to continue as it is then this would serve as a disincentive to the smallholder producers. In the case of a landlocked Malawi for example, the cost of sending a fully loaded 20-foot container to Mumbai India is $1,800 compared to $800–1,200 for Tanzania, and $500–800 for Kenya (Jones et al. 2002). Improvement of infrastructure in Mozambique has reduced the transportation costs below those of Kenya. For countries in eastern and southern Africa to remain competitive, productivity needs to increase, transaction costs have to be reduced, and quality standards improved.

Conditions in basic grains markets have clearly affected farmers' strategies in adopting legume based technologies. In Zambia, the parastatal maize and cash crop marketing systems with their stable outlets and price subsidies for maize producers – collapsed in the early 1990s, and private markets have not developed sufficiently to replace them (Scherr and Franzel 2002). In Kenya the Cereal Board collapsed and the farmers are no longer guaranteed that their maize will be bought after production. The instability common to markets for higher-value products also affects farmer adoption of legume based technologies. For instance, the milk industry in Kenya became unstable after the Kenya Cooperative Creameries lost its monopoly. Increased risk motivated farmers to substitute farm-produced calliandra for dairy meal to reduce cash costs. Financial returns to calliandra are not very sensitive to changes in the milk price itself, especially when calliandra is used as a substitute for dairy meal rather than as a supplement (Scherr and Franzel 2002).

5.5 Policy and How It Affects Legume Technology Development/Dissemination

A conducive policy environment is important for the adoption of any agricultural technology, but there are several reasons why policies are particularly critical for legume based technologies. Unfortunately, most policy measures used to support agriculture currently act as powerful disincentives against sustainability. In the short-term, this means that farmers switching from modern high-input agriculture to resource-conserving technologies such as the use of legumes can rarely do so without incurring some transition costs. In the long-term, it means that sustainable agriculture will not spread widely beyond the types of localized success. Some legume based practices tend to be more complex than many agricultural practices, because they involve mixing trees with crops and because several years are often required before they generate returns. Agroforestry practices may rely on supportive policies for their development and diffusion to a greater extent than do other agricultural practices (Franzel and Scherr 2002).

The previous section of this discussion indicated that there is considerable potential for many legume based technologies to increase farm income, meet household needs, contribute to local economic growth and possibly improve soil fertility status in many parts of the continent. However, there is need for specific types of policy action to achieve widespread adoption and significant aggregate impact from the technologies being put forward by the researchers. Of concern here are key areas such as legume germplasm supply, agricultural input supply, markets for the agricultural products, land tenure systems, technology design for extension and institutional arrangements for on-farm research and extension support (Place and Dewees 1999). Outdated national policies on new varietal testing and approval processes have been identified as a key constraint to the making of improved germplasm available to the farmers. This is further complicated by a general lack of regional coordination and policy harmonization, creating international barriers to the dispersion of, and trade in improved seed (Jones et al. 2002). Agricultural researchers have developed legume varieties that are both high-yielding and can meet the precise and varying quality requirements of end-user needs. However adoption of these new seed technologies is hampered by lack of access to improved seeds as well as the lack of established quality standards within local market systems that can reward farmers for producing a differentiated product.

Franzel and Scherr (2002) suggest that germplasm policy must focus more on reducing the cost of planting materials, especially important for technologies such as improved fallows, which require large numbers of seedlings per unit area and over time. Small-scale tree nursery promotion is needed for improved fallows in Zambia and for timber, pole and fodder trees in Kenya. Without the necessary market incentives, farmers are less likely to invest in improved seed and realize the productivity and quality gains that have been demonstrated from planting improved varieties.

5.6 Methods and Techniques Used in Legume Technology Dissemination-Strengths and Limitations

The adoption of agricultural innovations has been intensively researched since the seminal work of Grilliches (1957) on hybrid corn in the USA. Rogers and Shoemaker (1971) described adoption by individuals as an innovation-decision process consisting of four stages, which include, first, knowledge where individual is exposed to the existence of the innovation and gains some understanding of how it functions. Secondly, persuasion where individual forms a favourable attitude towards the innovation. Thirdly, decision where individual engages in activities that lead to a choice to adopt or reject an innovation. Lastly, confirmation stage where individual seeks reinforcement for the innovation decision with the option of reversing that decision based on increased experience with the innovation.

Legume based technologies for soil fertility replenishment are information intensive in the sense that they are not simple adjustments in the current or past farming systems (FAO 1996). They require much more learning and interaction. Evidence from pilot trials in central and western Kenya show that use rates are significantly higher in pilot villages, marked by greater farmer access to technical backstopping, than in nearby villages (Place et al. 2002). According to Matthews et al. (1993) it does not require formal education per se but the familiarity with a particular practice that determines the willingness to participate in a new technology. The more familiar farmers are with a technology the more likely they are to express an interest in its future use. Because pilot development projects cannot be replicated in many other areas and extension systems are often weak, the question of how to support farmers learning and sharing of information at wide scales is a difficult one.

In the case of African agricultural extension, attention has been increasingly turned towards a loosely defined collection of participatory approaches, none of which have asserted themselves in any form of operational dominance. The key to the success of these so-called Farmer Participatory Training (FPT) approaches is the empowerment of farmers with an understanding of the agro-ecology of their own fields. The approach enables those with a legitimate interest in the outcome of the research to be able to exert some influence on priorities in decision-making (Mbugua et al. 1997). This enables farmers to become informed decision-makers, making crop management decisions based on detailed observations of their own fields and the experience they build-up over the course of the training and thereafter during their own farming (Okali et al. 1994).

Participatory Rural Appraisal (PRA) tools are now identified as the most appropriate in facilitating information sharing, problem identification/ diagnosis, and identification of existing opportunities. In a Soil Management Project (SMP) in Kisii, South West Kenya, a PRA exercise was initiated before project development. Among the tools applied during the PRA were transect walks, semi-structured interviews (SSI), time and trend lines, daily routines for both women and men, social and economic resource maps, seasonal calendars and problem cause analyses

(Mbugua et al. 1997). Several major benefits have been accrued from the farmer participatory approach adopted by this project. First, the approach and in particular the use of the PRA allowed the research team to quickly familiarize themselves with the agroecological conditions, farming systems, and farming community at each site. Second, because of the emphasis on development of a collegial relationship between farmers and research teams, the approach has promoted information flow between the two groups – farmers and researchers. Third, the approach has encouraged a holistic or systems level view of constraints to agricultural productivity. These are some of the benefits identified by earlier researchers (Chambers et al. 1989; Gupta 1989; Stroud 1993; Okali et al. 1994) using farmer participatory approaches who observed that lack of prior consultation with farmers could lead to the failure of some projects.

Recently, interest has begun to coalesce around the potentials offered by the Farmer Field School (FFS) approach, based upon its tremendous success achieved among smallholder farmers in South and Southeast Asia (Simpson and Owen 2002). The Integrated Pest Management (IPM) FFS emerged out of a decade of experimentation in implementing participatory farming activities in the Philippines beginning in the late 1970s. Refinements in the Philippines program and a new major effort in Indonesia in the late 1980s led to the birth of the FFS movement (Pontius et al. 2000). The education philosophy of the FFS rests on the foundations of adult non-formal education, and reflects the four elements of the 'experiential learning cycle' proposed by Kolb (1984): concrete experience, observation and reflection, generalization and abstract conceptualization, and active experimentation. FFS puts greater emphasis on group and individual capacity building as well as indigenous knowledge in addition to the other components of participatory approaches. This approach was introduced to Africa through the Food Agriculture Organization (FAO) supported Global IPM Facility (GIF) and has gained widespread embrace in countries such as Ghana, Mali, Senegal, Zimbabwe, Kenya, Uganda and South Africa. FFS is now applied in the dissemination of legume based technologies as well as in IPM in legumes thus helping farmers cut down on cost of pesticide and fertilizer use, merits that have encouraged the adoption of legumes in smallholder agriculture.

Although participatory approaches have promised to be effective, there are also some problems. One major problem stems from the holistic viewpoint taken during the PRAs. Farmers' responses tend to include issues that are often beyond the mandate of the research team. Thus expectations in the community are raised that the project will help them deal with such issues as roads, wells, health care etc. Where the community feels that such issues have not been well addressed, the project may experience resentment. A second problem is that while specific procedures have been developed for the diagnostic stage of participatory research, such procedures have not been developed for the actual implementation of the participatory research agenda. As a result, the active involvement of the farmers in the research process tends to diminish after the diagnostic stage, which can eventually lead to researchers pursuing agendas that are of little value or interest to the farmers. This is an experience learnt in the Soil Management Project (SMP) in Western Kenya (Mureithi et al.

1997). A third problem is that the participatory approach, which requires frequent, regular contact between at least some members of the research team and the farming community over a period of several years (Bentley 1994), may be difficult to sustain given often limited resources in terms of vehicles, fuel and human.

The complementarities between on-station and on-farm research approaches as a means of effective technology dissemination has been echoed by numerous authors (Mettrick 1993; Franzel et al. 2001; Franzel and Scherr 2002; Place et al. 2002). In the strive to ensure exchange of information among the various stakeholders in the use of agroforestry and thus legume-based technologies, the International Centre for Research in Agroforestry (ICRAF) employs three types of trials to assess the feasibility, profitability and adoption of soil fertility technologies (Franzel et al. 2001): type 1 – researcher designed-researcher managed, type 2 – researcher designed-farmer managed, and type 3 – farmer designed-farmer managed. In an assessment of various case studies, Franzel and Scherr (2002) reflected the evolution of a new approach to accelerate the process of technology development and dissemination. The key elements of this approach are:

- On-farm trials, with differing levels of researcher and farmer involvement, as a central locus for collaborative technology development.
- On-farm trials greatly expand the potential contributions of research to technology development. The case studies confirm the importance and complementarity of the three types of on-farm trials, which are suited to different objectives.
- Farmer participation, adaptation and innovation as key inputs in technology development.

Indeed the case studies demonstrate the importance of the participatory research approach for improving the effectiveness of the design, implementation and evaluation of on-farm research. Participatory research strengthens the researcher-farmer partnerships as when farmers and researchers collaborate in the design, implementation and evaluation of type 1 or type 2 trial. Researchers contribute their scientific knowledge and farmers contribute their indigenous technical knowledge and information about their needs, preferences and circumstances. Benefits have also been gained from the inclusion of several institutions within the research team. First, scarce resources have been pooled. Second, the institutions involved have often found their skills and objectives complementary. In Kenya, for example, the national agriculture institution, the Kenya Agricultural Research Institute (KARI) researchers have gained valuable insights on methods of interacting with farmers from non-governmental organization personnel while the latter have learned basic techniques of experimentation from the researchers (Njue et al. 1997; Mureithi et al. 1997).

Conversely, there have also been constraints to collaboration between institutions arising from differences in operational procedures and objectives. Another problem has developed in terms of frontline staff. Some extension frontline staff have been unable to provide the projects with adequate support, either due to lack of time because of conflicting duties or because of lack of interest in the objectives of the project.

Although there are some problems with the farmer participatory approaches, the benefits far outweigh the constraints. The use of this approach promises a higher likelihood of increased adoption of improved legume technologies because they are developed with the participation of the end-users. In addition linkages are strengthened (Researcher/Extension/Farmer), easy communication flow between scientists and the farmers. This approach together with effective communication approaches such as radio, televisions, drama and poetry should be encouraged for the effective dissemination and adoption of legume based technologies.

5.7 Radio Communication

Knowledge, and more broadly, information, can enable farmers to bring about improvements to their environment and agricultural activities, and create new income and employment opportunities (Odame 2001). Therefore, more attention must be paid to the carrier of information and its accessibility. Experience in rural communications demonstrates that radio is a practical and creative medium for facilitating the education and empowerment of the poor, including women and youth (Balit 1999). In sub-Saharan Africa, radio has proven to be an effective media through which agricultural information can be communicated across long distances, in languages familiar to rural people. For many years, global plans of action for improved rural information and communication, supported by agencies like the FAO and World Bank, have highlighted the importance of rural radio.

Broadcasting-radio and television is a vital medium of public information. The most effective of them is television broadcasting because it combines picture and sound, which makes it more interesting and attractive. However, in developing countries television receivers and TV programme production is expensive. Lack of availability of electricity in most rural areas limits access to television, thus making it an urban-centred medium. The average rural dweller therefore has no access to it. On the other hand, the transistorised radio is comparatively cheap and can operate easily on ordinary dry cell batteries; therefore lack of electricity does not affect its use. In Africa there are estimated to be about 100 radio receivers per 1,000 people, whereas there are no more than ten television receivers for the same number, and even those are concentrated in the towns.

Different formats of radio programmes such as live talks show/call-in programmes, documentaries and short promos/announcements could be used based on the effect each format has on the audience. Experience from South African community radio indicates that live talk show/call-in programmes make a massive impact, because they give people an opportunity to voice their opinions during the discussion of a particular topic (Mosotho 2001). However the use of radio as a mass media has its limitations such as poor signals, limited reach in certain areas, top-down approach, limited airtime and in appropriate programming (Nakabugu 2001).

Case Study 1: How Radio Has Been Used for Agricultural Development in Uganda

Radio Uganda, the oldest and only public radio broadcaster, broadcasts to all regions in the country. It broadcasts in 25 of the country's 52 languages. Rural issues are extensively and substantively addressed. Agricultural programmes are produced by the Farm and Environment Department. The Farm and Environment Department produces programmes in English and 20 local languages. The Department sources its material from government, NGOs, local governments, research institutions, academic institutions and farmers.

Materials for programmes are collected through field visits and observations, field recordings and invitations to studios. The materials are presented as discussions, talkshows, features, interviews and magazines. A minimum of 15 min per language is produced and aired every week. The major setbacks the station experiences are that of inadequate facilitation for field recordings, distance from audiences and inappropriate programme timing (some programmes are missed as farmers are out in their fields).

Since government liberalised the airwaves over 20 commercial radio stations have come up in Uganda. The first community radio station was established on 20 July 1999 in Kibaale District known as Kagadi-Kibaale Community Radio (KKCR) station, an initiative of Uganda Rural Development Training project (URDT). It broadcasts in prevalent local languages. This station is distinctly different from public or commercial radios that are rural-based in that it is fully owned by the local villagers in the sub-county of Kagadi, in Kibaale District, Western Uganda. The villagers have selected a committee to represent the different interests and stakeholders groups in the management of the station.

By: Nakabugu (2001)

Case Study 2: Radio Communication in Ghana

In Ghana there are about 219 radio receivers to 1,000 people compared to 13 TV receivers for the same number of people; and they are mostly found in the cities and urban areas. Comparatively, there are over 3 million (3,078,000) radio sets with an estimated audience of 8.2 million compared to 199,000 television sets with estimated viewers of 4 million in Ghana. So far, radio is the most useful and efficient medium available to the rural population, which is scattered over a wide area and has no accessible roads. Radio broadcasting can therefore be seen, as one of the vehicles through which national aspirations can be pursued since it is financially accessible to both government and people.

(continued)

Case Study 2: (continued)

To enhance and promote rural broadcasting in the country, the Upper Regional Administration (URA) FM radio station at Bolga, Ho, Bawku, and another at Apam have been established. These stations broadcast a series of programmes to farmers and fishermen mostly in the language and dialects commonly used in the areas. They broadcast rural programmes based on policies set up by the Ghana Broadcasting Cooperation (GBC), but also take into consideration the economic, social cultural and religious lives of the people.

As a developing country with over 40 languages spoken, and facing language barrier, Ghana uses six languages (Akan, Dagbani, Ewe, Ga, Nzema and Hausa). Therefore, some rural programmes are broadcast in the evening when the targeted rural listener is likely to have retired to bed after a hard day's work. However, these issues can easily be solved if the plan to establish FM radio stations in the regions, throughout the country in place of the relay stations materialises. This is because such stations can broadcast rural programmes in languages commonly used in their area.

By: Abbey-Mensah (2001)

5.8 Drama

The African Radio Drama Association (ARDA) is a development communication organisation inaugurated after assessing the development communication needs of the continent at a symposium of African radio experts in Harare, Zimbabwe in November 1994.

At the forum, it was agreed that for the largely illiterate and low-income population, dependent on oral narrative communication, radio was a highly accessible and very effective medium through which information and education could be passed. Secondly, research findings shared revealed that drama programs were particularly effective in motivating behaviour change. After all, in African tradition, entertainment is a key element of educational activities. Folk tales, music, dance and drama always serve didactic purposes, but have the added advantage of capturing, engaging and motivating audiences. Drama imitates real life in characterisation, situations, conflicts and resolutions. It is this characteristic of drama which allows it to captivate individuals, provide the confidence needed to use new ideas and skills. Hence, delegates at the Harare symposium saw a need for regional organisations to draw attention to radio and create opportunities for networking among practitioners, whilst also setting higher professional standards for the development radio genre. The Nigerian chapter has since inception spearheaded as well as collaborated with other agencies like the Ford Foundation, UNICEF,

United States Information Service (USIS), MacArthur Foundation, John Hopskin University, World Wide Fund for Nature (WWF)'s African Communication Network (ACN) to communicate various development messages, ranging from population and reproductive health to environmental and civic rights issues through its radio drama serials and other related programs (Phido 2001). The following case in Kenya illustrates how drama can be used to disseminate agricultural information.

Case Study 3: Use of Drama in Agricultural Information Dissemination in Kenya

Drama continues to play a significant role in the dissemination of agricultural information in Kenya. The soap and magazine programme "Tembea na Majira" started in 1996 through the sponsorship of Overseas Development Agencies (ODA) in collaboration with Agricultural Information Center, a government department under the Ministry of Agriculture (MoA), is evidence of how drama can be used to convey agricultural information. Tembea na Majira is a rural radio "soap" programme (continuous drama) which is set in a small Kenyan roadside village. Weekly actors are heard re-enacting scripts written by highly trained Kenyan writers who bring to life the loves and traumas of a small farming village with carefully interwoven messages about sustainable rural livelihoods. The soap is 15 min long and has been running since January 1996. It now also includes a lively 15 min magazine programme broadcast immediately after the soap, which allows the audience to send letters, and comments on the soap to be rebroadcast and competitions are also run on issues developed in the soap. More technical information can be carried in the magazine programme with in depth interviews with experts, farmers, success stories, etc.

The programme has faced a number of financial setbacks where for example sponsorship of airtime has been stopped. Agricultural Information Center often finds it difficult to find funds and partners to share the production and airtime costs. However through the financial assistance of organizations/ companies such as East African Industries, Department for International Development (DFID), ODA, Mediae and the MoA, Tembea na Majira was able to be broadcast to an audience of over 6.9 million listeners in August 1996 and 4.5 million in August 1999 despite being off air for 6 months.

Other initiatives in the use of drama and theatrical skills in the dissemination of agricultural information in Kenya have also been demonstrated by the Kenya Wood Fuel Working Group (KWFWG) and International Centre for Research in Agroforestry (ICRAF) in western Kenya.

By: Campbell (2001)

5.9 Germplasm and Legume Based Technology Development and Dissemination

With the promotion and initial success of legume based technologies, comes the problem of seed supply. The improved technologies for example require large quantities of germplasm. Seed for preliminary screening experiments may be available from international genebanks, but quantities are usually small and the material needed for larger experiments of promising material, and subsequent on-farm testing needs to be produced in the respective countries. For most African countries, importation is costly, time consuming and may involve elaborate quarantine procedures. Large-scale commercial production of forage seed is often portrayed as a complex enterprise, requiring detailed knowledge of plant physiology with appropriate equipment for establishing, managing, harvesting, cleaning and storing the seed crops. In theory, farmers can provide for their seed needs by collecting seed or even saving a relatively few trees, since all the fallow species are relatively good seeders. But, in practice this is not easy to achieve. In Kenya, DeWolf et al. (2000) found that 131 of 150 improved fallow users did not have enough seed to plant the size of fallow desired. In Eastern Zambia, there were reports of farmers traveling over 20 km with ox-carts to find Sesbania seedlings (Kwesiga et al. 1999). The involvement of the formal private sector (e.g. retailers) is nonexistent and the relatively low value of the fallow seed would suggest a limited role. There is more potential for local informal initiatives in germplasm production and distribution. The increased farmer-to-farmer trade in seed and establishment of local seed orchards managed by groups or individuals are positive steps towards increasing supplies of tree fallow seed. However, experience in this project has shown that it is possible to produce reasonable quantities of basic seed with fairly limited resources and using mainly manual labour.

A number of germplasm evaluation and screening studies have been carried out on legumes (Byenkya 1988; Muyekho 1988; Drechsel et al. 1996; Dyck 1997; Maobe et al. 1998). A detailed screening and evaluation of 40 legume species (both exotic and traditional) was undertaken at 11 sites covering several agroecological zones in Kenya by the Legume Screening Network (LSN) in 1995 and 1996 (Dyck 1997; Gachene et al. 1997; Maobe et al. 1997; Otsyula and Okwuosa 1997; Saha et al. 1997). The legume cover crops (LCC) evaluated included materials suitable for multiple uses such as green manures, human consumption and livestock feed. The species were evaluated for establishment, biomass production, nodulation, N-fixation, ground cover, seeding abilities, persistence to drought, pests and diseases. Moreover the major species were screened on the major soil types such as acrisols, luvisols, nitisols and ferralsols (Byenkya 1988; Niang et al. 1996; Fischler 1997; Hanson and Lazier 1998; Maobe et al. 1998). Whereas the performance of most of these species under optimum conditions including rhizobium inoculation and P-fertilization was satisfactory, the evaluation did not consider the conditions and niches in which such species are to be found on-farm. As observed by Gachene et al. (1997) and Gachene and Haru (1998), intercropping such legume cover crops with food crops may result in competition for soil moisture, a condition that can greatly affect the performance of the legumes. Further, many of the soils under farm conditions in the regions are severely depleted in nutrients hence this could also result in reduced LCC performance and N-fixation.

In order to increase legume grain quality production requires first the introduction of new germplasm, then ensuring a regular seed supply to maintain grain quality. For non-hybrid crops such as legumes, farmers invariably use own-saved seed, and if for any reason seed is lost they can usually access it from a range of sources including social networks and local grain markets (Jones et al. 2002). Farmer seeds are both resilient and flexible even under extreme conditions, and experience shows that the farmers are reluctant to pay large premiums for certified seed except for hybrids where, because hybrid vigour declines in farm-saved seed. Test-marketing of small seed packs has shown that farmers are willing to purchase small quantities at prices several times higher than the grain price, but once they have access to the germplasm they will then save their own seed rather than returning to the seed market (Tripp 2001). For self-pollinated crops like beans and groundnuts recycling seed in this way poses few problems, but for crops like pigeonpea where there is more degree of out-crossing, varietal integrity will change over time. For markets of pigeonpea where grades and standards are important, as the case of the stringent standards demanded by European buyers, this may be a problem hence it calls for maintaining varietal integrity in breeding.

Barriers to seed dissemination result because of a number of socio-economic constraints. In Malawi just as in many other countries in sub Saharan Africa, poor distribution of inputs and produce in the region results from poor infrastructure that exists (Mkandawire et al. 2000). The road and bridge network are so vulnerable that vast areas are inaccessible particularly during the rainy season. Markets are not adequately established. The quasi-government marketing organizations have better networks of markets but are poorly organized. They offer the farmers the lowest prices and generally do not have ready cash to offer the farmer when the crop is ready for sale. Entrepreneurs are coming into the system due to market liberalization. Such entrepreneurs may have ready cash and transport but lack necessary knowledge to handle seeds differently from other inputs and produce. In many instances, farmers are not organized into groups or cooperative societies, which are powerful bodies that would assist them do combined efforts to break through the barriers of the marketing systems. In Malawi the credits system broke down as the country changed from dictatorship to a plural society.

5.9.1 Legume Germplasm Production in Malawi

There exists on-station and contract farmer systems in Malawi for seed production (Phiri et al. 2000), however the quantities of seed produced are far too small to meet the bean seed requirement for the nation. The national bean seed requirement was

Table 5.2 Amounts (metric tonnes) of certified seed distributed to and produced by NASSPA in the 1999/2000 crop season in Malawi

Crop	Seed (MT) distributed to associations	Seed (MT) produced by associations
OPV Maize	31.64	3,970
Groundnuts	53.14	410
Beans	8.48	60
Soybeans	0.74	20
Pigeon Peas	1.00	80

Source: European Union 2000

Table 5.3 Bean seed supplied to Seed Producing Groups (SPG's) between 1995 and 1998, and seed production by smallholder farmers under Actionaid Malawi Smallholder Seed Multiplication and Development Project (MSSMDP)

Bean variety	Number of groups	Seed supplied (kg)	Seed produced (kg)
Nasaka	22	1,141	5,705
Kalima	37	1,794	8,970
Kambidzi	2	140	700
Mkhalira	4	130	650
Nagaga	4	160	800
Maluwa	5	210	1,050
Napilira	1	40	200
Nyauzembe	1	20	100
Total	76	3,635	18,175

Source: Phiri et al. (2000)

projected at over 2,000 tonnes in 1999 alone. Thus there is need to produce large quantities of seed to meet the national requirement. In Malawi ActionAid; an NGO that has a food security-oriented seed scheme, developed a Malawi Smallholder Seed Development Project (MSSDP) after a flood disaster in 1992 when 1.1 million households benefited from 3,000 tonnes of relief seed issues. The objectives of this Project were to: (a) develop low cost mechanisms, which would improve and sustain availability and accessibility of appropriate, improved seed to resource poor farmers in participating communities, (b) establish sustainable self-motivating community based groups, which will manage seed multiplication and distribution in the communities, and (c) train selected extension staff and participating community groups in seed production, seed quality control, on farm seed selection and storage, group dynamics and in community participatory methodologies. In 1998 a total of 3.6 tonnes of certified bean seed was supplied and made available to 76 groups with a total membership of 1,500, who in turn multiplied more than 12 tonnes of first generation approved bean seed. Tables 5.2 and 5.3 is a summary of National Smallholder Seed Producers Association (NASSPA's) and the Seed Production Groups (SPG's) seed production in Malawi (Table 5.3).

5.9.2 Legume Germplasm Production in Tanzania

In Tanzania, until 1990 the government had been responsible for national seed production and distribution (Due and Jean 1988). The Tanzania Seed Company Ltd (TANSEED), a government parastatal, was established in 1973 and had the mission of producing sufficient certified seed for farmers at affordable prices, and of providing seed extension, dissemination, and advisory services. TANSEED had a monopoly on seed production, importation, distribution, and sale of certified maize, wheat, rice, and beans (Due and Jean 1988; Ministry of Agriculture and Co-operatives 1997). However, TANSEED was not able to perform these duties efficiently or effectively, resulting in poor performance of the national seed industry. Several problems contributed to poor performance and low impact of improved seeds that were handled by TANSEED, including insufficient transport and funding, lack of humidity-controlled warehouses, and inadequate seed drying equipment. TANSEED-produced seeds had low germination rates, and prices were considerably greater than prices of unimproved seed available through local markets. In the early 1990s, TANSEED began to concentrate their seed production activities on the most profitable seeds. Concentration on the most profitable crops meant dropping a number of less profitable but essential crops including grain legumes (Ministry of Agriculture and Co-operatives 1997). Thus, improved seeds of appropriate but unprofitable crop varieties became unavailable to smallholder farmers. Due to the deficiencies of TANSEED, and in recognition of the existence of a relatively large untapped market for improved seed in Tanzania, the government liberalised the seed industry in 1990. After liberalisation, a number of foreign and domestic private seed companies entered the seed sector to produce, distribute, and market improved seed (ICARDA 1994; Ministry of Agriculture and Co-operatives 1997). By 2000, the largest private seed company operative in Tanzania was Cargill Hybrid Seed. Other emerging private seed companies include Alpha Seed, INCOFIN Tanzania Ltd, East African Seed Company, and Pannar. All these companies concentrate on hybrid and composite maize seed, leaving a gap in the production and marketing of improved seeds of other essential crops such as grain legumes. Improved seeds of grain legumes, including beans, are one category of crops that is not handled by the private seed companies and is therefore not adequately supplied in most areas of Tanzania (Due and Jean 1988; Grisley 1992).

5.9.3 Legume Germplasm Production in Nigeria

The results from a simple on-station seed multiplication plots in Nigeria (Abuja Road seed multiplication plot- Kaduna) have demonstrated that seed multiplication of forage legumes is feasible using locally available inputs (Tarawali 1994). The results from this demonstration should provide an impetus for scientists of national research and extension programmes to develop in-country programmes for the production of basic quantities of forage seed for both research and subsequent dissemination to farmers. The importance of appropriate selection of material is apparent. Accessions with poor seed yields are likely to be very expensive to produce, and therefore would be of limited value to either researcher or farmer. The availability of such seed means that, as appropriate technologies are developed for specific farming systems, farmers themselves could become involved in seed production, possibly using contract growing approaches. In this instance, farmers would be provided with appropriate seed by the national programme, together with technical expertise and advice on seed production. The farmer would be mandated to provide a specified quantity of clean seed for the national programme. A more opportunistic approach is to adapt on-farm utilisation of forage species to allow for seed collection. With both approaches land and labour requirements and gender implications need to be taken into account when considering the development of on-farm forage seed production. Attention should also be paid to seed cleaning and storage, and here, a combination of modern technical knowledge and traditional methods may be the optimum approach.

5.9.4 Legume Germplasm Production in Kenya

Despite the success of the introduction of legume cover crops in western Kenya, the main challenge has been seed production for continuity of the trials. In a study by Mukalama (2000), women groups discussed the possibilities of providing preferred legume seed of black and white seeded mucuna, lablab cv Rongai, black seeded lablab and canavallia to the community. As a kick start, the groups were given small quantities of seed to bulk up and to produce seed for themselves and the other groups within the communities. The presence of NGOs such as the Kenya Woodfuel Working Programme (KWAP) in the surrounding districts has helped establish seed production schemes where the groups have been trained in quality seed production. Other efforts to bulk seed have been done by Tropical Soil Biology and Fertility (TSBF) where bulking of legume seed has been done on school land. During an on-farm trials conducted in western Kenya on the introduction of legume cover crops, non-government organizations (NGOs), farmers associations, and private individuals believe that the formation of a small-scale seed production sector is necessary to meet the needs of smallholder farmers. Through this means, seeds of new, improved varieties of traditional smallholder crops such as beans will be available in a timely fashion and at affordable prices to many smallholder farmers. Smallholder seed production programs are not in competition with the large commercial seed producers. They are intended to compliment large-scale seed production by being part of a seed production system that integrates small- and large-scale seed production systems (Ministry of Agriculture and Co-operatives 1997).

5.10 The Future of Seed Multiplication and Dissemination

Community empowerment and capacity building: Smallholder farmers who multiply seed need to be thoroughly trained in all aspects of seed multiplication and seed business. This would promote farmers' ability to understand and articulate the basic principles of seed multiplication and identify markets for their seed.

The concept of 'Para-inspectors' or 'Community Facilitators': Government and NGO's have limited capacity to effectively reach farmers throughout the country who are multiplying seed. The use of 'Para-inspectors' or 'Community Facilitators' is to complement some of the roles of government and NGO staff by delivering information to seed-producing farmers. This would promote seed multiplication and availability on the local market. This concept can well be initiated through communal seed production groups where farmers can be trained on effective seed handling.

Commercialising small-scale seed multiplication: Unless improved varieties are recognized by farmers as being superior over their traditional varieties, and that marketing seed of improved varieties can be a viable agro-enterprise, seed multiplication of improved varieties will remain a non-sustainable practice. Several factors can influence the commercialization of seed production, including: maintenance of seed quality and purity; effective processing and packaging; and timely supply of the seed on the market. Farmers' associations should be promoted to enable seed farmers to compete for better market prices.

Promoting adoption of improved varieties: Acceptance and use of improved varieties depends on several factors including: (1) availability of basic or certified seed from which farmers can readily replenish their seed supplies, and (2) access to markets for the harvested crops. On-farm research trials are a good way to advertise but are not sufficient to guarantee acceptance of the new varieties. These trials must be accompanied by a combination of variety demonstrations and farmer field days. NGO's are very capable of establishing variety demonstrations and organizing farmer field days.

The private sector including NGOs continue to play a key role of complementing government efforts to increase availability and accessibility to quality seed. However, the sustainability of small-scale seed multiplication and dissemination calls for concerted efforts from all parties concerned with food security, including the state, the public sector, the private sector, and NGO's. There is a need to facilitate effective flow of breeder- basic/foundation- certified seed to reach farmers who are multiplying seed. Farmers should be given access to credit facilities. By promoting linkages to lending institutions, farmers will gain access to capital funding for investing in their seed enterprise. And finally, the seed industry should offer various seed varieties for multiplication and farmers should be given the opportunity to choose the varieties that best fit their varying interests.

5.11 Scaling Out the Technologies

There is no legume based technology, no matter how technically current or elegant, will have a significant impact on the land scene unless it is adopted by a significant percentage of the intended users. According to Caveness and Kurtz (1993) adoption or non-adoption of legume based technology is a function of the practice proposed, perception of need, the availability and distribution of factors of production and perception of risks. Any technology introduced to farmers must appear practical and beneficial in both human and economic terms to warrant adoption. Bunch (1982) identifies several criteria that farmers use when choosing an appropriate technology for adoption. He argues that a technology must be able to use locally available resources, which must be obtained on a permanent, reliable basis and must be inexpensive. He also points out that the technology should fit in the existing farming systems, promise quick and recognizable success, and to some degree promise farmers substantial and dependable increase in either food supply or income at local prices. According to Raintree (1983), adoptability of a technology by farmers is greatly enhanced when the proposed technology holds potential to solve 'perceived problems' in a particular location. Therefore, an innovation must seem to address the felt needs of the community. In addition to the above, Bunch (1982), identifies other factors such as, a technology that is simple and easy to learn and understand, ecologically safe, one that requires minimum on-site supervision, relatively free of risk and one that is culturally acceptable.

On the basis of relatively few empirical studies directly related to legume based technologies, it is difficult to draw definitive conclusions about what are the most important factors affecting adoption and their implications for scaling up. According to recent reviews by Franzel (1999) and Place and Dewees (1999), several factors are most likely to affect adoption of a legume based technologies innovation: biophysical adaptation; profitability which considers the labor returns and land as well as financial returns; farmers awareness of the innovation; access to land, labor and water; access to social capital particularly where group action is needed; availability of essential inputs; access to financial capital; and degree of risk and uncertainty. Successful contribution of legumes grown in rotation with food crops has been reported in Rwanda (Raquet 1990), Uganda (Fischler and Wortmann 1999), Tanzania (Temu and Anne 1995) and Kenya (Niang et al. 1996) despite the presence of a number of constraints in other areas.

There is a general consensus among scientists, farmers and policy makers that low adoption of legume cover crop technologies is mainly caused by high opportunity costs, lack of farmers appreciation about the short term benefits of the legumes, poor understanding of the socio-economic factors dictating the decision of farmers and the problem in identification of niches for growing legumes without affecting the current farming systems. The success rate in achieving effective adoption of soil improving and forage legumes in sub- Saharan Africa has been low (Thomas and Sumberg 1995) as farmers prefer food legumes over forage or/legume cover crops in that the opportunity cost of green manure is so high to allocate part of the resources of food legumes to legume cover crops. Therefore, there is need to develop an effective guideline that targets different legume types to different niches in different agroecological zones and socio-economic strata.

The entry point for the adoption of legume-based systems in the farming systems of tropical Africa is based on the associated beneficial effects such as improved soil fertility and increased food production. However other benefits have also been identified. For example from 1939 until 1953, Marejea (*C. ochroleuca*) was cultivated at Peramiho, Tanzania to control couchgrass (Rupper 1987). Only later was it recognised as a means of improving soil fertility. In West Africa the suppression of the weed *Imperata cyclindrica* was the entry point for farmers' acceptance of Mucuna technology (Vissoh et al. 1998). In Rwanda (Raquet 1990; Drechsel et al. 1996) and Tanzania (Pfeiffer 1990), erosion, fodder and fuelwood problems led to the adoption of short-term fallows and macro-contour lines, respectively. Even for the same legume based technology farmers may have different reasons for adoption. For example in two villages in Benin the motivation of farmers to test *Mucuna* technologies was different in the two districts. Whereas in Sunyani more than two-thirds of the farmers wanted to improve or maintain their soil fertility through *Mucuna*, the main motivation in Atebubu was its weed suppressing potential. The main weed species occurring in Atebubu is *Imperata cylindrica*, which is more difficult to control than *Chromolaena odorata*, which is the dominant species in Sunyani. Although the soils in Atebubu are unfertile, weeds appear to be the major constraint to farming and restrict yields more than soil fertility. In Kenya some legume cover crop species are believed to have the additional advantage of controlling the weed striga (Oduwo and Onyango 1998) that results in large losses of maize production in the western part of the country. For wider adoption of the legume-based technologies to be achieved, first the legumes must be able to address an immediate problem of the community envisaged. The multiple uses of legumes are thus paramount to their adoption. For example, the macro-contour line studies in the Usambara Mountains indicated potential for integrated approach to livestock and crop production (Pfeiffer 1988, 1990). Adoption of legume cover crops in Rwanda was mainly due to lack of fire wood and stakes for fencing which became readily available after planting legume shrubs for short-term fallow duration in badly degraded areas. Also in Rwanda, the introduction of the high-yielding climbing beans proved to be an effective incentive for farmers to produce stakes (Graf and Ndorehayo 1988).

5.12 Research Needs and Way Forward

The incorporation of legumes into cropping systems has great potential for the maintenance of sustainable smallholder farming systems in the sub-Saharan region. The major disadvantage of incorporating legumes in smallholder agriculture is that it requires a sacrifice of time and/or space normally used for food production. This has tended to be a major disincentive to adoption and dissemination of legume based technologies especially considering the current and urgent land pressures and food insecurity. For widespread adoption of legume-based technologies therefore calls for the need to identify spatial and temporal niches in the farming systems (Woomer 1998) and all the time demonstrating the immediate benefits to be accrued by the adopters.

As observed by Giller and Cadisch (1995) there is perhaps sufficient understanding of both the processes and resources which are available for the enhancement of soil fertility to make an impact, and yet little direct application of this knowledge in smallholder farming systems in Africa. Legumes with alternative uses that are attractive to smallholder farmers need to be identified. Better mechanisms are required for sharing of knowledge between all those involved in trying to improve the productivity of smallholder agriculture (farmers, researchers, extension agents, NGO workers, etc.). Flow of knowledge cannot be achieved through a narrow prescriptive approach, but requires development and testing of a battery of possible interventions for soil fertility improvements suited to the specific agroecological environment together with the farmers. Multi-location screening network to screen legumes for a range of environments and uses is therefore imperative. Legume based technologies may increase labour requirements for certain operations such as planting and clearing of fallows, or collecting, transporting and spreading biomass. But such technologies may also reduce labour for other operations such as land preparation and weeding. Besides the potential for increased crop yields, farmers can also benefit from reduced soil erosion and soil nutrient loss by improved crop growth and establishment of rapid ground cover, fewer pests and diseases and diminished impact of drought crops (Rao et al. 1998).

Typically, farmers are looking for means of combining the legume inputs and employing them in ways that minimize the requirements for additional cash, labour, and land. Past or ongoing experimentation exists to guide them through the choices, and extension offers little counsel. From the previous discussion, it is clear that flexible soil fertility recommendations are required that better addresses actual nutrient deficiencies, take advantage of cropping system opportunities, are efficient under the highly variable rainfall regime faced by most smallholders, and are compatible with farmers' socio-economic circumstances. Transferring experience in fertility management from other regions of the world requires not only adaptation but also a much greater understanding of the processes through which fertility can be managed under African conditions.

References

Abbey-Mensah S (2001) Rural broadcasting in Ghana. Paper presented during the international workshop on farm radio broadcasting held between 19–22 Feb 2001, Rome, Italy

Amadalo B, Niang A, Gathumbi S, Jama B (1998) Using short-duration improved fallows in Western Kenya for soil fertility and crop yield improvement. National Agroforestry Research Centre, Maseno

Balit S (1999) Voices for change: rural women and communication. FAO communication for development group extension, education and communication service, Rome, Italy. In: Proceedings of the high-level consultation on rural women and information, Rome, Italy, 4 to 6 October 1999

Bentley JW (1994) Facts, fantassies and failures of farmer participatory research. Agr Hum Values 11:140–150

Bonnard P, Scherr S (1994) Within gender differences in tree management: is gender distinction a reliable concept? Agrofor Syst 25:71–93

Bunch R (1982) Two ear of corn: a guide to peopled-centered agricultural improvement. World Neighbors, Oklahoma City, pp 98–116

Byenkya, GS (1988) A review of forage cover germplasm introduction and evaluation in Uganda. In: Dzowela BH (ed.) African forage plant genetic resources evaluation of forage germplasm and extensive livestock production systems. Proceedings of the 3rd workshop held at the international conference centre, Arusha, Tanzania, 27–30 April. ILCA, Addis Ababa, pp 91–103

Campbell (2001) Food security – brief history to an initiative to provide sustainable radio programming and production in Kenya. Paper presented during the international workshop on farm radio broadcasting held between 19–22 Feb 2001, Rome, Italy

Caveness FA, Kurtz WB (1993) Agroforestry adoption and risk perception by farmers in Senegal. Agrofor Syst 21:11–25

Chambers R, Pacey A, Thrupp LA (1989) Farmer first: farmer innovation and agricultural research. Intermediate Technology Publications, London

CIMMYT (1993) CIMMYT Economics programme. The adoption of agricultural technologies: a guide for survey design. CIMMYT, Mexico

Dewees PA, Scherr SJ (1996) Policies and markets for non-timber tree products. Environment and production technology division discussion paper No. 16. International Food Policy Research Institute, Washington, DC

DeWolf J, Rommelse R, Pisanelli A (2000) Improved fallow technology in western Kenya: potential and reception by farmers. International Centre for Research in Agroforestry, Nairobi

Drechsel P, Steiner KG, Hagedorn F (1996) A review of the potential of improved fallows and green manure in Rwanda. Agrofor Syst 33:109–136

Due, Jean M. (1988) Tanzania cultivar release, seed multiplication and distribution systems. Technical Report No. 7. Department of Rural Economy. SUA, Morogoro, Tanzania, pp 46

Dyck E (1997) Screening legumes for potential soil productivity improvement in Kenya: Kenyan legume research network. Poster presented at Green Manure Cover crops conference in Santa Catarrina, Brazil, 6–12 April 1997

Eicher CK (1999) Institutions and the African farmer. Third Distinguished Economist Lecture. International Maize and Wheat Improvement Centre, Mexico

European Union (2000) 1999/2000 Annual report of maize productivity task force – action group 2. Ministry of Agriculture and Irrigation Development, Lilongwe

Fischler M (1997) Legume green manures in the management of maize-bean cropping systems in Eastern Africa with special reference to Crotalaria (*C. ochroleuca*). Ph. D. thesis No. 12099, ETH Zwich, Switzerland

Fischler M, Wortmann CS (1999) Green manure research in eastern Uganda — a participatory approach. Agrofor Sys 47(1/30000000):123–138

Food Agriculture Organization (FAO) (1996) Sowing the seeds of knowledge. World Food Summit held in Rome, 13–17 Nov 1996

Franzel S (1999) Socioeconomic factors affecting the adoption potential of improved tree fallows in Africa. Agrofor Syst 47(1–3):305–321

Franzel S, Scherr SJ (eds.) (2002) Trees on the farm: assessing the adoption potential of agroforestry practices in Africa. CAB International, Nairobi

Franzel S, Phiri D, Place F, Mwangi M (2000) The uptake of improved fallows in eastern Zambia. International Centre for Research in Agroforestry, Nairobi

Franzel S, Coe R, Cooper P, Place F, Scherr SJ (2001) Assessing the adoption potential of agroforestry practices in sub-Saharan Africa. Agr Sys 69(142):37–62

Gachene CKK, Haru R (1998) Soil moisture extraction in maize-mucuna based cropping system. Paper presented at the 16th conference of soil science society of East Africa held at Tanga, Tanzania, 13–19 Dec 1998

Gachene CKK, Makau M, Haru R (1997) Soil moisture extraction by different legume cover crops. Paper presented at the Legume screening Network Workshop in Kanamai, Mombasa, March 1997

Gachene CKK, Palm CA, Mureithi JG (1999) Legume cover crops for soil fertility improvement in the East African Region. Report of an AHI workshop TSBF, Nairobi, 18–19 Feb 1999

Gachene KKC, Palm CA, Mureithi JG (2000) Legume cover crops for soil fertility improvement in the eastern Africa region. Report of the African highland initiative workshop, 18–19 February 1999, No. 11. TSBF, Nairobi, Kenya

Giller KE (2001) Nitrogen fixation in tropical cropping systems, 2nd edn. CAB International, Wallingford, Oxon

Giller KE, Cadisch G (1995) Future benefits from biological nitrogen fixation: an ecological approach to agriculture. Plant Soil 174:255–277

Giller KE, McDonagh JF, Cadisch G (1994) Can biological nitrogen fixation sustain agriculture in the tropics? In: Syers JK, Rimmer DL (eds.) Soil science and sustainable land management in the tropics. CAB International, Wallingford

Graf W, Ndorehayo V (1988) A review of research methodologies and results. In: Wortmann C (ed.) Proceedings of a workshop on soil fertility for bean cropping systems in Africa, Addis Ababa, Ethiopia, 5–9 September 1988. CIAT African Workshop Series, No. 3, pp 123–137

Grilliches Z (1957) Hybrid corn: an exploration into the economics of technological Change. Econometrica 25:501–523

Grisley W (1992) Seed for bean production in sub-Saharan Africa: issues, problems and possible solutions. Agr Syst 43:19–33

Gupta AK (1989) Scientists' view of farmers' practices in India: barriers to effective interaction. In: Chambers R, Pacey A, Thrupp LA (eds.) Farmer first: farmer innovation and agricultural research. Intermediate Technology Publications, Westminster, 218 pp

Hanson J, Lazier JR (1998) Forage germplasm at the ILCA for Africa. In: Dzowela BH (ed.) African forage plant genetic resources, evaluation of forage germplasm and extensive livestock production systems. ILCA, Addis Ababa, pp 69–73

Haque I, Powell JM, Elwis SK (1995) Improved crop-livestock production strategies for sustainable soil management in tropical Africa. In: Lal R, Stewart BA (eds.) Soil management: experimental basis for sustainability and environmental quality. CRC Press, Boca Raton

Hartemink AE, Buresh RJ, Jama B, Janssen BH (1996) Soil nitrate and water dynamics in sesbania fallow, weed fallow, and maize. Soil Sci Soc Am J 60:568–574

Hoang Fagerstrom MH, van Noordwijk M, Phien T, Vinh NC (2001) Innovations within upland rice-based systems in northern Vietnam with Tephrosia candida as fallow species, hedgerow, or mulch: net returns and farmers' response Agriculture. Ecosyst Environ 86:23–37

Hoekstra DA (1985) Choosing the discount rate for analysing agroforestry systems/technologies from private economic viewpoint. Forest Ecol Manag 10:177–183

Honlonkou AN, Manyong VM, Tchetche N (1999) Farmers' perceptions and the dynamics of adoption of a resource management technology: the case of mucuna fallow in southern Benin West Africa. Int Forest Rev 1(4):228–235

ICARDA (1994) Organisations and management of national seed programmes. In: Proceedings of follow-up seminar/workshop, 12–24 Nov 1994, Aleppo, Syria. ICARDA, pp 283

Inaizumi H, Singh BB, Sanginga PC, Manyong VM, Adesina AA, Tarawali SA (1999) Adoption and impact of dry-season dual-purpose cowpea in the semiarid zone of Nigeria. IITA, Ibadan

Jama B, Buresh RJ, Ndufa JK, Shepherd KD (1998) Vertical distribution of roots and soil nitrate: tree species and phosphorus effects. Soil Sci Soc Am J 62:280–286

Jones R, Ade Freeman H, Lo Monaco G (2002) Improving the access of small farmers in eastern and southern Africa to global pigeonpea markets. Agricultural Research and Extension Network, Network Paper No. 120

Kang BT, Sipkens GF, Wilson GF, Nangju D (1981) Leucaena leococephala prunings as a nitrogen source for maize (*Zea mays*). Fert Res 2:279–287

Kindu M, Buresh RJ, Bashir J (1997) Root and inorganic nitrogen distribution in sesbania fallow, natural fallow and maize fields. Plant Soil 188:319–327

Kolb D (1984) Experiential learning. Prentice Hall, Englewood Cliffs

Kwesiga FR, Franzel S, Place F, Phiri D, Simwanza CP (1999) Sesbania sesban improved fallows in eastern Zambia: their inception, development and farmer enthusiasm. Agrofor Syst 47:49–66

Leaky RB, Newton AD (1994) Domestication of tropical trees for timber and non-timber products. MAP Digest 17 UNESCO, Paris

Mafongoya PL, Nair PKR (1997) Multipurpose tree prunings as a source of nitrogen to maize under semiarid conditions in Zimbabwe. Agrofor Syst 35:31–46

Maobe SN, Dyck E, Ondicho AR (1997) Legume screening for soil fertility improvement in a high potential area in south west Kenya. Paper presented at the LSN Workshop held at Kanamai, Mombasa, Kenya, 24–26 March 1997

Maobe SN, Dyck EA, Mureithi JG (1998) Screening of soil-improving herbaceous legumes for inclusion into smallholder farming systems in Kenya. In: Waddington et al. (Eds.) Soil fertility research for maize-based farming systems in Malawi and Zimbabwe. Proceedings of the soil fert net results and planning workshop held from 7 to 11 July 1997 at Africa University, Mutare, Zimbabwe. Soil Fert Net and CIMMYT-Zimbabwe, Harare

Matthews S, Pease SM, Gordan AM, Williams PA (1993) Land owner perceptions and the adoption of agroforestry practices in Southern Ontario, Canada. Agrofor Syst 21:159–168

Mbugua DM, Maobe SN, Kwach JK, Oigo AB (1997) Participatory evaluation, management and utilization of improved fodders in south west Kenya. In: Mureithi et al (eds.) Participatory technology development for soil management by smallholders in Kenya. Special publication of Soil Management and Legume Research Network Projects. Kenya Agricultural Research Institute, Nairobi

Mettrick A (1993) Development oriented research in agriculture: an ICRA textbook. ICRA, Wageningen

Ministry of Agriculture and Co-operatives (1997) The agricultural policy of Tanzania. Government Printers, Dar es Salaam, pp 77

Mkandawire ABC, Masangano C, Magayane F, Nchimbi-Msolla S, Mabagala R (2000) Bean seed multiplication and dissemination in East Africa. Bean/Cowpea collaborative research support program. In: East Africa proceedings: bean seed workshop Arusha, Tanzania, 12–14 Jan 2000

Mosotho S (2001) Community radio and farming in South Africa. Paper presented during the international workshop on farm radio broadcasting held between 19–22 Feb 2001, Rome, Italy

Mugendi DN, Nair PKR, Mugwe JN, O'Neill MK, Swift MJ, Woomer P (1999) Alley cropping of maize with calliandra and leucaena in the sub humid highlands of Kenya. Part 2: Biomass decomposition, N mineralization and N uptake by maize. Agrofor Syst 46:51–64

Mugendi DN, Nair PKR, Graetz DA, Mugwe JN, O'Neill MK (2000) Nitrogen recovery by alley-cropped maize and trees from ^{15}N- labelled tree biomass in the sub humid highlands of Kenya. Biol Fertil Soils 31:97–101

Mukalama J (2000) Introduction of legume cover crops to communities in western Kenya. Legume Research Network Project Newsletter: Issue No. 4. Kenya Agricultural Research Institute (KARI), Nairobi, Kenya

Mureithi JG, Waanyonyi M, Farley C, Dyck E (1997) Sustaining farmer participation in agricultural research in western Kenya. In: Mureithi et al (eds.) Participatory technology development for soil management by smallholders in Kenya. Special publication of Soil Management and Legume Research Network Projects. Kenya Agricultural Research Institute, Nairobi

Muyekho FN (1988) A review on evaluation and agronomy of promising legume varieties in Kenya. In: Dzowela BH (ed.) African forage plant genetic resources, evaluation of forage germplasm and extensive livestock production systems. International Livestock Centre for Africa (ILCA), Addis Ababa, pp 192–195

Nakabugu SB (2001) The role of rural radio in agricultural and rural development translating agricultural research information into messages for farm audiences. Paper presented during the international workshop on farm radio broadcasting held between 19–22 Feb 2001, Rome, Italy

Niang A, Gathumbi S, Amadalo B (1996) The potential of short-duration improved fallow for crop productivity enhancement in the highlands of Western Kenya. In: Mugah JO (ed.) People and institutional participation in agroforestry for sustainable development. Kenya forestry Research Institute, Nairobi, pp 218–230

Njue EK, Makelo MN, Mbugua DM, Makini FW (1997) Gender analysis of the soil management project activities. In: Mureithi et al (eds.) Participatory technology development for soil management by smallholders in Kenya. Special publication of Soil Management and Legume Research Network Projects. Kenya Agricultural Research Institute, Nairobi

Odame HH (2001) Making waves in the CGIAR: rural radio and agricultural science. Paper presented during the international workshop on farm radio broadcasting held between 19–22 Feb 2001, Rome, Italy

Oduwo A, Onyango C (1998) Green manuring to improve soil fertility and reduce striga infestation by smallholder farmers in lower potential South Nana. C-MAD, Kenya, Annual Report

Okali C, Sumberg J, Farrington J (1994) Farmer participatory research: rhetoric and reality. Overseas Development Institute

Otsyula RM, Okwuosa EA (1997) Legume screening in the mid altitude area in West Kenya, Kakamega. Paper presented at the LSN workshop held at Kanamai, Mombasa, Kenya, 24–26 March 1997

Peterson J (1999) The Zambian integrated agroforestry project baseline survey: lessons learned in Chadiza, Chipata, Katete and Mambwe districts. Eastern Province, International Centre for Research in Agroforestry, Nairobi

Pfeiffer R (1988) Fodder production potential for zero grazing systems in the bimodal West Usambara Mountains of Tanzania. In: Dzowela BH (ed.) African forage plant genetic resources, evaluation of forage germplasm and extensive livestock production systems. ILCA, Addis Ababa, Ethiopia, pp 192–195

Pfeiffer R (1990) Investigating possibilities of combining fodder production with erosion control and agroforestry in the West Usambara Mountains of Tanzania. In: Kotsch J (ed.) Ecofarming practices for tropical smallholdings, vol 5, Tropical Agroecology Margraf, Weikersheim, pp 55–80

Phido DA (2001) The African Radio Drama Association, ARDA, Nigeria. Paper presented during the international workshop on farm radio broadcasting held between 19–22 Feb 2001, Rome, Italy

Phiri MAR, Chirwa RM, Kandawole S, Tripp R (2000) Introducing new bean varieties with small seed packs: experience from Malawi. CIAT, Cali, Occasional Publication Series, No. 32

Pisanelli A, Franzel S, DeWolf J, Rommelse R, Poole J (2000) Adoption of improved tree fallows in western Kenya: farmer practices, knowledge and perception. International Centre for Research in Agroforestry, Nairobi

Place F (1995) Cost-benefit analyses of selected agroforestry technologies in the southern AFRENA. ICRAF, Nairobi

Place F, Dewees P (1999) Policy and incentives for the adoption of improved fallows. Agrofor Syst 47:323–343

Place F, Franzel S, De Wolf J, Rommelse R, Kwesiga F, Niang A, Jama B (2002) Agroforestry for soil fertility replenishment: evidence on adoption processes in Kenya and Zambia. In: Barrett CB, Place F, Aboud AA (eds.) Natural resource management in African agriculture: understanding and improving current practices. CAB International, Wallingford

Pontius J, Dilts R, Bartlett A (eds.) (2000) Ten years of building community: from farmer field schools to community IPM. FAO Community IPM Programme, Jakarta

Raintree JB (1983) Strategies for enhancing the adoptability of agroforestry innovations. Agrofor Syst 1:173–187

Raintree J, Warner K (1986) Agroforestry pathways for the intensification of shifting cultivation. Agrofor Syst 4:39–54

Rao MR, Mathuva MN (2000) Legumes for improving maize yields and income in semi-arid Kenya. Agr Ecosyst Environ 78:123–137

Rao MR, Niang A, Kwesiga F, Duguma B, Franzel S, Jama B, Buresh R (1998) Soil fertility replenishment in sub-Saharan Africa: New technologies and the spread of their use on farms. Agrofor Today 10:3–8

Raquet K (1990) Green manuring with fast-growing shrub fallow in the tropical highland of Rwanda. In: Kotschi J (ed.) Ecofarming practices for tropical smallholding, vol 5, Tropical Agroecology. Margraf, Weikersheim, pp 55–80

Rizvi SJH, Tahir M, Rizvi V, Kohli RK, Ansari A (1999) Allelopathic interactions in Agroforestry systems. Critical Rev Plant Sci 18:773–796

Rogers EM, Shoemaker FF (1971) Communication of innovations. Free Press, New York

Rupper G (1987) Cultivation of marejea (*Crotolaria ochroleuca*): experience of Peramiho. In: Proceedings of a writers workshop held at Peramiho, Songea, Tanzania. Benedictine Publication, Ndanda Peramiho, pp 9–12

Saha HM, Njunie MN, Tsanje NM (1997) Legume screening for soil fertility improvement in the coastal lowland of Kenya. Paper presented at the LSN workshop held at Kanamai, Mombasa, Kenya, 24–26 March 1997

Simpson MB, Owens M (2002) Farmer field schools and future of agricultural extension in Africa. Sustainable Development Department (SD), Food and Agriculture Organization, Rome

Sperling L, Loevinsohn ME (1993) The dynamics of adoption; distribution and mortality of bean varieties among small farmers in Rwanda. Agr Syst 41:441–453

Sperling L, Scheidegger U, Buruchara R, Nyabyenda P, Munyanesa S (1994) Intensifying production among smallholder farmers: the impact of improved climbing beans in Rwanda. Occassional Publications Series, No. 12. CIAT Network on Bean Research in Africa, Kampala

Stahl M (1993) Land degradation in East Africa. Ambio 22:505–507

Stroud A (1993) Conducting on-farm experiments. Centro Internacional de Agricultura Tropicale (CIAT), Cali

Swinkles RA, Franzel S, Shepherd KD, Ohlsson E, Ndufa JK (1997) The economics of rotation improved fallows: evidence from areas of high population density in Western Kenya. Agricultural Systems 55:99–121

Tarawali SA (1994) Evaluation of *Stylosanthes hamata* (L.) Taub. Accessions for livestock and crop enterprises in subhumid Nigeria. Aust J Exp Agr Sci Camb 127:175–182

Temu AEM, Anne JB (1995) Effect of green manuring and rotations on yield in the Southern Highlands of Tanzania. Nor J Agr Sci Suppl 21:93–98

Thomas D, Sumberg JE (1995) A review of the evaluation and use of tropical forage legumes in sub Saharan Africa. Agr Ecosyst Environ 54:151–163

Tripp R (2001) Seed provision and agricultural development: the institutions of rural change. James Currey, Oxford

Vissoh P, Manyong VM, Carsky JR, Osei-Bonsu P, Galiba M (1998) Experiences with Mucuna in West Africa. In: Buckles D, Ateka A, Osiname O, Galiba M, Galiano G (eds.) Cover crops in West Africa. Contributing to sustainable agriculture. IDRC/IITA, Ibadan, pp 1–32

Wilson GF, Kang BT, Mulongoy K (1986) Alley cropping: trees as sources of green manure and mulch in the tropics. Biol Agr Hortic 3:251–267

Woomer PL (1998) Legume niches in East African Highlands smallhold agriculture: an approach to legume characterization. Department of Soil Science, University of Nairobi, Nairobi

Chapter 6
Comparative Analysis of the Current and Potential Role of Legumes in Integrated Soil Fertility Management in West and Central Africa

A. Bationo, J. Kimetu, B. Vanlauwe, M. Bagayoko, S. Koala, and A.U. Mokwunye

Abstract Legumes form an important component of sustainable food production systems in West and Central Africa. The region is characterized by low and erratic rainfall distribution, low soil fertility, high temperature. Legumes being fast maturing and able to withstand some of these constraints emerge as a strategic investment by the farmers. Legume technologies provide opportunities to contribute to the conservation of the natural resource base while maintaining inter-annual yield and generating increased output. Grain legumes are a key source of nitrogen-rich edible seeds, providing a wide variety of high-protein products and constituting the major source of dietary protein in the diets of the poor in most parts of sub-Saharan Africa. Largely grown as subsistence food crops, they are predominantly crops grown by women and used within the family. The most common grain legumes grown include groundnut (*Arachis hypogaea*), cowpea (*Vigna unguiculata*) and other vigna types (*Vigna radiata,*

A. Bationo (✉)
Alliance for a Green Revolution in Africa (AGRA),
Soil Health Program, PMB KIA 114, Airport-Accra, Ghana
e-mail: abationo@agra-alliance.org

J. Kimetu
Institute for Sustainable Energy, Environment and Economy (ISEEE), University of Calgary, Alberta, Canada
e-mail: jmkimetu@ucalgary.ca

B. Vanlauwe • S. Koala
Tropical Soil Biology and Fertility Inst. of CIAT (TSBF-CIAT), Nairobi, Kenya
e-mail: b.vanlauwe@cgiar.org; s.koala@cgiar.org

M. Bagayoko
Centre Régional de Recherche Agronomique de Niono, Mali
e-mail: minamba.bagayoko@yahoo.fr

A.U. Mokwunye
Mokwunye Consulting, Accra Ghana
e-mail: uzo.mokwunye@alumni.illinois.edu

A. Bationo et al. (eds.), *Fighting Poverty in Sub-Saharan Africa: The Multiple Roles of Legumes in Integrated Soil Fertility Management*,
DOI 10.1007/978-94-007-1536-3_6, © Springer Science+Business Media B.V. 2011

Vigna unbellata), common bean (*Phaseolus vulgaris*), Soybean (*Glycin max.*), pigeon pea (*Cajanus cajan*) bambara groundnut (*Voandzeia subterranean*). Cowpea biomass is a major source of animal fodder especially during the dry season. The grain and biomass of most of the above legumes are marketable, thus proving the farmers with the much needed cash for domestic use.

Intensification of low-input agricultural production has led to a rapid increase in soil degradation and nutrient depletion in many parts of West and Central Africa, constituting serious threat to food production and food security. Nitrogen depletion in maize-based systems in some farmers' fields in West African savanna is estimated to be 36–80 kg N ha^{-1} per year. Despite the recognized need to apply chemical fertilizers for high yields, the use of mineral fertilizers in West Africa is limited by lack of capital, inefficient distribution systems, poor enabling policies, and other socioeconomic factors. The ability to fix atmospheric nitrogen makes legumes excellent components within the various farming systems because they provide residual nitrogen and reduce the needs for mineral nitrogen fertilizers by associated non-legumes. Legumes also play a key role in improving other soil properties such as soil organic matter (SOM), water infiltration and workability.

Over the years, considerable efforts have been made to introduce and promote legume technologies in West and Central Africa. However, with the exception of grain legumes, the adoption of legume technologies has been disappointingly low. The increased production observed in legume grain production has mainly been due to increased area (extensification) exposed to cultivation as opposed to increased per unit (intensification) production. For increased adoption, research and development efforts have to be more farmer-oriented and more emphasis should be given to the development of technologies that are capable of delivering benefits to farmers in the short term and, at the same time, contribute to improved soil fertility.

The importance of legumes for improved soil fertility management is recognized; however, legumes alone are not a solution to the problem of persistent food shortages and land degradation. There is need for research and development efforts to be directed at improving access to improved germplasm, judicious use of organic and inorganic fertilizer, and improved crop-livestock integration.

6.1 Introduction

Western and central Africa are ancient centers of agriculture and domestication of many forage and grain legumes species. The most frequently grain legumes encountered are: groundnut (*Arachis hypogaea*), cowpea (*Vigna unguiculata*) and other vigna types (*Vigna radiata, Vigna unbellata,*), common bean (*Phaseolus vulgaris*), Soybean (*Glycin max.*), pigeon pea (*Cajanus cajan*) bambara groundnut (*Voandzeia subterranean*), the potatoes of Mossi (*Sphenostylis stenocarpa*). Forage legumes include *Mucuna pruriens, Dolicos lab lab, Canavalia ensiformis, Crotalaria Juncea, Crotalaria espectabilis, Crotalaria breviflora, and Sesbania rostrata.*

In West Africa, plant growth conditions are characterized by low and erratic rainfall distribution, low soil fertility, high temperature. During most of the year, potential evapotranspiration is much higher than rainfall so that moisture deficit generally limit crop yields. Sunshine hours and potential evapotranspiration are fairly similar from year to year but the amount of rainfall and its distribution within a year vary greatly between seasons (Sivakumar 1989). Changing yearly pattern of soil water availability is a dominant aspect for plant growth. Under such harsh environmental conditions increasing the use efficiency of available soil moisture through cropping strategies is a critical issue to secure food production.

Soils of Western and central Africa are of different origin and besides climatic constraints, most of the soils have low organic matter, nitrogen content and low phosphorus (P) availability (Bertrand et al. 1972; Traoré 1974; Lombin and Singh 1986; Christianson et al. 1990). Since N and P are very limited in many soils, raising their supply levels will substantially increase the productivity of agricultural soils. In Western and central Africa, grain legumes are traditionally grown in rotation or in intercrop with cereals to secure food production. Improvement of the traditional cereal legume cropping systems can be a good opportunity to increase soil N balance. Specific problems in grain legume production by African farmers are yield instability, drought susceptibility, changing product prices and varying fertilizer responses. Because legumes are important part of daily protein for human and animals, strategies to develop improved production practices are urgently needed.

Population pressures, intensive cultivation, shortening fallow periods and increasing nutrient losses have resulted in declining soil fertility and hence threatened sustainable agricultural production in the densely populated areas of West and Central Africa. For decades these regions relied on shifting cultivation practices as their main farming systems. These traditional slash-and-burn systems operated efficiently in the past, when land for cropping was abundant (Manyong et al. 1999). However, with increasing human population, these agricultural practices are under severe pressure. Marginal lands are now being brought into cultivation and smallholder farmers are compelled to change from extensive to intensive use of the land. The bush fallows practiced by smallholders, which used to be long enough (10 years or more) to allow the replenishment of soil fertility, have been drastically shortened or simply abandoned (Vissoh et al. 1998). Inadequate agricultural practices have led to a dramatic decline in soil fertility, invasion by noxious weeds, and extremely low crop yields. Thus, achieving food security for a rapidly growing population will require intensification of food production on existing crop land through improving soil fertility and agronomic practices.

Various uses are attached to legumes which include: improving human nutrition, fodder situation and to increase crop production. All these gains are indirectly through soil fertility replenishment by the leguminous plants. These benefits are of special interest especially for regions with nitrogen deficiency as is the case in most regions of sub-Saharan Africa. Two main reasons why legumes are preferred to mineral nitrogen fertilizers are: the high cost of buying fertilizer and the danger of making farmers market dependent (Breman and van Reuler 2002).

The limited availability of nitrogen can be improved by exploiting the biological fixation of N by leguminous plant species. However, for legumes to obtain a comparative advantage over other species, they have to benefit from conditions that enable them to exploit the advantage of their biological nitrogen fixation mechanism. Legumes will do well in N deficient soils and where available P is relatively abundant, the right rhizobium strain are present, and limited pest and disease pressure.

N deficiency is a common phenomenon in West Africa soils that legumes could be useful in most parts of this region. However, the only limitation is the low levels of available P in the soil. Thus legume contribution to the annual production of the herb layer of natural vegetation and rangelands is only about 5%. Some legumes like *Zornia glochidiata* are able to fix 15–55 kg N ha^{-1} annually (Penning de Vries and Djiteye 1982).

In the northern Sahel, several constraints to legume development have been identified which include: limited water availability, inherent limited P solubility and lack of enough time for mycorrhizae to work effectively (Breman and De Ridder 1991). Although there is lower pressure of legume pests and diseases due to the short single growing season in the semi-arid lands, the extreme high temperatures and relatively reliable rainfall favour other plant species which offer steep competition to slow germinating prostrate C_3 legumes. Thus only leguminous tress seems to benefit from the relatively favorable soil and climate conditions of the semi-arid zones (Breman and van Reuler 2002).

Nitrogen deficiency in the soils of the sub-Saharan Africa region is such that the minimum values are reached soon after germination and establishment of plants after which crop growth is retarded leading to low yields (Penning de Vries and Djiteye 1982). The use of leguminous plants can be considered as one option to counteracting this deficiency.

6.2 Climate, Soil and Biophysical Characteristics

West Africa has several distinct agroecological zones. The growing season varies from about 300 days along the coast from Sierra Leone to Cote d'Ivoire to about 60 days in the Sahel. The arid one has 60–120 days, and the semi-arid zone has 120–150 days. Subhumid (150–270-days growing season) zones are found in Senegal, the Gambia, Guinea Bissau, Guinea, Mali, Burkina Faso, Cote d'Ivoire, Ghana, Togo, Benin, Gambia and Nigeria. The humid forest zone has a length of growing period of 271–365 days. The humid forest zone is divided into different subzones following gradients in degradation of the original forest: forest margins, forest pockets and degraded forest (Keatinge et al. 2001). The rainfall pattern encountered in this region is bimodal where the length of the growing period is eight or more months. Soil conditions determine the intensity of cropping, the need for fallowing, the varieties of crops that can be grown, and the risk of drought. Major soils are the Alfisols in the savanna areas and Oxisols or Ultisols in the humid areas (Jones and Wild 1975). Using the Food Agriculture Organization (FAO) of the United Nations Manyong et al. (1996) described the soils of West Africa into two categories – that is, those fertile enough to support crops – occupy only 24% of West Africa. The remaining 76% of soil can be characterized as poor, requiring judicious use of inorganic fertilizers, organic matter management and soil erosion control.

Weeds are a major constraint to crop growth in the humid and subhumid zones and weeding requires a major labour input. Two major weeds in this region include the spear grass (*Imperata cylindrica*) in the humid zones and the witchweed (*Striga* spp.) in the dry zones (Akobundu and Agyakwa 1987). Spear grass is common in the intensively used lands and is considered one of the most difficult weed to control in the tropics. The weed has caused many farmers in southern Benin to abandon their fields to fallow. The most serious species of witchweed is *Striga hermonthica*, which is parasitic to maize, sorghum and millet. The weed is difficult to control because it does much damage beneath the soil surface (Weber et al. 1995) to flowering plants.

6.3 Legumes in Context – Importance in Relation to Other Crops

There is a wide range of technological options for improving soil conservation and land management that are economically viable, ecologically sound, and socially acceptable. These include inorganic fertilizers, crop residue management, green manure, composting, farmyard manure, agroforestry technologies, alley farming, planted fallow, cover crops, and cereal legumes intercropping or rotation (IITA 1998; Buckles et al. 1998). Legumes form an important component of sustainable food production systems in West and Central Africa. Supplementing inorganic fertilizers with legume green manuring may add nutrients, like N, S, and P to the soil. Green manuring may also help maintain or build soil organic matter (OM), which will in turn help improve the soil's structure, pore size, water holding capacity and inorganic-fertilizer-use efficiency.

6.4 Mucuna

One of the most promising legume cover crops to improve soil fertility in intensified cropping systems is *Mucuna* fallow. *Mucuna* is indeed a legume cover crop that is an efficient, low-cost source of nitrogen with considerable potential to improve soil fertility in intensified cropping systems (Sanginga et al. 1996; Carsky et al. 1998; Buckles et al. 1998; IITA 1998). It is also efficient in controlling noxious weeds such as *Imperata cylindrica* (Akobundu and Udensi 1995) and nut grass (*Cyperus rotundus*), two of the most difficult weeds to control in the tropics (Carsky et al. 1998). There are reports that *Mucuna* can be used for reducing nematode populations. *Mucuna* has also been used in cattle grazing, and can provide animal feed and human food (Carsky et al. 1998; Versteeg et al. 1998). Two different management systems have been developed for the integration of *Mucuna* into cropping systems.

One is a sole cover crop fallow for severely degraded fields. The other is a maize/ *Mucuna* relay crop for fields requiring less rehabilitation. For severely degraded and *Imperata*-infested fields, *Mucuna* should be planted in a pure stand at the onset of the rainy season. Production of 7–9 t ha^{-1} of dry matter is commonly observed in the bimodal rainfall zone (Vissoh et al. 1998). In the dry season, *Mucuna* ends its life cycle, leaving a thick mulch free of weeds. This allows for a subsequent maize crop during the major rainy season with little or no land preparation or weeding. Maize can be seeded directly through the mulch using a stick, hoe, or cutlass. *Mucuna* can also be intercropped with maize when the *Imperata* infestation is not severe. The technology is to plant *Mucuna* in a relay cropping system with food crops such as maize. After the maize harvest, the land is left to *Mucuna* fallow, which allows groundcover to develop fully for biomass accumulation and nitrogen fixation while weeds, such as *Imperata*, are smothered. During the following dry season, the *Mucuna* dies off and the farmer can farm the field in the next main cropping season (Versteeg and Koudokpon 1990).

Mucuna's ability to suppress spear grass was the major reason for its adoption in the bimodal rainfall zones of Benin. In Mono province for example, Versteeg and Kaudokpon (1990) indicated that Mucuna reduced Imperata to less than 10% of its initial density on farmers fields. Although most benefits from Mucuna use have not been extensively measured, Versteeg and Koudokpon (1990) reported maize grain yield increases on farmers fields of about 500 kg ha^{-1} for local maize variety and about 800 kg ha^{-1} for an improved variety, following 1 year of fallow with Mucuna. Osei-Bonso and Buckles (1993) estimated that Mucuna as an intercrop or as a sole crop provided an equivalent of more than 100 kg N ha^{-1} to the following maize. This is similar to an amount Sanginga et al. (1996) estimated for bimodal rainfall zone of southern Nigeria. In another study in Nigeria, Codjia (1996) observed 98% higher maize yields after a Mucunas short fallow without chemical fertilizer application and a 179% increase with 51 kg N, 46 kg P and 28 kg K ha^{-1}. These results indicate that for high maize yields to be attained, Mucuna residue should be supplemented with moderate amounts of inorganic fertilizer.

Mucuna fallowing has additional benefits, such as erosion control and maintenance or improvement of soil physical, chemical and biological properties. In an onstation study at IITA by Hulugalle et al. (1986), porosity and infiltration rates increased and penetrometer resistance decreased with amount of Mucuna biomass produced. Azontonde (1993) studied soil erosion under a maize-Mucuna-relay intercropping system in southern Benin and observed soil losses of 3–7.5 t soil ha^{-1} in the maize-Mucuna plot as compared with 30 t soil ha^{-1} in the flat sole-maize plot and 10 t soil ha^{-1} when maize was planted on contour ridges.

During the evaluation of the Mucuna technology in southern Benin, farmers observed that the legume was very effective at suppressing one of their most serious weeds, *Imperata cylindrica*. Imperata takes hold as the length of fallows increases and soil fertility declines and requires a substantial amount of labour to deal with, often forcing farmers to abandon their fields. This was obviously a serious constraint to crop production in a highly densely populated area as is southern Benin. Evaluation with farmers also revealed their reluctance to lose a second season food crop because

Mucuna did not yield a marketable or consumable product (Manyong et al. 1999). Farmers were calculating that the immediate opportunity cost of the lost crop was higher than the future benefits of a Mucuna cover crop. Efforts to deal with the constraint identified focussed around the creation of markets for Mucuna seeds, using its residues as fodder for livestock, and enhancing the edibility of Mucuna seeds for humans and livestock by removing toxic L-Dopa (Carsky et al. 2001).

6.5 Pueraria

Pueraria phaseoloides (Roxb.) Benth (Pueraria), used as a cover crop in association with food crops can improve soil nutrient availability, stimulate soil biological activity and control soil erosion, leading to high soil productivity (Tian et al. 2001). In West Africa, Pueraria has been shown to accumulate 150–250 kg N ha^{-1} of which 68% of this N is derived from the atmosphere on P and K poor soil, and 87% on soils with sufficient P and K.

Pueraria has been shown to have the ability to maintain soil physical properties due to its good ground cover and improvement in soil faunal activities (Tian et al. 2001). When intercropped with food crops, Pueraria covers the ground completely before harvesting of the crops. Salako et al. (1999) reported enhanced soil aggregate stability in Pueraria fallow system more than in either bush fallow or in leucaena alley cropping. In a study by Salako et al. (2000), lower soil loss was noted in Pueraria cropping system as compared to cowpea system (Fig. 6.1). The reduction in soil loss was attributed to greater residue supply from Pueraria and the increased water infiltration after the inclusion of Pueraria (Gijsman and Thomas 1996).

Significantly higher yields of maize, cowpea and cassava under Pueraria fallow (compared to natural fallow) were reported in Ibadan, Nigeria (Lal et al. 1978; Tian et al. 2001). Tian et al. (2000a) in a study at Ibadan 1995 reported doubling of maize grain yields after Pueraria compared to grass fallow. The implication was that 1 year Pueraria fallow replaced 50–60 kg ha^{-1} of fertilizer nitrogen for maize. In a different study in on-farm trials, Tian et al. (2000b) demonstrated that Pueraria fallow systems resulted in higher maize grain yield compared to natural regrowth controls (Fig. 6.2).

6.6 Leguminous Trees

Due to their many uses, environmental and socioeconomic benefits, trees and shrubs feature prominently in traditional farming systems in the tropics. Some of the notable systems include bush fallow or improved fallow systems, home gardens, parkland systems, alley cropping systems and the agroforests in the humid zone. Most traditional farmers use woody species to suppress weeds and pests as well as for soil fertility restoration. Soil enrichment is mainly through nutrient cycling and plant litter fall and biological nitrogen fixation by legume species (Kang and Shannon 2001).

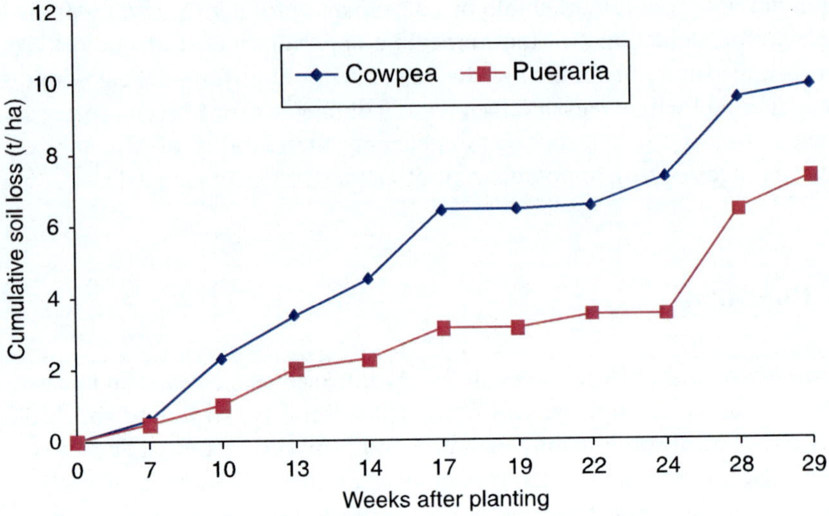

Fig. 6.1 Reduction of soil erosion by pueraria residues in southern Nigeria (Source: Salako et al. 2000)

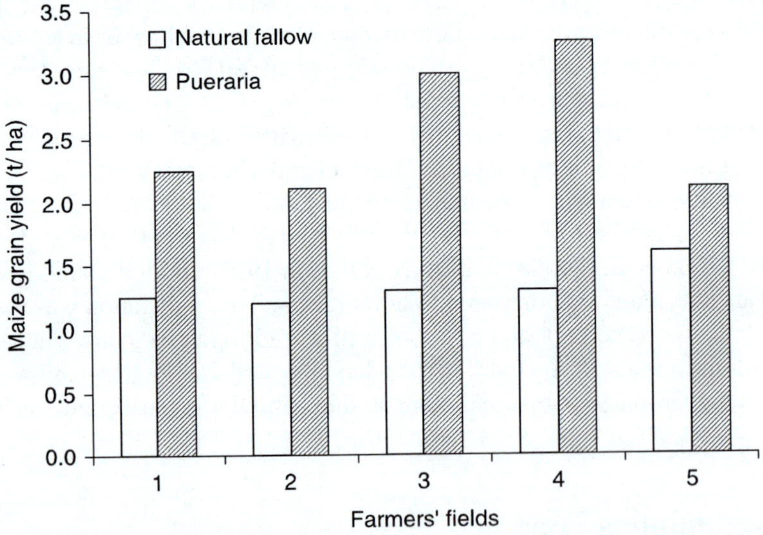

Fig. 6.2 Maize grain yields after pueraria cover crop fallow and traditional natural fallow in 5 farmers' fields at Alabata, Nigeria (Source: Tian et al. 2000b)

In the humid zone, the effect of woody species in fallow systems on soil fertility is highly dependent on species and soil factors (Kang and Shannon 2001). In acidic and low base status soils, the effect is less pronounced as compared to high base status soils.

6.7 Grain Legumes in Central and West Africa

There are many other crops, including legumes (e.g., cowpeas, beans, groundnut), and roots and tubers (e.g., cassava, yams, potatoes), which are immensely important food crops in Africa, both in terms of production and caloric contribution to African diets. However, the available evidence of research impacts on these food crops is scanty, not only in Africa but also in other parts of the developing world. Some of these crops have been labeled "orphan crops," since little interest was shown by colonial powers and, initially, by the international donor community to improve them. However, this has changed in the last few decades when they became the mandated research crops of many International Agriculture Research Centres (IARCs) (e.g., IITA, ICRISAT, CIAT, CIP) and a focus of some donor-supported collaborative research programs in Africa (e.g., Bean – Cowpea CRSP).

6.8 Cowpea

Cowpea, *Vigna unguiculata* (L.) Walp., is a grain legume grown in savanna regions of the tropics and subtropics, especially West and Central Africa. Cowpeas, which most likely originated in West Africa, are now found throughout the world. The principal production areas are Nigeria and Niger, with approximately one half of world cowpea production (Sanders et al. 1994). FAO estimates that 3.3 million tonnes of cowpea dry grains were produced worldwide in 2000. Nigeria produced 2.1 million tonnes, making it the world's largest producer, followed by Niger (650,000 t) and Mali (110,000 t). Total area grown to cowpea was 9.8 million hectares, about 9.3 million hectares of these in West Africa. World average yield was 337 kg ha^{-1}; average yield in Nigeria was 417 kg ha^{-1}, and in Niger was 171 kg ha^{-1}. Even though the cowpea area and production in other West African countries are relatively small, cowpeas are very important in this region, both as a nutritional food source for the farm household and for improving the cropping system by providing the fertility and other benefits of a cereals-grain legumes rotation. The value of cowpeas lies in its high protein content, its ability to tolerate drought, and the fact that it fixes atmospheric nitrogen which allows it to grow on, and improve, poor soils. In fresh form, the young leaves, immature pods, and peas are used as food.

Traditionally in West and Central Africa, cowpea is grown on small farms, often intercropped with cereals such as millet and sorghum. Pesticides are generally not used, because they are too expensive or not available. The cowpea and cereal are usually planted in alternating rows, although recent research at International Institute for Tropical Agriculture (IITA) has shown that planting four rows of cowpea to two rows of cereal is more productive. The cereal is planted first, followed by the cowpea. The fast growth and spreading habit of traditional cowpea varieties suppress weeds, and supply soil nitrogen which improves cereal growth. The two crops are harvested at different times, spreading out demand for available labor.

Research on cowpeas and the establishment and maintenance of cowpea germplasm was probably first carried out in the early 1950s at the Federal West African Agricultural Research Station at Bambey, Senegal (Bingen et al. 1988). At present, the largest cowpea research program in the region is in Nigeria. However, several National Agriculture Research Systems (NARS) in Senegal, Mali, Burkina Faso, Niger, Ghana and Cameroon also have active cowpea research programs. In addition to the efforts of NARSs, IITA, donor supported collaborative networks (e.g., Collaborative Research Support Programs - CRSP) and non governmental organizations (NGOs) play an important role in developing and promoting the use of improved cowpea varieties in this region. Cowpea breeding research in this region is mainly focused on developing resistance for *striga*, field insects (aphids, thrips), storage insects (bruchids) and diseases. This research has had substantial successes in the introduction of new early maturing cowpea varieties in Western and Central Africa. In Ghana a new variety, Vallenga, released in 1987, has been introduced on more than 20,000 ha in the north, raising farmers' yields to approximately 1 t ha^{-1}. The introduction of an early maturing cowpea variety in northern Senegal in 1985 doubled the area sown to cowpeas and increased the national production from the previous 15 year average of 17,800 metric tons to 66,000 t in that year (Schwartz et al. 1990). Diffusion of eight new cowpea varieties in Burkina Faso was estimated to reach almost 100% of the cowpea area by 1992 (Sanders et al. 1994). The estimated yield differentials between improved and traditional cowpeas vary greatly depending on the production techniques. In Niger and Cameroon, the yield advantage was estimated to range from 25% to 46% depending on the region (Sterns and Bernsten 1994; Mazzucato and Ly 1994). The yield advantage of the most recent Institut Sénégalais de Recherches Agricoles (ISRA) releases in Senegal, is about 40% over older improved varieties in on-farm trials. Increases in cowpea production are larger with phosphate application and pest management. The available evidence on economic rate of returns to cowpea research show that improvement research on this crop has ranged from as low as 3% in northern Cameroon to as high as 32–92% in Senegal, suggesting that research and technology transfer activities in cowpeas have contributed to increased productivity of the agricultural sector in several countries. The benefits of cowpea research are also in the form of gains due to maintaining yields over time in an increasingly difficult environment. It was estimated that, as a result of the development and diffusion of improved resistant varieties, about 25–50% in yield loss was averted (Sanders et al. 1994). For Burkina Faso and Mali, (Sanders et al. 1994) estimated that the annual benefits to maintaining farmers' yields ranged from $0.8 to $4.8 million per year between 1984 and 1992, using the most conservative assumption of 25% yield decline in the absence of new varieties. There are, however, other impacts of cowpea improvement research that are not quantifiable with the rates of returns measure. Notable among these is the impact on household food security by increasing the supply of fresh green-pods in the "hungry season." The early harvest of cowpea grain and fresh green pods made possible by improved, early maturing varieties helps to alleviate the household food constraint faced early in the season. In Senegal, for example, it was estimated that about 17.5% of total cowpea production was consumed as fresh green pods.

Table 6.1 Millet grain and total dry matter yield at harvest as influenced by millet/cowpea cropping system at Sadore (Niger) (Source: Bationo and Ntare 2000)

	Grain yield (kg ha^{-1})			Total dry matter yield (kg ha^{-1})		
Cropping system	1996	1997	1998	1996	1997	1998
Continuous millet	937	321	1,557	4,227	2,219	6,992
Millet after cowpea	1,255	340	1,904	5,785	2,832	8,613
P>F	<0.001	0.344	<0.001	<0.001	<0.001	<0.001

Another benefit of cowpea research not included in most rates of return studies is the decline in storage losses of dried cowpeas. Recognizing the importance of post-harvest losses of cowpeas, the research agenda in several countries have shifted the breeding strategy to develop varieties resistant to storage pests, such as bruchids, but also to develop new storage methods and techniques to reduce grain losses. An impact assessment of progress in Cameroon shows an approximately 6% ROR for cowpea storage research if the technology is used only in northern Cameroon. However, if these technologies are extended and adapted to other Western African countries the rates of return would be almost 50%.

Despite its importance, the yields of cowpea are very low due to several constraints including poor soil, insect pests, and drought. The soils in semiarid West Africa are inherently low in nitrogen and phosphorus. Soil, water, and nutrient management practices are inadequate to sustain food production and to meet the food requirements of the fast growing population. Research results show that proper management of organic amendments such as crop residues and manure, which are essential complements to mineral phosphorus fertilizers, can increase yields of cowpea and associated cereals more than three fold. Direct application of indigenous phosphate rocks can be an economical alternative to the use of imported, more expensive soluble phosphorus fertilizers for cowpea production in the region. The agronomic effectiveness of indigenous phosphate rock is about 50% compared to the imported single super-phosphate. Furthermore, when the unreactive phosphate rocks are partially acidulated at 50%, their agronomic effectiveness can increase to more than 70%. Studies on cereal – cowpea rotation revealed that yields of cereals succeeding cowpea could, in some cases, double compared to continuous cereal cultivation. Table 6.1 shows the effect of cowpea-millet rotation on millet grain and total biomass production. In a period of 3 years, there was an increase of about 3 t ha^{-1} of total dry matter production when millet was grown in rotation with cowpea. With efficient soil fertility management, cowpea can fix up to 88 kg N ha^{-1} and this results in an increase of nitrogen use efficiency on the succeeding cereal crop from 20% in the continuous cereal monoculture to 28%when cereals are in rotation with cowpea. Furthermore, the use of soil nitrogen increased from 39 kg N ha^{-1} in the continuous cereal monoculture to 62 kg N ha^{-1} in the rotation systems. There is also a need to quantify the below-ground nitrogen fixed by different cowpea cultivars. The increase of cowpea productivity in the cropping systems in this region will improve the nutrition of people, increase the feed quantity and quality for livestock, and contribute to soil fertility maintenance. This should contribute to reduction in poverty and environmental degradation.

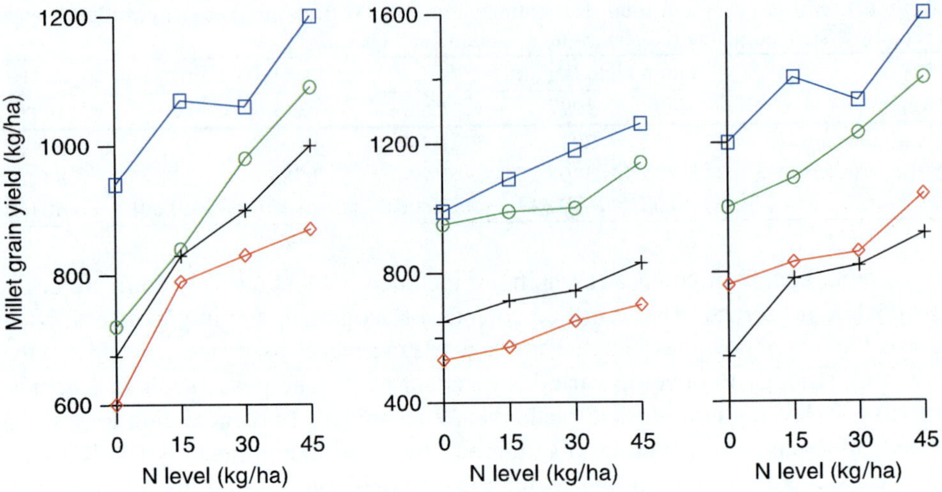

Fig. 6.3 Effects of nitrogen and rotation on pearl millet grain yield (kg ha^{-1}, average of 4 years) at Sadore, Tara and Bengou

In a study by Bationo (unpublished data) at three sites (Sadore, Bengou and Tara), rotation of pearl millet with groundnut and cowpea resulted in higher significant pearl millet grain yields than in the monoculture cropping of pearl millet on average over a 4-year period (Fig. 6.3). Grain yield of millet following cowpea was higher than millet following groundnut at the three sites. Similarly, millet total dry matter was higher in the cowpea/millet rotation, except at Bengou where the total dry matter was higher in the groundnut/millet rotation system. With no application of N fertilizer millet grain yield after cowpea increased by 57% at Sadore, 18% at Bengou and 87% at Tara. Even at higher levels of N, continuous cropping of millet produced the lowest yields of both millet grain and total dry matter. The response of legumes to rotation with millet was also apparent at Tara and Bengou where legume yields were consistently lower in monoculture than when rotated with millet (Fig. 6.4). This suggests that factors other than N alone contributed to the yield increases in the millet-legume rotations. It has been assume by many workers that the positive effect of rotations arise from the added N from legume in the cropping system. Some workers, however, have attributed the positive effects of rotations to the improvement of soil biological and physical properties (Hoshikawa 1990) and the ability of some legumes to solubilize occluded P and highly insoluble calcium bounded phosphorus by legume root exudates (Arihara and Ohwaki 1989). Other advantages of crop rotations include soil conservation (Stoop and Staveren 1981), organic matter restoration (Spurgeon and Grimson 1965) and pest and disease control (Curl 1963).

Bagayoko et al. (1997) noted that cowpea yield after maize and after legumes converged at higher N levels (Fig. 6.5), suggesting that only N benefits were realized. At Cinzana, Mali (latitude 13°N), millet was grown after cowpea for 4 years with a continuous millet control and it was found that 40 kg N ha^{-1} applied to the continuous millet gave a yield similar to cowpea rotation. Thus, it was estimated that the

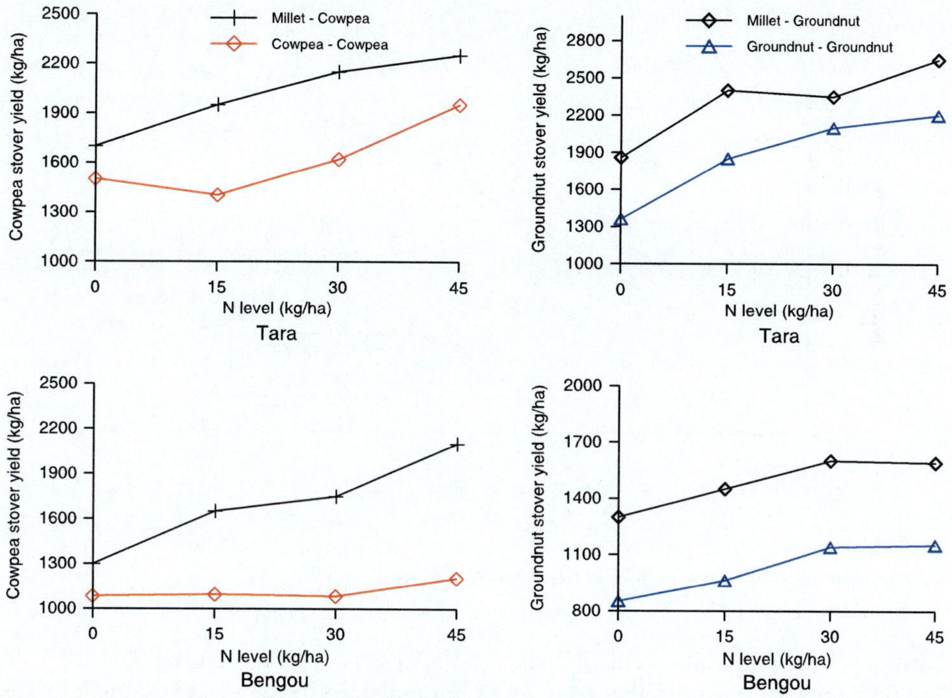

Fig. 6.4 Effects of nitrogen and rotation on legume stover yield (kg ha^{-1}, average of 4 years) at Tara and Bengou

Fig. 6.5 Effect of cowpea rotation on response of millet to N fertilizer in Mali (Source: Bagayoko et al. 1997)

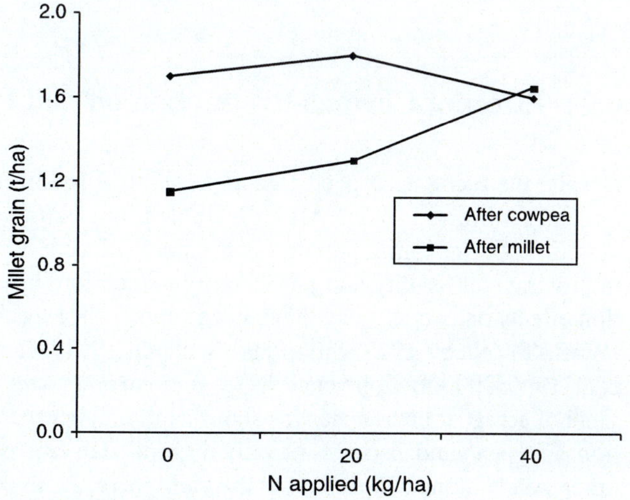

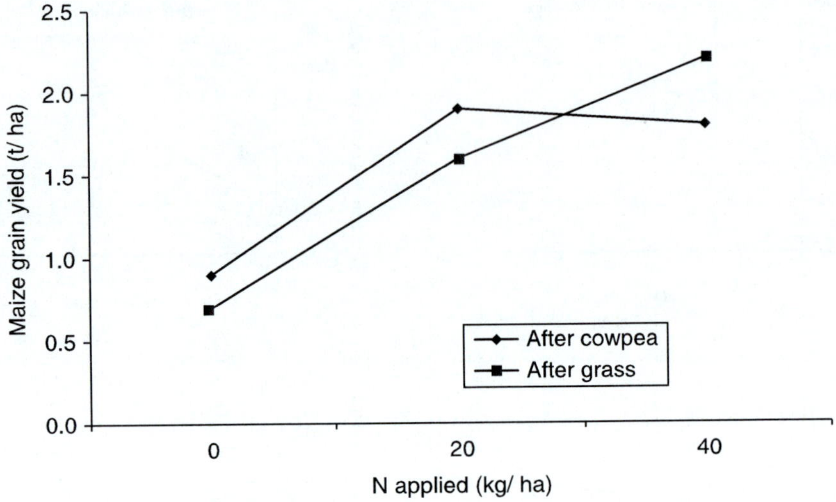

Fig. 6.6 Response of maize to N after cowpea and after natural grass fallow at two northern Guinea savanna sites in northern Nigeria (Source: Carsky et al. 1999)

Nitrogen Fertilizer Replacement Value (NFRV) was approximately 40 kg ha^{-1}. It should be noted, however, that soil N was not higher in the cowpea system than the continuous millet system after 4 years. The two previous estimates of Nitrogen Fertilizer Replacement Value (NFRV) are substantial, approaching 40 kg ha^{-1}. In contrast, the mean Nitrogen Fertilizer Replacement Value (NFRV) from two sites in northern Nigeria was only 9 kg ha^{-1} for one season of cowpea in the first year followed by one season of maize in the second year in the Guinea savanna of northern Nigeria at latitude 11°N (Carsky et al. 1999; Fig. 6.6). In this case, the cowpea effect was compared to native fallow rather than a continuous cereal control. A continuous cereal control is likely to give higher estimates of NFRV than fallow because of higher N export by the cereal.

6.9 Effect of Cowpea Production on Soil Fertility Improvement

Despite the recognized need to apply chemical fertilizers for high yields, the use of mineral fertilizers in West Africa is limited by lack of capital, inefficient distribution systems, poor enabling policies, and other socioeconomic factors. Cheaper means of improving soil fertility and productivity is therefore necessary. Cereal – legume rotation effects on cereal yields have been reported for the West Africa semi arid tropics (WASAT) (Nicou 1977; Stoop and Staveren 1981; Klaij and Ntare 1995; Bagayoko et al. 1996, 2000; Bationo et al. 1998; Bationo and Ntare 2000). In all these studies, the yield of cereal after cowpea was significantly higher than in continuous cereal cultivation. Cowpea yield also significantly responded to crop rotation, indicating that factors other than N alone contributed to the yield increases in the cereal – legume rotations.

A synthesis of results from the savanna zone of West Africa done by Carsky et al. (2002) suggest that cowpea rotation can be considered to be an effective resource management technology in cereal-based systems. Part of the N requirement of cereal crops can be satisfied by cowpea crop rotation. Furthermore, benefits of cowpea rotation are sometimes higher than expected based on the N content of the cowpea crop alone. Reasons for this have been cited to be substantial root biomass and N, substantial N-sparing by the legume, and other benefits such as reduction in *Striga hermonthica*, or pests and often diseases, and possibly access to sparingly soluble P. The characteristics to be encouraged to maximize the N benefit are the amount of nitrogen derived from the atmosphere and the amount of N returned in the residues. These researchers also suggested that (1) the maturity class of the cowpea variety should be as late as possible, (2) the cereal should be planted as soon as possible after cowpea has been harvested, and (3) minimum soil requirements for optimum cowpea growth should be respected.

6.10 Bean Production in Central and West Africa

Common beans (*Phaseolus vulgaris*) are grown on more than 3.5 million hectares in Africa. Beans are cultivated for subsistence and, increasingly, as a cash crop by many farmers, most of them women. For many rural and urban consumers in Africa, beans provide the least expensive source of calories and protein. Although bean yields in Africa have increased modestly during recent years, the rate of increase in production still lags behind population growth. The full realization of this crop's potential to combat hunger and poverty requires a major research effort aimed at overcoming key constraints.

Although a precise figure on research resources devoted to beans is not available, bean improvement research is becoming increasingly important at both the national and international levels. For several years, the national bean research programs in Africa have been developing various technologies, including: high yielding, stress tolerant varieties; pest management strategies; and soil fertility amendments to address bean production constraints and increase yields. As a result of these increased efforts, the national programs in Africa have released 42 new bean varieties since 1984. Nineteen of these were breeding lines developed at CIAT headquarters, and another 23 were selected in Africa based on CIAT introductions.

A review of the dissemination and adoption of new bean technologies in eight Eastern and Central African countries in 1992–1996 revealed that approximately 69 improved bean varieties were widely available to farmers in these countries through formal and informal seed channels (David 1997). No comprehensive information is yet available on the adoption and impact on crop production due to the adoption of new bean varieties. Recent adoption studies by Sperling et al. (1994) and Hoogendijk and David (1997), however, report wide adoption of climbing bean varieties in Rwanda and Uganda. In terms of area, 10–20% of the total bean area in Rwanda in the early 1990s was planted to 20 improved climbing beans introduced 5 years earlier. The study concluded that "the use of improved climbing beans annually brings 31,000–66,000 additional tons of beans for Rwanda, equivalent to an extra US$ 8 to 15 million in income to Rwandan farmers" (Sperling et al. 1994, p.8).

6.11 Groundnut

Groundnuts (*Arachis hypogaea L.*) or peanuts are of South American origin where they were grown by the Indian communities. The crop was first introduced to West-Africa in the Senegambia area by the Portuguese in the sixteenth century. Then it spread quickly in the interior of Africa. With industrialization of Europe, cultivation on a larger scale began when the potential of the crop for oil was realized (need for lubricant, production of soap, and for cooking).

Today, groundnut is one of the main food and cash agricultural products of western and central Africa. It is a high value crop that can be marketed with little processing and can be used in a wide range of products. The oil can be used for cooking, as a base for confectioneries and the grain can be used to make peanut butter.

In West Africa (Niger, Mali, Sénégal) the crop is widely grown on acid sandy soil as sole crop or intercrop with sorghum and millet. The use of mineral fertilizer is not a common practice but local farmers sometimes use farm yard manure to fertilise the crop. Although the crop has a high production potential, the average pod yields in the Sahelian zones are only about 500–700 kg ha^{-1}. Because phosphorus appears as the most yield limiting factor in West Africa (Traoré 1974; Christianson et al. 1990) with application of P fertilizers, good yield can be achieved with groundnut (Bertrand et al. 1972; Lombin and Singh 1986). Molybdenum deficiency has been observed in several producing areas of West Africa (Martin and Fourier 1965; Heathcote 1970) and has been found to impair symbiotic N_2 fixation of groundnut in Niger (Hafner et al. 1992).

Groundnut generally grows very poorly in the infertile sandy soils of West Africa. Although the crop can get most of its N requirement symbiotically, (Rebafka 1993). Increased N_2 fixation and N uptake were observed by Hafner et al. (1992) and Rebafka (1993) with addition of crop residues to soil. Since nitrogenase activity in groundnut is depressed by high concentration of Mn (Nambiar and Anjaiah 1989), improved N_2 fixation and dry matter production of legume by crop residue may also be related at least in part to a decrease in the availability of Mn as reflected in the low concentration in the soil and shoot (Rebafka 1993).

In the poor sandy soils of West Africa, the symbiotic N_2 fixation is substantially depressed by low molybdenum (Mo) availability (Hafner et al. 1992). This is because these soils have high free iron oxides concentrations and Mo solubility is depressed by the free iron and the low pH.

To alleviate the depressive effect of Mo deficiencies on groundnut growth and N_2 fixation, Hafner (1992) proposed seed pelleting with molybdenum trioxide at the rate of 100 g Mo ha^{-1}. According to Hafner (1992), the use of this technique may increase the yield potential by 25–80% depending on some other growth limiting factors such as water stress and diseases.

6.12 Bambara Groundnut

Bambara groundnut [*Voandzeia subterranea* (L.)] has many common names such as Congo groundnut, Congo goober, Madagascar groundnut, earth pea, baffin pea, njugo bean (S. Africa), voandzou, nzama (Malawi), indhlubu, and underground bean. It is a leguminous plant which is grown for its underground seeds. The entire plant is similar to the common peanut, with compound leaves of three leaflets. Like the peanut, it forms pods and seeds on or just below the ground. The pods which are round, wrinkled, contain one or two seeds that are round, smooth, and very hard when dried. The seeds may be white, brown, red, or black-eyed. Many nitrogen-fixing nodules can be observed on the roots. Bambara nut, produces a nutritious food under cultivation throughout Africa.

The crop grows best in climates suitable for peanuts. It needs bright sunshine, high temperatures, at least 4 months and regular rains to grow best. The plant, however, is highly adaptable and tolerates harsh conditions better than most crops. It yields under conditions too dry for peanuts and is known to grow in rain forest areas, and the cool moist highlands of Rhodesia. In soils high in nitrogen, it produces mostly above ground leafy biomass and only a few pods and seeds. When grown like the peanut, Bambara groundnuts may take about 3–6 months to mature, depending on climatic conditions and the cultivar.

The seeds are often roasted and may be eaten raw when immature, but become too hard when mature. Bambara nut are sweet when roasted and can be used to make flour. Seeds contain 14–24% protein and about 60% carbohydrate. The protein is reported to be higher in the essential amino acid methionine than other grain legumes. Bambara groundnuts contain 6–12% oil, which is less than half the amount found in peanuts, making them not useful as an oilseed crop.

6.13 Soybean

Soybean has been identified as an alternative source of less expensive high quality protein for improving nutrition and health and livelihoods of Africa's rural communities (IITA 2002). This has led to the initiation of agronomic and breeding research by IITA with the main aim of developing management practices to optimize yields and varieties that are better adapted to the African environment. Within a period of about 7 years (1987–1993), IITA implemented an IDRC/CIDA-funded project in Nigeria which led to an increased local consumption of soybean and stimulated production and industrial processing of the beans in Nigeria. For the past 12 years, breeders in IITA have focused on developing improved varieties of soybean that can fix a high proportion of nitrogen from the atmosphere as well as kill *Striga hermothica.* According to these researchers at IITA, these varieties produce large amounts of biomass. In addition, when these new improved varieties are planted in rotation with maize or sorghum, the productivity and sustainability of crop production is enhanced. A researcher-managed on-farm experiment in West Africa was conducted by Jagtap and Adeleye (1999) on soil with low available P (1.90 mg kg^{-1}) to evaluate nutrient carryover from soyabeans grown under different P management options and productivity of maize grown in rotation. Promiscuously nodulating 110 day maturing soyabean variety TGX 1479-1E was sown under eight fertilizer treatments comprising of two levels of N (0 and 60 kg ha^{-1}) and four levels of P (0, 15, 30 and 60 kg ha^{-1}). In the following year, maize variety DMLSR-Y was sown without fertilizer on soyabean fallow plots with two bush fallow treatments at 0 and 60 kg N ha^{-1} as checks for system efficiency evaluation. With 0 kg N, P fertilization resulted in 66–75% of total soyabean N uptake in pods while that in the stover decreased from 29% to 19%. Conversely, with 60 kg N, P fertilization caused 11% reduction in pod N and a proportional increase in stover. An increase of N in the stover resulted in higher N carryover in litter fall and roots to the subsequent cereal crop. Soil N mining by soyabean was reduced by application of P at 0 kg N, while it increased with P at 60 kg N. Of the N estimated to have been retained in the field only 10% was recovered in the maize and the remainder was lost to leaching, wind removal and animal grazing. Soil N monitored during non-cropped and cropped periods showed greater nutrient concentrations in the soyabean fallow plots than in the bush fallow plots. Maize following soyabean was productive, soil nutrient conserving and more nutrient efficient than the maize grown on the bush fallow plots. Economic analysis of the two production systems showed that gross returns from the soyabean-maize rotation were $275–617 higher than from the maize-maize rotation with no N inputs at all levels of P. Farmer adoption and profitability may be further enhanced with strategies that improve nutrient recovery in subsequent maize and retain maximum stover as litter fall in the field.

6.14 Opportunities and Challenges for Grain Legume Production and Legume Adoption in Western and Central Africa

The production of the main grain legumes in these regions has shown some substantial increase over the last 13 years (Fig. 6.7). Groundnut production has grown by about 3 million metric tons between 1990 and 2002 while cowpea production grew from 1.5 to more than 3 million metric tons over the same period of time. Nevertheless, this increased production has only been due to increased area exposed to cultivation which increased from about 4 million hectares to approximately 7 million for groundnuts and from less than 5 million to about 9.5 million hectares for cowpea (Fig. 6.8). This has led to a constant per area productivity over the years as indicated in Fig. 6.9 (FAO 1999).

Programs like the Sasakawa Global 2000 have been on the forefront in promoting the adoption legume technologies in West Africa. For example between 1994 and 1996 the adoption rate of Mucuna in southern Benin tripled largely due to the intensified effort of Sasakawa Global 2000 (SG2000) who bought about 15 t of seed in 1995 (Manyong et al. 1999). The decline observed after 1996 relates to the reduced

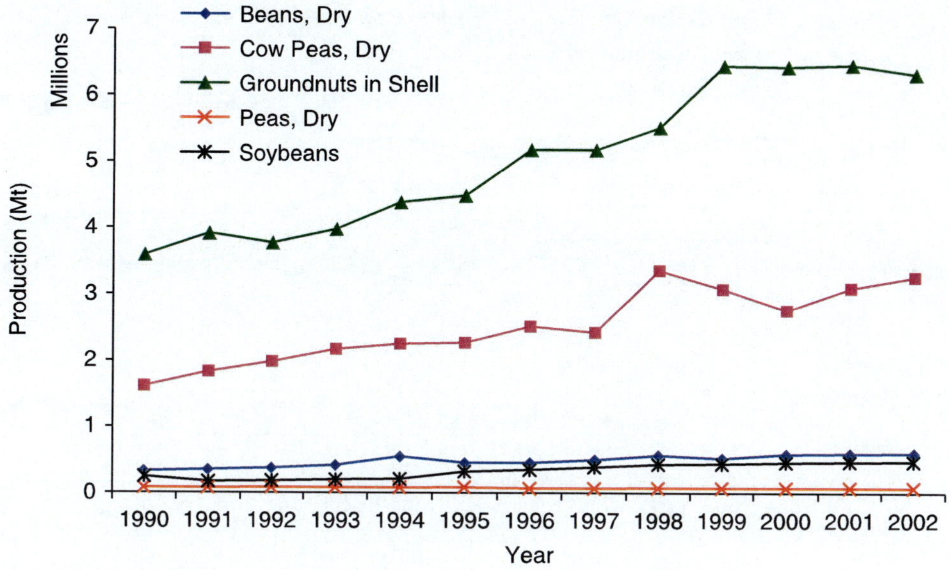

Fig. 6.7 Grain legume production in West and Central Africa (1990–2002)

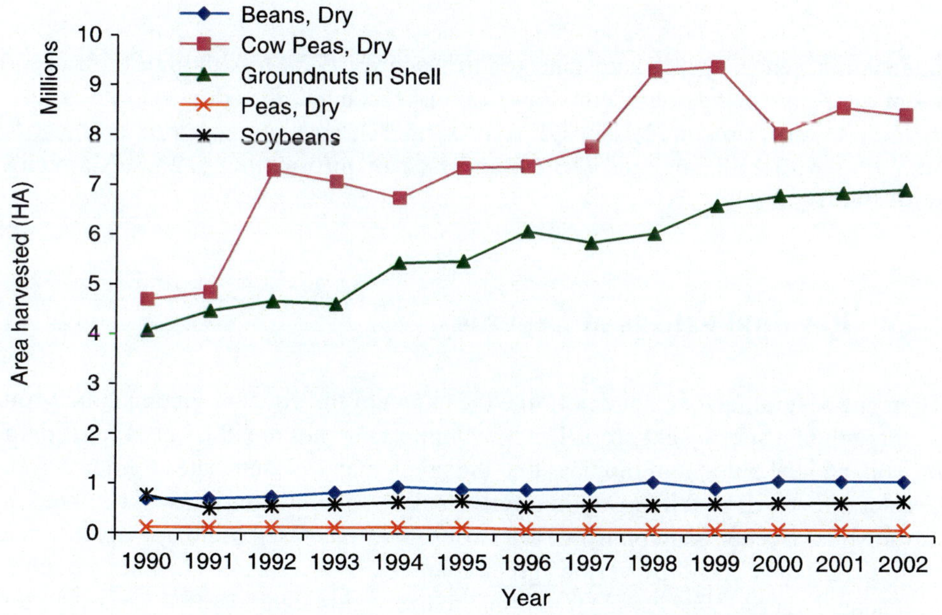

Fig. 6.8 Grain legume production (Area) in West and Central Africa (1990–2002)

effort of SG2000 to buy seeds and a collapse in the market (Doughtwaite 2002). The major drivers for adoption were need for weeding (39% predicted probability) and cash income (41%) (Manyong et al. 1999), the latter likely largely driven by the market created by SG2000. Other factors include degraded fields (25%), access to

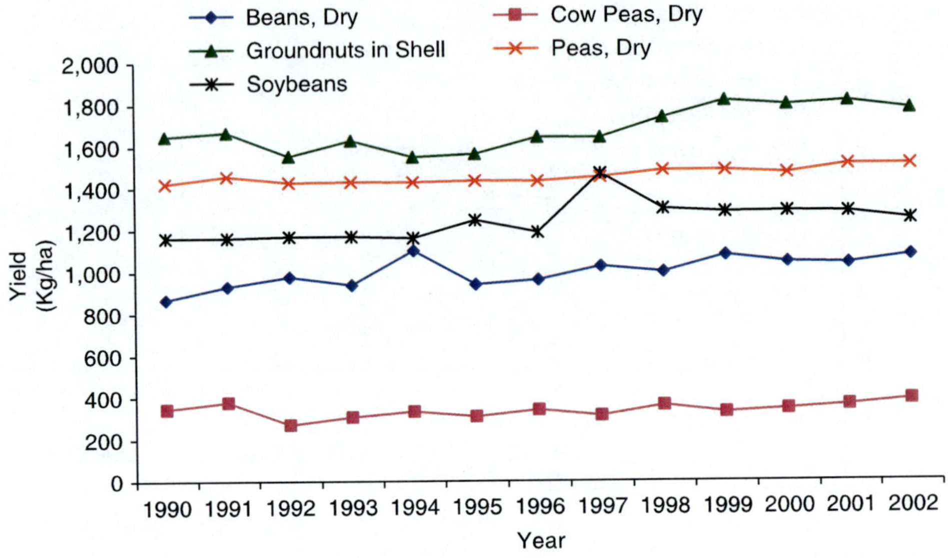

Fig. 6.9 Grain legume yield in West and Central Africa

extension services (24%), and land tenure security (22%). Although the rate of adoption is promising, many constraints are likely to halt the adoption rate. These include: loss of a second season (42% of farmers), insecure land property rights (19%), unavailability of seed (16%), and lack of information (12%) (Manyong et al. 1999).

6.15 Residual Effects of Legume

There are several factors that determine the extent of the effect of green manures on the growth of a subsequent crop. The most important ones are: the initial fertility of the soil, and the amount of biomass that the green manure contributes (Giller 2001). Greater effect will be felt mostly in infertile soils while little impact will be noted in fertile soils. Becker and Johnson (1998) observed increased yields of upland rice with increased N added by various green manures in several sites in Cote d'Ivoire. In Ibadan, Nigeria, Vine (1953) found that maize yields significantly increased after mucuna fallow especially where there has been continuous cropping. In southern Benin the fertilizer value of mucuna was as high as 66 kg N ha^{-1} (Houngnandan et al. 2001).

Generally, green manures do not have a long-term effects on crop yields and on soil organic matter but in the short-term, significant effects have been noted. This has been attributed to the rapid decomposition of most legume residues (Giller 2001).

Table 6.2 Estimates of nitrogen fixation by grain legumes

Legume species	Estimates N_2 fixed
Cowpea	73–80
Groundnut	72–240
Soybean	20–260

Sources: Yamada 1974

6.16 Biological Nitrogen Fixation

The measurement of the biologically fixed nitrogen is an important step to access the contribution of leguminous crops to plant nutrition and total soil N balance. Although there exist several methods for quantifying the atmospheric fixed N by plants, none of them measures actual N_2 fixation with absolute accuracy. The symbiotic performance and yield enhancement of inoculated and non inoculated grain and forage legumes have been reported in a number of studies in West Africa (Awonake et al. 1993; Ganry and Gueye 1992; Awonake and Hardson 1992; Danso et al. 1987; Hardarson et al. 1984; Senaratne et al. 1987). Depending on experimental conditions, estimates of N_2 fixed by legumes may be as high as shown in Table 6.2. For this reason, the beneficial effects of grain or forage legumes on the succeeding crops is assumed to be almost exclusively due to the increased soil N. However, this may not be true unless legumes are used as a green manure or as pasture components. Positive N balance may be expected when much of legume biomass is returned to the soil. However, leguminous crops may not significantly contribute to soil N when grains and straw are removed for human and animal nutrition. Some leguminous crops species are reported to deplete soil N. One such well known legume that deplete soil N is soybean (Hoshikawa 1991). In West Africa, large beneficial effects of legumes in mix cropping or in rotation with or without fertilizer have been observed (Bationo and Ntare 2000; Bagayoko et al. 1997; Bagayoko et al. 2000). It has been suggested that factors other than symbiotic N_2 fixation may be responsible for the high residual effects of legumes because these effects could not be compensated by addition of N fertilizers (Bationo and Ntare 2000).

6.17 Factors Influencing Legumes N_2 Fixation

Increased exploitation of the biological N is constrained by various environmental and nutritional factors, including the cropping patterns used.

Leguminous crops depending on symbiotic nitrogen fixation are usually more sensitive to soil acidification than legumes using bound N (de Carvalho et al. 1981; Foy 1988). Generally in acid soils, the legume rhizobium symbiosis is impaired by high concentration of H+ and particularly Al and Mn and by low concentration of available Mo (Reisenawer et al. 1962; Burmeister et al. 1988) and P (Casman et al. 1980). High H+ concentrations depress the attachment of Rhizobium at the root surface (Caetano-Anolles et al. 1989) and the ability of root exudates to induce the expression of nodulation genes (Richardson et al. 1988). Graham and Donawa (1981) and Bell and Edwards (1986) showed that high concentration of Al^{3+} inhibit root infection with rhizobium and nodule initiation even in the presence of sufficient numbers of rhizobia (Whelan and Alexander 1986).

An evaluation of traditional cropping systems in Africa shows that crop rotation involving legume and cereal monocultures is by far more sustainable than intercropping, the most dominant cultural practice in the continent. Tree legumes also fix about 43–581 kg N ha^{-1} per year, making their leaf prunings an important component of sustainability in agroforestry and alley cropping systems. In a single year, the prunings of *Sesbania sesban* can provide up to a hectare of cereal crop, up to 448 kg N, 31.4 kg P, 125 kg K, 114 kg Ca and 27.3 kg Mg, thus making the foliage of this legume the 'ideal' fertilizer. Clearly, achieving sustainable yields in Sub-Saharan Africa would require a deeper understanding of how fixed N in legume residues is managed in the soil environment, in addition to expanding the use of neglected African food legumes, which exhibit considerable drought resistance and nitrate tolerance. In Africa, where soil moisture often limits yields, research on neglected, symbiotic native legumes with NO_3^- and drought-tolerant traits would constitute a sound basis for increased sustainable production (Dakora and Keya 1997).

6.17.1 Nitrogen in Legume Biomass

Quantifying the biologically fixed N in legume biomass is challenging. Legume biomass method appear to be the easiest method for N accumulation. The method measures the N concentrations of both the grain and straw. A simple calculation results in a measure of the quantity of N produced by the legume, available for incorporation into the soil and potentially usable by the following crop. Experiments using this method through out the world have resulted in N contribution values between 18 and 390 kg of N ha^{-1}. However, not all the legume N produced is mineralised and available to the subsequent crop. Some studies suggest that only 10–30% of the N incorporated in legume material is absorbed by the following crop (Ladd et al. 1983). There exists large variability in the estimate of N contribution using this method which may be attributed to legume species, legume stand and age and the harvest management employed during the year of measurement. Maximum and minimum N accumulation in legume biomass reported in Table 6.3 show large variability in experimental data for West Africa.

Becker and Johnson (1998) studied N accumulation, N derived from the atmosphere (Ndfa), weed suppression, and the effects on rice yield in 50 legumes, grown at four sites in Cote d'Ivoire with contrasting climate, soils, and rice production systems. The results indicated that biomass was in most instances significantly greater in the legume fallow than in the 'weedy' fallow control, and several legume species suppressed weed growth. N accumulation by legumes varied between 1–270 kg N ha^{-1} with 30–90% Ndfa (Table 6.4). Across sites, *Mucuna spp.*, *Canavalia spp.*, and *Stylosanthes guianensis* showed consistently high N accumulation. Grain yields

Table 6.3 Minimum and maximum N concentration (g kg^{-1}) in straw, grain and pod shells of cowpea, groundnut and Bambara groundnut in West Africa

	Straw		Grains		Pods	
Crops/country/sources	Min	Max	Min	Max	Min	Max
Cowpea						
Mali						
Haverman 1986	21.0	37.3	38.1	55.0	10.7	23.4
Nigeria						
Evans and Boulter 1974			36.7	53.7	–	–
Nnadi et al. 1976	14.8	18.3	40.5	45.4	–	–
Sénégal						
Jacquinot 1967	12.0	17.9	33.9	43.0	6.3	24.6
Groundnut						
Nigeria						
RFMC 1980	13.8	21.1	–	–	–	–
Bromfield 1973	15.2	–	39.1	–	8.1	–
Sénégal						
Gillier 1964	13.2	15.2	44.5	46.6	9.6	10.8
Ndiaye 1978	15.6	27.4	–	–	–	–
Bambara groundnut						
Nigeria						
Nnadi et al. 1976	19.0	–	39.4	–		

Table 6.4 N accumulation and % N derived from biological nitrogen fixation in selected fallow legumes in Cote d'Ivoire, 1995/1996 dry season

	Guinea savanna		Derived savannah		Bimodal forest		Monomodal forest	
	Kg N ha^{-1}	%Ndfa	Kg N ha^{-1}	%Ndfa	Kg N ha^{-1}	%Ndfa	Kg N ha^{-1}	%Ndfa
Fallow vegetation								
Aeschynomene histrix	27	35	82	78	159	86	66	68
Cajanus cajan	45	60	125	61	221	71	18	55
Canavalia ensiformis	77	58	256	68	191	69	96	57
Canavalia rosea	41	54	180	77	95	78	70	66
Crotalaria juncea	49	59	185	77	120	86	1	–
Crotalaria anagyroides	23	48	37	41	270	67	10	36
Mucuna pruriens	81	69	125	74	257	83	30	60
Tephrosia villosa	27	67	119	60	102	80	39	57
Stylosanthes guianensis	40	63	124	68	201	79	69	65
Weedy fallow	12	0	22	0	44	0	17	0

Source: Becker and Johnson 1998

Values for N accumulation were based on maximal legume biomass during a 6 month growing period while %Ndfa was determined by natural abundance of rice which had been preceded by a legume fallow were on average 30% greater than that preceded by a natural weedy fallow control. At the savanna sites where fallow vegetation was incorporated *Canavalia ensiformis* showed relatively higher N accumulation with 234 kg N accumulation per hectare higher than in weedy fallow. In the bimodal forest zone, the highest N accumulation was noted with *Crotalaria anagyroides fallow*. In general, the effects of legume fallows on N accumulation were most significant in environments with favourable soil and hydrological conditions.

Agricultural production in the densely populated highlands of Rwanda is subject to serious soil fertility constraints. As the use of imported mineral fertilizers is beyond the economic means of resource-poor farmers, research and extension efforts of several projects, institutes and organizations have concentrated during the last 15 year on the development and promotion of improved fallows with woody and herbaceous nitrogen fixing legumes (such as Tephrosia, Cajanus, Crotalaria, Sesbania, Mucuna and Mimosa) planted over one or more seasons as pure green manure, in hedgerows (alley cropping), or on fields as seasonal inter-or relay-crops. Green manuring proved to be a risky enterprise, due to highly variable biomass production and residual effects. Yield increments on-farm of up to 74% in the first season and 46% in the second season did not compensate loss of yields and labour investments during green manuring. Even where biomass production was sufficient, residual effects were in most cases unsatisfactory, due to rapid nutrient leaching (N, K) or inappropriate foliage incorporation on-farm. In research-managed trials, residual effects were in general somewhat higher, but more than mere compensation of lost yields was not possible and farmers' adoption of these labour-intensive technologies was rather low. Due to acute land shortage, farmers were also reluctant in allocating land to fallows or hedgerows, with the exception of fields already out of production. Consequently, the concept of improving soil fertility and crop yields with the help of planted fallows or green manure in rotation failed. Woody legumes might have a future on abandoned fields and in wide spaced contour hedges, mainly for the production of firewood and bean stakes. For soil fertility management, the production and availability of farmyard manure and country-own mineral fertilizers, such as travertine and volcanic ashes should be supported. The question is raised as to whether sustainable agricultural development is possible without a credit system for small farmers, reallocating land and creating off-farm employment (Drechsel et al. 1996).

In field experiments in 1995–1996 in Ibadan, Nigeria, soyabeans cv. TGX 1740-2F and TGX 1448-2E were seed inoculated with Bradyrhizobium strain IRj 284A or not inoculated and given 0 or 40 kg P ha^{-1}. Inoculation and P fertilizer led to increases in nodulation in both cultivars in both years. Dry matter accumulation at podding increased due to inoculation and P fertilizer in TGX 1448-2E in both years and in TGX 1740-2F in 1995. At seed formation, these treatments led to increases in dry matter accumulation of TGX 1740-2F in 1996. In both years, all the treatments led to increases in N and P in plant tissue when compared with the control. Grain yield increased in response to the combination of inoculation and P in both years (Taiwo et al. 1999).

In a study by Carsky et al. (1997), maize grain yield increase following soyabeans was variable but the main effect of the previous soyabean crop was positive even though the soyabeans were not inoculated with rhizobia and the aboveground soyabean residues, except litter fallen before harvest, were exported from the fields, following the present farmer practice. The yield increase following the medium duration soyabean cv. TGX 1660-19F was similar to that from 40 kg N ha^{-1} applied 4 weeks after sowing to maize preceded by maize. The increase following the early soyabean cv. TGX 1456-2E was significantly ($P<0.05$) smaller than that following the medium duration soyabean cultivar. Total N in the 0 to 10 cm depth of the previous TGX 1660-19F plots (0.063%) was significantly greater ($P<0.05$) than in the previous maize plots (0.058%) when analysed for the combined sites. Thus, the effect of previous soyabean crops on maize grain yield was due mostly to residual N availability.

Legumes in general and biological N fixation specifically have a potentially important role to play in Integrated Soil Fertility Management (ISFM) strategies. Alley cropping with leguminous hedgerows has been proven to be technically sound and generated a lot of process work in agroforesty systems in the West African savanna. As with alley cropping systems, inclusion of Mucuna in a cropping system was observed to significantly enhance crop yield. This role of mucuna is rather associated with its ability to suppress *Imperata cylindrica* weeds than with improving the soil fertility status, thereby also limiting its niche for adoption. Dual purpose grain legumes – cereal rotations using improved legume germplasm that provides net N benefits to the cropping system besides grains, significantly enhance cereal yield and supply the farmer with immediate products that can be consumed or sold. This technology shows a lot of promise and is currently spreading in the Northern Guinea savanna zone of Nigeria, but require the creation of local processing skills and/or markets for the grains and intensive interaction between breeders, soil fertility specialists, and farmers.

As with soybean, improved germplasm of cowpea can also be used to enhance the soil fertility status while yielding immediate benefits to farmers.

A field experiment was conducted at Bel Air station, in Dakar using ^{15}N isotope dilution technique and the non-nodulating soyabean variety m129 as reference plant to estimate nitrogen fixation in common bean (*Phaseolus vulgaris*) (Diouf et al. 1999). Dry matter production and nitrogen accumulation were increased in plants inoculated with Rhizobium strain ISRA 355 compared to uninoculated plants (+31% and +50%, respectively) and in 20 N-fertilized plants compared to non-N-fertilized plants (+36% and +39%, respectively). In addition, %Ndfa was increased in common bean inoculated with strain ISRA 355 (+32% higher than non inoculation) or supplied with 20 kg N ha^{-1} (+37% higher than non N fertilization). However, the amount of fixed nitrogen was increased by the interaction inoculationX20 N-fertilization only (+155% over the value averaged across all uninoculated plants). From all treatments, estimates of fixed nitrogen ranged between 63 and 139 kg N ha^{-1}. Beside this estimation, nitrogen fixation was inhibited in common bean supplied with 80 kg N ha^{-1}. Thus, 20 kg N ha^{-1} should be considered as starting rate of nitrogen fixation in common bean cultivated at this site.

6.18 Partitioning Non-N Benefits of Legume Cropping Systems

6.18.1 *Soil Organic Matter*

In low activity clay soils, soil organic matter (SOM) decline rapidly under continuous cropping (Kang 1993). Due to the rapid rate of oxidation (Giller et al. 1997), SOM content of tropical soils cannot be maintained under cultivation. Vine (1953) indicates that fallow with legume cover crops could improve SOM because of increased shading on which build up of humus largely depends especially in tropical conditions. The litter fall from the cover crops contribute to increase in soil organic matter stocks. In their study in Ibadan, Tian et al. (1999) reported rapid soil organic carbon (SOC) decline under continuous cropping. However, there was nearly no decline in SOC under pueraria cover crop fallow. The improvement of SOM quality by pueraria fallow has been noted to increase N concentration in particulate organic matter (POM).

Bationo et al. (2002) noted a higher level of organic carbon in the rotation systems with cowpea compared to the continuous cropping systems, this was partially attributed to the fallen cowpea leaves (Fig. 6.10).

6.18.2 *Legume Interaction with Mineral Fertilizer*

In a study by Bationo et al. (2002) on the interaction between phosphorus fertilizers and different cropping systems, they observed that application of P had a significant effect on yield of cowpea and pearl millet. Higher millet yields were obtained in a rotation whereby millet followed cowpea than continuous cultivation (Fig. 6.11). There was increasing millet yields with increased P supply in all cropping systems with highest millet yields being obtained from millet-cowpea cropping system. This was an indication that P fertilizer could be of more benefit in a cropping system whereby cereal is alternated with legume.

6.18.3 *Soil Physical Properties*

Mucuna has also been reported to help protect the physical properties of soil from raindrop impact thereby reducing water erosion through runoff (Carsky et al. 2001). Thus helping to maintain soil physical properties. Soil erosion was reduced to about 10 times in maize/ mucuna relay intercropping as compared to sole maize cropping system in southern Benin (Azontonde 1993). In addition, Hulugalle et al. (1986) reported increased porosity and infiltration rates due to increased biomass production

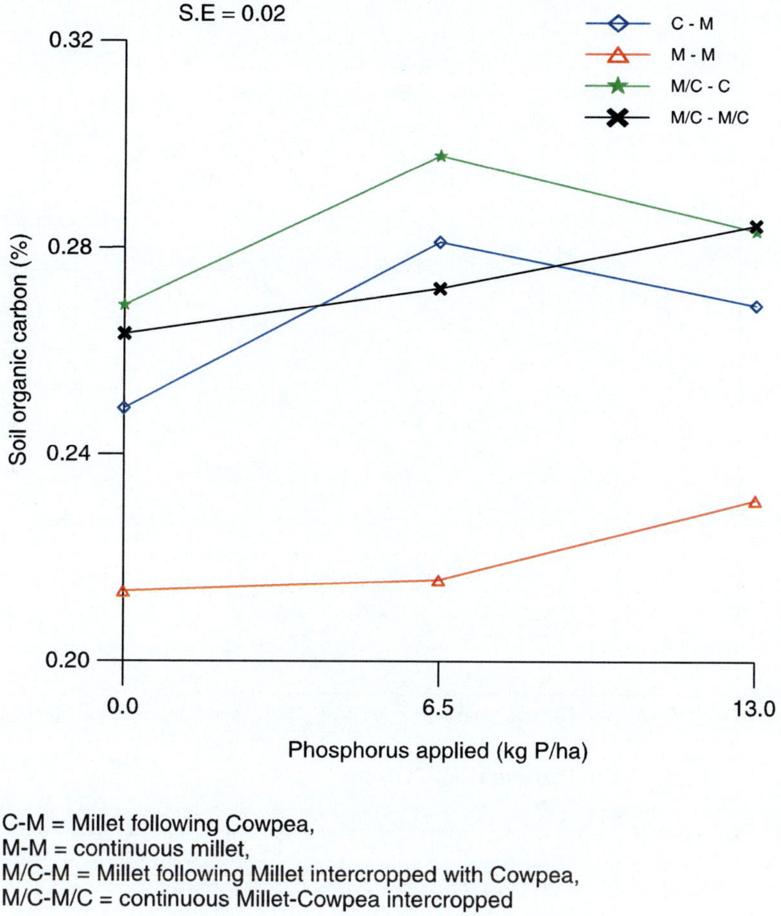

Fig. 6.10 Effect of phosphorus and cropping system on soil organic carbon, Sadore, Niger 1995

by mucuna. In a study in southern Cameroon, Hauser (unpublished data 2000) noted significantly lower soil bulk density after mucuna relay fallow compared to natural regrowth fallow.

Obi (1999) carried out a study to assess the physical and chemical responses of a degraded sandy clay loam Ultisol to two leguminous and four grass cover crops in southern Nigeria. Relative increases of 26% and 112% in soil organic carbon and phosphorus levels were observed as well as appreciable improvements in the cation exchange capacity (CEC) and Calcium levels under vegetative covers compared with the initial conditions. The improvements were more pronounced with legume covers than with grass covers. Furthermore, the vegetative covers improved mean organic carbon level by 28% and appreciably improved mean CEC, Ca, and Mg levels over the values for the bare soils. The percentage of water-stable aggregates >1.0 mm was significantly reduced under bare fallow (x=27.7%) compared with soils under vegetative cover (x=79.3%). The correlation between water-stable aggregates >1.0 mm

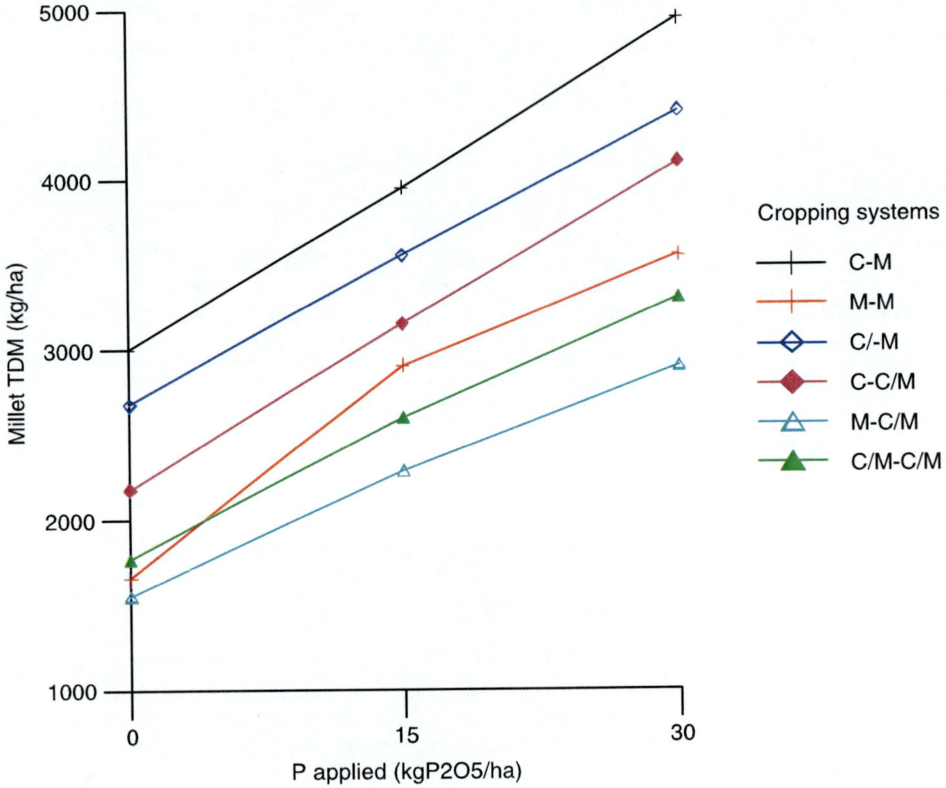

Fig. 6.11 Effects of phosphorus and different cropping systems on millet yield over 4 years, Sadore, Niger

and exchangeable aluminium was negative and significant (r=–0.7050) at P=0.05. There were highly significant treatment effects (P=0.01) for penetrometer resistance, pore size distribution, water infiltration, water retention and saturated hydraulic conductivity. Grass and legume fallows which protect the soil and guarantee regular additions of organic materials are ecologically sound and socially acceptable components of sustainable agricultural production. Indications, however, are that this degraded tropical Ultisol would require a period exceeding 5 years under vegetative covers for restoration of its fertility to acceptable productive status.

Generally, the beneficial effects of legumes are better noted in heavy clay soils than light soils (Hoshikawa 1991). Larger effects of soybean cultivation on maize yields have been noted on poorly drained heavy soil as compared to well-drained light soils (Griffith et al. 1988; Hoshikawa 1991). This has also been noted on soils with low-tillage systems than under conventional tillage (Edwards et al. 1988; Hoshikawa 1991).

Armbrust et al. (1982) and Gantzer et al. (1987) have reported the ability of Soybean to loosen the soil hence improving soil tilth. However, the direct benefit on the succeeding crop might only be realized in the long term.

6.19 Conclusion

Intensification of low-input agricultural production has led to a rapid increase in soil degradation and nutrient depletion in many parts of West and Central Africa, constituting serious threat to food production and food security. Well adapted and highly productive grain, forage, and green manure legume species are available for the West and Central African agroecological zones. The different legume types vary in terms of their potential contribution to soil fertility improvement. Past and ongoing research show great potential of legumes to contribute to the conservation of the natural resource base while maintaining inter-annual yield and generating increased outputs. As discussed in this and elsewhere in this book, legumes alone are not a solution to the problem of persistent food shortages and land degradation. There is need for research and development efforts to be directed at improving access to improved germplasm, judicious use of organic and inorganic fertilizer, input-output access and improved crop-livestock integration.

References

Akobundu IO, Agyakwa CW (1987) A handbook of West African weeds. International Institute of Tropical Agriculture, Ibadan, Nigeria, pp 521

Akobundu IO, Udensi UE (1995) Effect of *Mucuna* species and fertilizer levels on the control of speargrass (*Imperata cylindrica* L.) [abstract]. 22nd Annual Conference, Weed Science Society of Nigeria, 6–10 Nov 1995, Ibadan, Nigeria. International Institute of Tropical Agriculture, Ibadan, Nigeria

Arihara J, Ohwaki Y (1989) Estimation of available phosphorus in vertisol and alfi sol in view of root effects on rhizosphere soil in XI Colloquium on plant nutrition-physiology and applications, 30 July to 4 August 1989. Wageningen, The Netherlands

Armbrust DV, Dickerson JD, Skidmore EL, Russ OG (1982) Dry soil aggregation as influenced by crop and tillage. Soil Sci Soc Am J 46:390–393

Awonake KO, Hardson G (1992) Biological nitrogen fixation of *Glyricidia Sepium* / Rhizobium Symbiosis as influenced by plant genotype, bacterial strain and their interactions. Trop Agric 69:381–385

Awonake KO, Danso SKA, Zapata F (1993) The use of a double isotope (^{15}N ^{34}S) labelling technique to assess the suitability of various reference crops for estimating nitrogen fixation in *Glyricidia sepium* and *Leucaena Leucocephala*. Plant Soil 155(156):325–328

Azontonde HA (1993) Degradation et restauration des terres de barre au Sud-Benin. In: Presentation to 10th erosion workshop, ORSTOM, Montpellier, 15–18 Sept 1993

Bagayoko M, Mason SC, Traore S, Eskridge KM (1996) Pearl millet/cowpea cropping systems yield and soil nutrient levels. Afr Crop Sci J 4:453–462

Bagayoko M, Mason SC, Traore S (1997) The role of cowpea on pearl millet yield, N uptake,and soil nutrient status in millet cowpea rotation in Mali. In: Renard G, Neef A, Becker K, von Oppen M (eds.) Soil fertility management in West African land-use systems. Margraf-Verlag, Weikerseim, pp 109–114

Bagayoko M, Buerkert A, Lung G, Bationo A, Romheld V (2000) Cereal/legume rotation effects on cereal growth in Sudano. Sahelian West Africa: soil mineral nitrogen, mycorrhizae, and nematodes. Plant Soil 218:103–116

Bationo A, Lompo F, Koala S (1998) Research on nutrient flows and balances in West Africa: state-of-the-art, pp 19–36. In: Smaling EMA (ed.) Nutrient balances as indicators of production and sustainability in sub-Saharan African agriculture. Agric Ecosyst Environ 71:1–3

Bationo A, Ntare BR (2000) Rotation and nitrogen fertilizer effects on pearl millet, cowpea, and groundnut yield and soil chemical properties in a sandy soil in the semi arid tropics, West Africa. J Agric Sci (Cambridge) 134:277–284

Bationo A, Ntare BR, Tarawali SA, Tabo R (2002) Soil fertility management and cowpea production in the semiarid tropics. In: Fatokum CA, Tarawali SA, Singh BB, Kormawa PM, Tamo M (eds.) Challenges and opportunities for enhancing sustainable cowpea production. Proceedings of the world cowpea conference III held at the International Institute of Tropical Agriculture (IITA), Ibadan, 4–8 Sept 2000

Becker M, Johnson DE (1998) Legumes as dry season fallow in upland rice-based systems of West Africa. Biol Fertil Soils 27:358–367

Bell LC, Edwards DG (1986) The role of aluminum in acid soil infertility. In: Latham M (ed.). Proceedings of the first regional seminar on soil management under humid conditions in Asia and the Pacific, Phitsanulok, Thailand, pp 201–223

Bertrand R, Nabo J, Vicaire R (1972) Exportations minérale par le mil et l'arachide, conséquenses sur la définition d'une fumure d'entretien d'un sol ferrugineux tropical dévéloppé sur matériau éolien à Tarna (Niger). L'agronomie Tropicale 12:1287–1303

Bingen RJ, Hall AE, Ndoye M (1988) California cowpeas and food policy in Senegal. World Dev16: 857–865

Breman H, van Reuler H (2002) Legumes: when and where an option ? (No panacea for poor tropical West African soils and expensive fertilizers). In: Vanlauwe B, Diels J, Sanginga N, Merckx R (eds.) Integrated plant nutrient management in sub-Saharan Africa: from concept to practice. CAB International, Wallingford

Breman H, Ridder N (eds.) (1991) Manuel sur les paturages des pays Saheliens. ACCT, Paris/ CTA, Wageningen/ KARTHALA, Paris, 485 pp

Bromfield AR (1973) Uptake of sulphur and other nutrients by groundnuts (Arachis hypogaea) in northern Nigeria. Exp Agric 9:55–58

Buckles D, Etéka A, Osiname O, Galiba M, Galiano G (1998) Cover crops in West Africa contributing to sustainable agriculture. IDRC, Ottawa, Canada; IITA, Ibadan, Nigeria; Sasakawa Global 2000, Cotonou, Benin

Burmeister CH, Adams JF, Odom JW (1988) Response of soybean to lime and molybdenum on vertisols in Northern Alabama. Soil Sci Soc Am J 52:1391–1394

Caetano-Anolles G, Lagares A, Favelukes G (1989) Adsorption of rhizobium meliloti to alfalfa roots: dependence on divalent cations and pH. Plant Soil 117:67–74

Carsky RJ, Abaidoo R, Dashiell K, Sanginga N (1997) Effect of soybean on subsequent maize grain yield in the Guinea savanna zone of West Africa. Afr Crop Sci J 5(1):31–38

Carsky RJ, Oyewole B, Tian G (1999) Integrated soil management for the savanna zone of W. Africa: legume rotation and fertilizer N. Nutr Cycl Agroecosyst 55:95–105

Carsky RJ, Becker M, Hauser S (2001) Mucuna cover crop fallow systems: potential and limitations. In: Tian G et al (eds.) Sustaining soil fertility in west Africa. SSSA Special Publication Number 58. SSSA, Madison, pp 111–136

Carsky RJ, Vanlauwe B, Lyasse O (2002) Cowpea rotation as a resource management technology for cereal-based systems in the savannas of West Africa. In: Fatokum CA, Tarawali SA, Singh BB, Kormawa PM, Tamo M (eds.) Challenges and opportunities for enhancing sustainable cowpea production. Proceedings of the world cowpea conference III held at the International Institute of Tropical Agriculture (IITA), Ibadan, Nigeria, 4–8 Sept 2000. IITA, Ibadan

Carsky JR, Tarawali SA, Becker M, Chikoye D, Tian G, Sanginga N (1998) *Mucuna*: a herbaceous cover legume with potential for multiple uses. International Institute of Tropical Agriculture, Ibadan, Nigeria. RCMD Research Monograph No. 25

de Carvalho MM, Edwards DG, Andrew CS, Asher CJ (1981) Aluminum toxicity, nodulation and growth of stylosanthes species. Agron J 73:261–265

Casman KG, Whitney AS, Stokinger KR (1980) Root growth and dry matter distribution of soybean as affected by phosphorus stress, nodulation and nitrogen source. Crop Sci 20:239–244

Christianson CB, Bationo A, Baethgen WE (1990) The effect of soil tillage and fertilizer use on pearl millet yields in Niger. Plant Soil 123:51–58

Codjia XC (1996) Etude de l' action fertilisante sur terre de barre du Mucuna noir (Mucuna pruriens var. cochinchinensis) et de leur interaction avee les engrais. Ag. Eng. thesis, Universite National du Benin, Cotonou, Benin

Curl EA (1963) Control of plant diseases by crop rotation. Bot Rev 29:413–479

Dakora FD, Keya SO (1997) Contribution of legume nitrogen fixation to sustainable agriculture in Sub-Saharan Africa. Soil Biol Biochem 29(5–6):809–817

Danso SKA, Zapata F, Hardson G, Fried M (1987) Nitrogen fixation in fababeans as affected by plant population density in sole or intercropping systems with barley. Soil Biol Biochem 19:411–415

David S (1997) Dissemination and adoption of new technology: a review of experiences in bean research in Eastern and Central Africa, 1992–1196. Network of Bean Research in Africa. Occasional Publications Series, No. 21. CIAT, Kampala, Uganda

Diouf A, Ndoye I, Spencer MM, Nef-Campa C, Gueye M (1999) Need for inoculation of common bean (Phaseolus vulgaris L.) in Senegal and assessment of nitrogen fixation using ^{15}N isotope dilution technique. Biological nitrogen fixation in Africa: linking process to progress. Eighth Congress of the African Association for biological nitrogen fixation, Cape Town, South Africa, 23–27 Nov 1998. Symbiosis-Rehovot 1999, 27:3–4, 251–257

Doughtwaite B, Manyong VM, Keatinge JDH, Chianu J (2002) The adoption of alley farming and Mucuna: lessons for research, development and extension. Agrofor Syst 56:193–202

Drechsel P, Steiner KG, Hagedorn F (1996) A review on the potential of improved fallows and green manure in Rwanda. Agroforest Syst 33(2):109–136

Edwards JH, Thurlow DL, Eason JT (1988) Influence of tillage and crop rotation on yields of corn, soybean and wheat. Agron J 80:76–80

Evans IM, Boulter D (1974) Amino acid composition of seed meals of yam bean (Sphenostylis stenocarpa) and Lima bean (Phaseolus Iunatus). J Sci Food Agr 25:919–922

Food and Agriculture Organization (1999) FAOSTAT Agriculture data, Rome. http://apps.fao.org/cgi-bin/nph-db.pl?subset=agriculture

Foy CD (1988) Plant adaptation to acid, aluminum toxic soils. Comm Soil Sci Plant Anal 19:959–987

Gantzer CJ, Buyanovsky GA, Alberts EA, Remley RA (1987) Effects of soybean and corn residue decomposition on soil strength and splash detachment. Soil Sci Soc Am J 51:202–206

Ganry F, Gueye F (1992) La mise en valeur des bas-fonds de la zone soudano-sahélienne par *Sesbania rostrata* est elle possible ? Agron Trop 46:155–159

Gijsman AJ, Thomas RJ (1996) Evaluation of some physical properties of an Oxisol after conversion of native savanna into legume-based or pure grass pastures. Trop Grassl 30:237–248

Gillier P (1964) Les Coques d'Arachides: Leur Utilisation en Agriculture. Oleagineux 19: 473–475

Giller KE, Cadisch G, Ehaliotis C, Adams E, Sakala WD, Mafongoya PL (1997) In: Buresh RJ et al (eds.) Replenishing soil fertility in Africa. SSSA Special Publication 51. SSSA, Madison, pp 151–192

Giller KE (2001) Nitrogen fixation in tropical cropping systems, 2nd edn. CAB International, Wallingford

Graham RA, Donawa AL (1981) Effect of soil pH and inoculum rate on shoot weight, nitrogenase activity and competitive nodulation of groundnut (*Arachis Apogaea* L.). Trop Agric (Trinidad) 58:337–340

Griffith DR, Kladivko EJ, Mannering JV, West TD, Parsons SD (1988) Long-term tillage and rotation effects on corn growth and yield on high and low organic matter, poorly drained soils. Agron J 80:599–605

Hafner H, Ndunguru BJ, Bationo A, Marshner H (1992) Effect of nitrogen, phosphorus and molybdenum application on growth and symbiotic N2-fixation of groundnut in an acid sandy soil in Niger. Fert Res 31:69–77

Hardarson G, Zapata F, Danso SKA (1984) Field evaluation of symbiotic nitrogen fixation by rhizobial strains using ^{15}N methodology. Plant Soil 82:369–375
Haverman PP (1986) A characterization of the growth of the cowpea crop in the Sahelian zone. MSc thesis Department of the Theoretical Production Ecology, Agricultural University, Wageningen
Heathcote RG (1970) Soil fertility under continuous cultivation in northern Nigeria. II. Response to fertilizer in the absence of organic manures. Exp Agric 6:345–350
Hoogendijk M, David S (1997) Bean production systems in Mbale District, Uganda with emphasis on the adoption of new climbing varieties and varietal diversity. Network on Bean Research in Africa. Occasional paper Series, No. 20, CIAT, Kampala
Hoshikawa K (1990) Significance of legume crops in improving the productivity and stability of cropping systems. International symposium on the use of stable isotopes in plant nutrition, Soil fertility and environmental studies, Vienna, Austria. 1-5 October, 1990
Hoshikawa K (1991) Significance of legume crops in improving the productivity and stability of cropping systems. In: Johansen C, Lee KK and Sahrawat KL (eds.) Phosphorus nutrition of grain legumes in the semi-arid tropics. ICRISAT, India, pp 173–181
Houngnandan P, Sanginga N, Okogun JA, Vanlauwe B, Merckx R, Van Cleemput O (2001) Residual nitrogen benefits of mucuna previously inoculated with bradyrhizobia to maize in farmers' fields in the derived savannah of Benin, Field Crops Research submitted
Hulugalle NR, Lal R, Ter Kuile CHH (1986) Amelioration of soil physical properties by mucuna after mechanized land clearing of a tropical rain forest. Soil Sci 141:219–224
IITA (International Institute of Tropical Agriculture) (1998) Towards sustainable development in Africa. Medium-term Plan 1999–2001. IITA, Ibadan, Nigeria
IITA (International Institute of Tropical Agriculture) (2002) Soybean crop and farming system. IITA Annual Report 2002, IITA, Ibadan, Nigeri
Jacquinot L (1967) Croissances et alimentations minérales comparées de quatre variétés de niébé. Agron Trop 22:575–640
Jagtap SS, Adeleye O (1999) Nutrient recovery and productivity of a soybean-maize rotation in the derived savanna ecology of West Africa. J Sust Agric 15(1):75–85
Jones MI, Wild A (1975) Soils of West African savanna. The maintenance and improvement of their fertility. Commonwealth Bureau of the Soils Technical Communication of the Harpenden, U.K. Commonwealth Agricultural Bureau (CAB), Farnham Royal, UK. 55, 246
Kang BT (1993) Change in soil chemical properties and crop performance with continuous cropping on an Entisol in the humid tropics. In: Mulongoy K, Merckx R (eds.) Soil organic matter dynamics and sustainability of tropical agriculture. Wiley, Chichester, pp 297–305
Kang BT, Shannon DA (2001) Agroforestry with focus on alley cropping. Sustaining Soil Fertility in West Africa. SSSA Special Publication No. 58. Soil Science Society of America and American Society of Agronomy, Madison, Wisconsin
Keatinge JDH, Breman H, Manyong VM, Vanlauwe B, Wendt J (2001) Sustaining soil fertility in West Africa in the face of rapidly increasing pressure for agricultural intensification. In: Tian G, Ishida F and Keatinge JDH (eds.) Sustaining Soil Fertility in West Africa, 1–22. Soil Science Society of America Special Publication 58, Soil Science Society of America/American Society of Agronomy, Madison, Wisconsin
Klaij MC, Ntare BR (1995) Rotation and tillage effects on yield of pearl millet (*Pennisetum glaucum*)and cowpea (*Vigna unguiculata*), and aspects of crop water balance and soil fertility in semi-arid tropical environment. J Agric Sci (Cambridge) 124:39–44
Ladd JN, Amato M, Jackson RB, Butler JHA (1983) Utilisation bu wheat crops of nitrogen from legume residues decomposing in soil in the field. Soil Biol Biochem 15:231–238
Lal R, Wilson GF, Okigbo BN (1978) No-tillage farming after various grasses and leguminous cover crops in tropical Alfisols I. Crop performance. Field Crops Res 1:71–84
Lombin G, Singh L (1986) Fertilizer response of groundnut (Arachis hypogaea L.) under continuous intensive cultivation in the Nigerian savannah. Fert Res 10:43–58
Martin G, Fourier P (1965) Les oligo-elements dans la culture de l'arachide du Nord Sénégal. Oléagineux 20:287–291

Manyong VM, Houndekon V, Sanginga PC, Vissoh P, Honlonkou AN (1999) Mucuna fallow diffusion in southern benin. Impact Series. International Institute of Tropical Agriculture, Ibadan, Nigeria

Mazzucato V, Ly S (1994) An economic analysis of research and technology transfer of millet, sorghum and cowpeas in Niger. (International Development Working Paper No. 40). Michigan State University

Nambiar PTC, Anjaiah V (1989) Effects of manganese toxicity on growth and N_2 fixation of groundnut *Arachis Hypogaea*. Ann App Biol 115:361–366

Ndiaye JP (1978) Enquête sur la fertilité en milieu paysan dans la région du Sine-Saloum. Rapport CNRA, Bambey, Sénégal, 17 p. + 174 p. annexes

Nicou RN (1977) Le travail du sol dans les terres exondees du Senegal. Motivations constraintes. ISRA,CNRA, Bombay, 51pp

Nnadi LA, Balasubramanian V, Singh L (1976) Yield and mineral composition of grain legumes. Ghana J Agr Sci 9:211–215

Obi ME (1999) The physical and chemical responses of a degraded sandy clay loam soil to cover crops in southern Nigeria. Plant Soil 211(2):165–172

Osei-Bonsu P, Buckles D (1993) Controlling weeds and improving soil fertility through the use of cover crops: experience with *Mucuna* spp. in Benin and Ghana. West African Farming Systems Research Network Bulletin 14: 2–7

Penning de Vries FWT, Djiteye MA (1982) La productivite des paturages saheliens. Une etude des sols, des vegetations et de I'exploitation de cette resource naturelle. Agricultural Research Report 918. Pudoc, Wageningen, 523 pp

Rebafka F-P (1993) Deficiency of phosphorus and molybdenum as major growth limiting factors of pearl millet and groundnut on an acid sandy soil in Niger, West Africa. Hohenheimer Bodenkundliche Hefte 9. Universität Hohenheim, 155 p

Reisenawer HM, Tabikh AA, Stout PR (1962) Molybdenum reaction with soils and the hydrous oxides of iron, aluminium and titanium. Soil Sci Soc Am Proce 26:23–27

Richardson AE, Djorjevic MA, Rolfe BG, Simpson RJ (1988) Effects of pH, Ca and Al on the exudation from clover seedlings of compounds that induce the expression of nodulation genes in *Rhizobium trifolii*. Plant Soil 109:37–47

RFMC (1980) Mémento de l'agronome. Paris, 1 600 p

Salako FK, Babalola O, Hauser S, Kang BT (1999) Soil macroaggregate stability under different fallow management systems and cropping intensities in southwestern Nigeria. Geoderma 91:103–123

Salako FK, Kirchhof G and Tian G (2000) Conservation of tropical Alfisol using legume cover crop intercropping and residue mulch. In: Proceedings of tillage at the threshold of the 21st century: looking ahead, 5th conference of the International Soil Tillage Research Organization (ISTRO), Fort Worth, 2–7 July, 2000

Sanders J, Bezuneh T, Schroeder A (1994) Impact Assessment of the SAFGRAD Commodity Networks. Washington, DC: USAID/AFR/ARTS/FARA

Sanginga N, Ibewiro B, Hougnandan P, Vanlauwe B, Okogun JA (1996) Evaluation of symbiotic properties and nitrogen contribution of *Mucuna* to maize growth in the derived savannas of West Africa. Plant Soil 179:119–129

Schwartz L, Sterns J, Oehmke J (1990) Impact study of the bean/Cowpea CRSP for Senegal. Michigan State University, East Lansing

Senaratne R, Amorripimol C, Hardarson G (1987) Effect of combined nitrogen on nitrogen fixation of soybean (*Glycine max. L. Merrill*) as affected by cultivar and rhizobial strain. Plant soil 103:45–50

Sivakumar MVK (1989) Agroclimatic aspect of rainfed agriculture in the Sudano-Sahelian zone. In: Renard C, Van Den Beldt RJ, Paar JF (Eds.) Soil, crop, and water management in the Sudano-Sahelian Zone. Proceedings of an international workshop, 11–16 Jannuary 1987, pp 17–38. International Crop Research Institute for Semi-Arid Tropics (ICRISAT), Patancheru

Sperling L, Scheidegger U, Buruchara R, Nyabyenda P, Munyenesa S (1994) Intensifying production among smallholder farmers: the impact of improved climbing beans in Rwanda. Network

on Bean Research in Africa, Occasional Publications Series, No. 12, CIAT/RESAPAC, Butare, Rwanda
Spurgeon WI, Grimson PH (1965) Influence of cropping systems on soil properties and crop production. Mississippi Agricultural and Forestry Experiment Station Bulletin No. 710
Sterns JA and Bernsten RH (1994) Assessing the Impact of Cowpea and Sorghum Research and Extension in Northern Cameroon, MSU International Development Working Paper No. 43. Department of Agricultural Economics, Michigan State University, East Lansing
Stoop WA, Staveren JPV (1981) Effects of cowpea in cereal rotations on subsequent crop yields under semi-arid conditions in Upper Volta. In: Graham PC, Harris SC (eds.) Biological nitrogen Fixation technology for tropical agriculture. Centro International de Agricultura Tropical (CIAT), Cali, pp 653–657
Taiwo LB, Nworgu FC, Adatayo OB (1999) Effect of Bradyrhizobium inoculation and phosphorus fertilization on growth, nitrogen fixation and yield of promiscuously nodulating soybean (Glycine max (L.) Merr.) in a tropical soil. Crop Res Hisar 18(2):169–177
Tian G, Kolawole GO, Salako FK, Kang BT (1999) An improved cover crop-fallow system for sustainable management of low activity clay soils of the tropics. Soil Sci 164:671–682
Tian G, Kolawole GO, Kang BT, Kirchhof G (2000a) Nitrogen fertilizer replacement indexes of legume cover crops in the derived savanna of West Africa. Plant Soil 224:287–296
Tian G, Kolawole GO, Ishida F, Carsky RJ (2000b) On-farm study of Pueraria cover crop-fallow system in the derived savanna of West Africa. In: Carsky RJ et al. (eds.) Cover Crop for Natural Resources Management in West Africa, Proceedings of a workshop organized by IITA and CIEPCA, Cotonou, Republic of Benin, 26–29 Oct 1999, pp 185–193
Tian G, Hauser S, Koutika LS, Ishida F, Chianu JN (2001) Pueraria cover crop fallow systems: benefits and applicability. In: Tian G et al (eds.) Sustaining soil fertility in West Africa. SSSA Special Publication Number 58. SSSA, Madison, pp 137–156
Traoré MF (1974) Etude de la fumure minérale azotée intensive des céréales et du rôle spécifique de la matière organique dans la fertilité des sols au Mali. Agron Trop 29:567–586
Vanlauwe B, Bationo A, Carsky RJ, Diels J, Sanginga N, Schulz S (Unpublished) Enhancing contribution of legumes and biological nitrogen fixation in cropping systems: experiences from West Africa
Versteeg MN, Koudokpon V (1990) *Mucuna* helps control Imperata in southern Benin. West African Farming Systems Research Network Bulletin, 7:7–8
Versteeg MN, Amadji F, Etéka A, Houndékon V, Manyong VM (1998) Collaboration to increase the use of Mucuna in production systems in Benin. Pages 33–43 In: Buckles D, Etéka A, Osiname O, Galiba M, Galiano G (eds.) Cover crops in West Africa contributing to sustainable agriculture. IDRC, Ottawa, Canada; IITA, Ibadan, Nigeria; Sasakawa Global 2000, Cotonou, Benin
Vine H (1953) Experiments on the maintenance of soil fertility at Ibadan, Nigeria, 1922–1951. Emp J Exp Agric 21:65–85
Vissoh P, Manyong VM, Carsky JR, Osei-Bonsu P, Galiba M (1998) Experiences with Mucuna in West Africa. Pages 1–32 In: Buckles D, Etéka A, Osiname O, Galiba M, Galiano G (eds.) Cover crops in West Africa contributing to sustainable agriculture. IDRC, Ottawa, Canada; IITA, Ibadan, Nigeria; Sasakawa Global 2000, Cotonou, Benin
Weber GK, Elemo K, Lagoke STO (1995) Weeds in intensive cereal-based cropping systems of the northern Guinea savanna. International Institute of Tropical Agriculture, Ibadan, Nigeria. Resource and Crop Management Research Monograph No. 20. 28 pp
Whelan AM, Alexander M (1986) Effects of low pH and high Al, Mn, and Fe on the survival of Rhizobium trifolii and the nodulation of subterranean clover. Plant Soil 92:363–371
Yamada N (1974) Biological nitrogen fixation – limitless resource supporting agriculture. (In Japanese) Nettai Noken Shuho 25:20–28

Chapter 7
Comparative Analysis of the Current and Potential Role of Legumes in Integrated Soil Fertility Management in East Africa

D.N. Mugendi, B.S. Waswa, M.W. Mucheru-Muna, J.M. Kimetu, and C. Palm

Abstract Legumes play an important role in farming systems in East Africa. Legumes are not only a source of food and fodder but also play a key role in improving soil fertility. As a result of the increasing pressure on land, the traditional fallow periods needed for land regeneration have shortened. Similarly, most farmers in the region do not use fertilizer. As a result of this, the soil fertility is low and declining. Whereas the nutrient deficiencies can be resolved by use of inorganic fertilizers, most smallholder farmers are poor hence cannot afford to purchase fertilizers. In other instances fertilizer is not accessible due to poor infrastructure. The use of farm yard manure on the other hand is limited since most farmers do not have livestock and where the manure is available, it is of poor quality. The growing of legumes which fix nitrogen and improve other soil properties emerges as a potential solution to the soil fertility problem among the smallholder farmers. Research in the region has identified a wide range of legumes that can be incorporated in the cropping systems. These include grain and herbaceous legumes

D.N. Mugendi (✉) • M.W. Mucheru-Muna
Department of Environmental Sciences, Kenyatta University, Nairobi, Kenya
e-mail: dmugendi@yahoo.com; moniquechiku@yahoo.com

B.S. Waswa
Tropical Soil Biology & Fertility (TSBF),
International Centre for Tropical Agriculture (CIAT),
P.O. Box 30677, 00100 Nairobi, Kenya
e-mail: bswaswa@cgiar.org

J.M. Kimetu
Institute for Sustainable Energy, Environment and Economy (ISEEE), University of Calgary, Alberta, Canada
e-mail: jmkimetu@ucalgary.ca

C. Palm
The Earth Institute, Columbia University, USA
e-mail: cpalm@ei.columbia.edu

A. Bationo et al. (eds.), *Fighting Poverty in Sub-Saharan Africa: The Multiple Roles of Legumes in Integrated Soil Fertility Management*,
DOI 10.1007/978-94-007-1536-3_7, © Springer Science+Business Media B.V. 2011

either grown in rotation or together with cereals. The legumes however, vary in their rate of growth, susceptibility to diseases and pest, adaptability to given agro-ecological conditions, ability to fix nitrogen and general acceptance by the farmers. Althugh the benefits of legumes are recognized and promoted in many parts of the world,, their use in East Africa is still low. This can be attributed to among other reasons diminishing land holdings, low and declining soil fertility, unavailability of improved germplasm, low and lack of proper agronomic knowledge on legume production, pest and diseases, poorly developed input-output markets among others. Strategies to increase legume production should aim at identifying niches for these legumes in both space and time; assessing their ability to fix N under on-farm conditions and to demonstrate immediate benefits such as substantial increase of food crops, fodder and wood products and cash to the farm. This chapter discusses the trends in legume production in East Africa and the role played by legumes in integrated soil fertility management (ISFM).

7.1 Introduction

The fertility and productivity of the soils in most East African countries is low and on the decline. Deficiencies of nitrogen (N), phosphorus (P) and potassium (K) are widespread. Stoorvogel and Smaling (1990) and Stoorvogel et al. (1993) calculated N, P and K balances for 38 African countries, and found that most East African countries experience high nutrient depletion rates (Table 7.1). Maize yields (the staple food crop) are generally less than 1 t ha^{-1} in a season (Swinkels et al. 1997). With declining soil N, the build up of *Striga hermonthica* – a parasitic weed in many cereals – increases (Oswald et al. 1996). The net effect of all these has been a decline in production of crops and food shortages.

Integration of legume cover crops into farming systems may offer feasible options for maintaining and improving soil fertility in smallholder farming. Legumes have numerous advantages, which include improved soil productivity through increased soil organic matter (SOM) content, improved soil physical and microbial properties, suppression of weeds and pests, and erosion control. Their contribution of nitrogen via nitrogen fixation to the soil and succeeding crops reduces the need for inorganic

Table 7.1 Average nutrient balances of N, P and K (kg ha^{-1} $year^{-1}$) of the arable land for some East African countries

	N		P		K	
Country	1982–1984	2000	1982–1984	2000	1982–1984	2000
Ethiopia	−41	−47	−6	−7	−26	−32
Kenya	−42	−46	−3	−1	−29	−36
Rwanda	−54	−60	−9	−11	−47	−61
Tanzania	−27	−32	−4	−5	−18	−21

Source: Stoorvogel et al. 1993

fertilizer (Palm et al. 1997). Recent studies have shown that incorporation of herbaceous N-fixing legume cover crops into crop production systems has an important role in the maintenance and improvement of soil fertility.

7.2 Legume Production and Fertilizer Use in Eastern Africa

Fertilizer consumption in eastern Africa continues to be low despite recognition that fertilizer use can greatly improve food production. Most of the fertilizer is used on cash crops such as tea, coffee and sugarcane. Legume production per hectare basis is low in most east African countries. According to the FAO (2003) statistics, the yields from dry peas, groundnuts, cowpeas, chickpeas, dry beans and pigeon peas hardly exceed 1 t ha^{-1} (Fig. 7.1). Yields for soybeans range between 0.9 and 1.5 t ha^{-1} while that of green beans is about 3 t ha^{-1}. The yield of green peas varies from 3.1 to 5.6 t ha^{-1} but has tended to be highly variable over the years. In terms of total legume production, the yields of dry beans were between 1 million metric tones and 1.7 million metric tones per year while that of groundnuts was between 0.5 and 0.8 m Mt year (Fig. 7.2). Although showing a general increase in total yields over the years, the yield may be as a result of the increase in land under cultivation (Fig. 7.3) not necessarily an increase in legume production per unit area as discussed above. For example, the area under dry beans production increased from the lowest observed in 1994 of 1.6 to 2.4 mha observed in 2002 while the area under groundnuts increased from 0.9 mha in 1993 to 1.3 mha recorded in 2002. As for the other legumes there

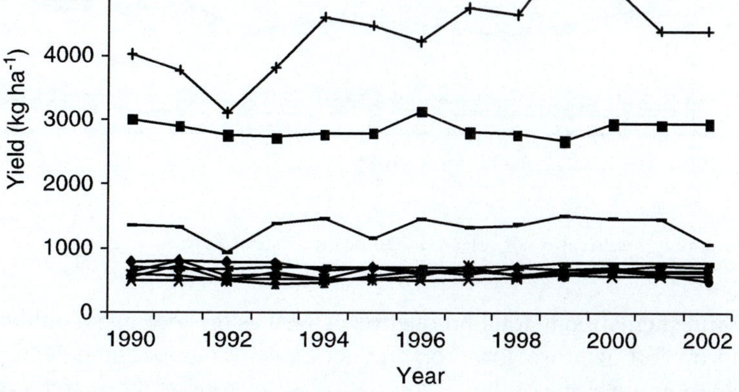

Fig. 7.1 Unit production trends for some selected legumes in eastern Africa

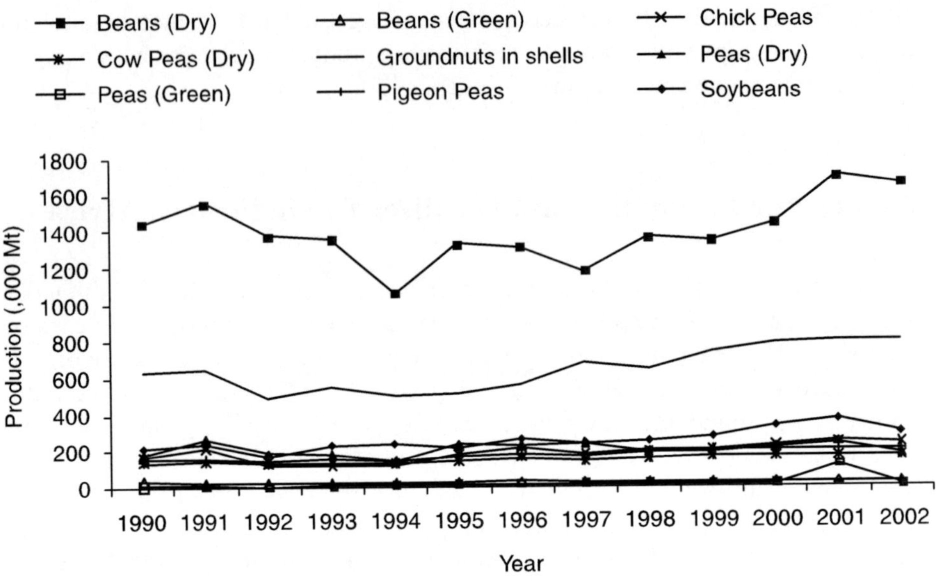

Fig. 7.2 Total production trends of selected legumes in eastern Africa

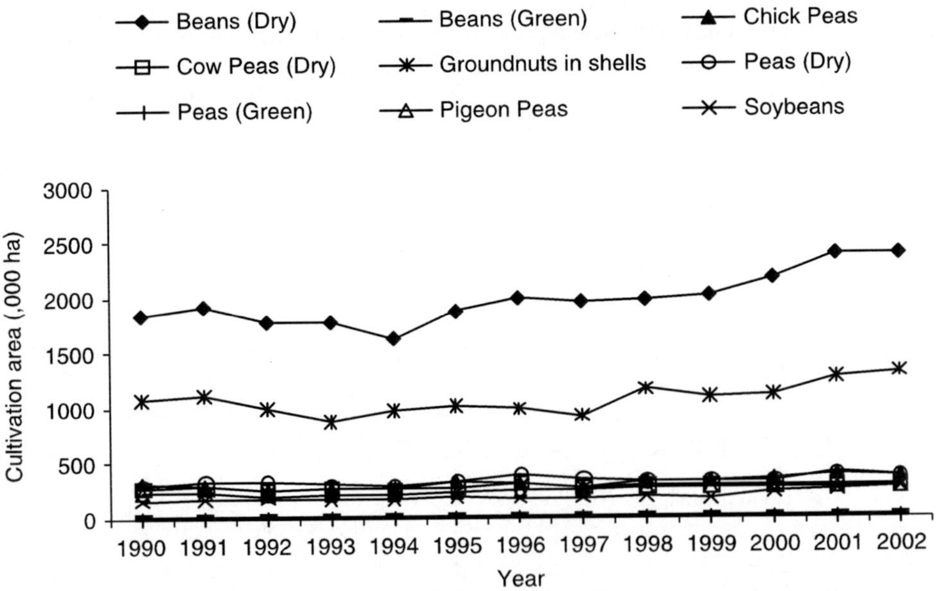

Fig. 7.3 Area under cultivation of selected legumes in eastern Africa

was a general stagnation in total production as well as the area under cultivation over the years. Low fertilizer use, low crop production per unit area and potential multiple roles of legumes are some of the reasons that have encouraged research in the use of legumes in integrated soil fertility improvement technologies.

7.3 Biomass Production by Legumes

The use of legumes as green manures basically depends on their biomass production and nutrient contents. In Kenya the Legume Research Network (LRNP) was initiated in 1994 to primarily screen suitable green manure legumes in 11 sites representing major agroecological zones in Kenya. The sites are spread across the country from 15 to 1900 m above sea level where about 40 legume species were screened. Among the best bet legume cover crops, *Mucuna pruriens*, *Crotalaria ochroleuca*, *C. grahamiana*, *Lablab purpureus* and *Canavalia ensiformis* were identified as the most outstanding in terms of dry matter production. At legume screening network sites, biomass production by mucuna was 4–7 t dry matter (DM) ha^{-1} at 3 months after planting (Dyck 1997). In Uganda, Tanzanian sunnhemp (*Crotalaria ochroleca*) performed well in all sites attaining DM of 3–5 t ha^{-1} at 3 months, while early sown *C. ochroleuca* produced as much as 13.8 t DM ha^{-1}. In Kitale, Kenya, the average DM yield for lablab ranged from 3.4 to 7.4 t DM ha^{-1}, depending on the severity of the dry season (Kirungu et al. 1998). Average green manure yield of *Vicia sativa* and *V. dasycarpa* after 3 months was 2.75 and 8.89 t DM ha^{-1} in Kabale, Uganda (Siriri 1999), while in high and medium rainfall areas of Kenya, purple vetch (*V. benghalensis*) accumulated over 8 t DM ha^{-1} after 6 months (Maobe et al. 1998). In a study by Amede (2002) in Ethiopia, dry matter production among legumes was significant regardless of the length of growing period. For short term fallows of 3 months or less, crotalaria gave significantly higher biomass yield (4.2 t ha^{-1}) followed by vetch and mucuna (2 t ha^{-1}), while for a medium-term fallow of 6 months, tephrosia was the best performing species (13.5 t ha^{-1}), followed by crotalaria (8.5 t ha^{-1}). In general, biomass production from legumes will depend on whether the legume is intercropped or sole cropped. Sole cropped green manure cover crops species produce greater yields of biomass as there is no competition for water, nutrient or light (Fischler 1997).

Selection of legumes for integration into farming systems continues to be a challenge to most research. This is due to the fact that whereas the legumes may play a significant role towards increasing soil fertility, they may not be preferred by farmers (Rao and Mathuva 2000) due to the farmers' varied socio-economic status. In a study by Amede (2002), farmer criteria in selection of legumes for incorporation into their farming systems included among others, firm root system, early soil cover, biomass yield, decomposition rate, soil moisture conservation, drought resistance and feed value. Overall, farmers would adopt the use of legumes based on land productivity, farm size, land ownership, access to markets and other uses for the legumes other than soil fertility such as animal feed.

7.4 Quality of Legume Organic Resources

A review of the TSBF Organic Resource Database gives the chemical properties of some legumes (Table 7.2) that can be integrated into farming systems by smallholder farmers.

Table 7.2 Chemical properties of selected legumes used for soil fertility improvement in East Africa

Genus	Nitrogen (%)	Phosphorus (%)	Lignin (%)	Total soluble polyphenol (%)
Canavalia ensiformis	3.5	0.16	5.5	2.4
Crotalaria juncea	3.9	0.16	6.8	1.3
Crotalaria ochroleuca	4.5	0.16	4.9	2.0
Desmodium intortum	3.4	0.15	8.8	5.5
Desmodium uncinatum	3.3	0.16	11.0	5.0
Lablab purpureus	4.0	0.18	5.6	2.5
Mucuna pruriens	3.6	0.17	7.5	3.3
Vicia benghalensis	3.7	0.16	5.0	1.1
Crotalaria grahamiana	3.2	0.13	6.8	2.00
Tephrosia vogelii	2.9	0.18	8.0	5.2
Dolichos lablab	1.2	0.18	17.6	0
Calliandra calothyrsus	3.4	0.15	17.6	9.9
Arachis hypogaea	2.2	–	6.3	1.3
Leucaena diversifolia	3.9	0.25	10.4	5.4
Cajanus cajan	3.1	0.13	14.7	4.9
Phaseolus sp.	0.8	0.06	11.3	0.3

The quality of the leaves of most of the 'best-bet' LCCs is high (high %N, low % lignin and low % polyphenols), characteristics that favor rapid decomposition and nitrogen release (Palm et al. 1997). Using these legumes singly or in combination with inorganic nutrient sources can help supply required crop nutrients and improve on soil organic matter aspects that can help boost food production in smallholder agriculture.

7.5 Effects of Legumes on Crop Yield

A number of studies are presently being undertaken to determine the short-term and long-term effects of legumes on crop response (TSBF, 1987). Emerging results from the Tropical Soil Biology and Fertility African Network (AfNet) work indicate that about 5–10 t dry matter per hectare of crop residues may supply enough nitrogen for a 2 t maize grain crop but cannot meet P requirements and hence P must be supplemented by inorganic P sources (Palm 1995). The availability and supply of 5–10 t of crop residues directly as nutrients sources may be a problem especially on resource poor farmers where there are competing uses such as livestock feed or fuel sources (Nandwa and Bekunda 1998). Legumes can be intercropped or relay cropped with food grains or planted as a sole crop in rotation with food crops. There has been mixed effect from these combinations in terms of the contribution of the legumes to the soil fertility status or to the resultant crop yields. Understanding these interactions has been basis for devising appropriate integration sequences that ensure minimal competition between the food crop and the legume cover crop. In legume cover crop/cereal intercropped systems, the yields of the initial intercrop food crop such as maize may be reduced by as much as 30% (Giller et al. 1997; Fischler and Wortmann 1999). However, this trend is reversed for the subsequent grain food crop where in most cases there are increases (20–100%) in grain yields in comparison to the continous maize crop (Giller et al. 1997). Yield increases of 75–86% were recorded in Rwanda after a one-year fallow (Raquet 1990). In Uganda, maize grain yields were consistently higher following sole legume cover crop than following sole maize (Fischler and Wortmann 1999). Increases in grain yields were 35%, 51%, 56% and 62% following sole crops of Lablab, Mucuna, Canavalia and Crotalaria, respectively. In Tanzania, Temu and Anne (1995) indicated that yield of maize grown after *C. zanzibarica* was >100% higher than when grown after maize and without fertilizers. In Western Kenya, maize yields were 70–80% higher after one season fallow of *Sesbania sesban* and *Tephrosia vogelii* than yields under continous maize (Niang et al. 1996b).

Gitari et al. (2000) studied the performance of green manure legumes under different combinations with crops in Mount Kenya region. In the long rain season, maize grain yield was 6.48 and 3.14 t ha^{-1}, where only legume residue was used as a source of N at Karurina and Gachoka sites respectively. This is compared with farmer practice (use of different combinations of inorganic fertilizers) grain yield of 3.49 and 2.71 t ha^{-1} for Karurina and Gachoka respectively. In general, maize grain yields at both sites were highest in plots where legume green manure was used alone or in combination with either animal manure or mineral fertilizers. In a similar study carried out by Kamidi et al. (2000) at Matunda in Western Kenya, legume green manures tended to boost maize and bean grain yields. The green manures planted were velvet bean (*Mucuna pruriens*), soyabeans (*Glycine max*), dolichos (*Lablab purpureus*), sunnhemp (*Crotalaria ochroleuca*) and cowpeas (*Vigna unguiculata*). The maize grain yields from plots under legume cover and half the recommended rates of inorganic fertilizers recorded significantly higher maize grain yields 7.2, 6.9, 7.4, 7.1, 6.6 t ha^{-1} for velvet bean, soyabeans, sunnhemp, cowpeas and dolichos respectively as compared to the farmer practice of 4.8 t ha^{-1}.

Gachene et al. (1999) reported that as a result of resource competition, maize yield was reduced by 30–45% when intercropped with green manure cover crops. However, there was an increase in maize yield (16–58%) during the subsequent seasons after incorporating the green manure cover crop residues. Although there is a loss of crop in one of the seasons, the increase in yields after the short duration fallow can compensate for this loss especially where labour for weeding and inputs such as seeds and fertilizers are taken into consideration.

The potential of legumes to improve food production continues to be the stimulus for the effort to encourage adoption by researchers. Several studies in Western Kenya suggest that improved fallows of a one-to-three season can increase soil fertility and improve yields. In one on-station study at an NPK-deficient site, a 6-month fallow Sesbania, tephrosia, crotalaria and *Cajanus cajan* increased maize yield by 35–128% compared with continous maize with no fertilizer application. This indeed has exciting prospects for many farmers in the area with fewer other options for improving crop yields. In this study maize yields were higher after fallows of sesbania (3.5 t ha^{-1}) and T*ephrosia vogelii* (3.6 t ha^{-1}) and similar for fallows of crotalaria, cajanus and natural fallow (2.7 t ha^{-1}). Maize yield was lowest for the continous cropping system (2.0 t ha^{-1}) although it was higher than the typical yield of 1.0 t ha^{-1} or less for such a system (Swinkels et al. 1997). The fallows showed considerable residual effects on the subsequent crops, which ranged from 36% for natural fallow to 44% with sesbania.

The potential of short-duration improved fallows has also been assessed on farm, under farmers own management conditions. In a study on N and P deficient sites, six-month fallow of crotalaria and tephrosia increased maize yields by 31–35% compared with continous cropping without addition of N and P. Maize yield from non-P addition soil was not significantly different for the legume fallows (crotalaria – 2.5 and tephrosia – 2.3 t ha^{-1}) but higher than the farmers no input control (1.6 t ha^{-1}) (Jama et al. 1999).

7.6 Nitrogen Fixation

One of the major reasons for the planting of legumes in farming systems continues to be their contribution to nitrogen fixation. As much as 30–60 kg N ha^{-1} $year^{-1}$ is reported to be added to the soil by legumes (Reijntjes et al. 1992). Sanginga et al. (1996) reported that across all cropping systems (mixed and mono cropping) *Mucuna pruriens* derived an average of 70% of its N from atmospheric nitrogen (estimate made by the ^{15}N isotope dilution method), representing 167 kg N ha^{-1} per 12 weeks in the field. Mucuna accumulated in 12 weeks about 313 kg N ha^{-1} as either a sole crop or 160 kg N ha^{-1} when mixed/intercropped with maize respectively (Sanginga et al. 1996).

Some early reports on alley cropping claimed that enormous large quantities of N were fixed by some fast-growing tree species used as hedgerows, especially *Leucaena leucocephala* and *Gliricidia sepium*. For example, Sanginga et al. (1995), in their review, cited N_2 fixation levels of 100–300 kg N ha^{-1} $year^{-1}$ and sometimes up to 500 kg N ha^{-1} $year^{-1}$. But such estimates are subject to a number of variables such as soil, climate, and plant management conditions. Furthermore, it has lately been found that high variability exists among provenances or isolines of nitrogen fixing trees (NFTs) in the percentage of total plant N derived from atmospheric N_2 (Sanginga et al. 1995, 1996). It is estimated that up to 50 kg N ha^{-1} $year^{-1}$ can be contributed by grain legumes to an associated or subsequent non-legume crop; but total N_2 fixation is two or three times this value (Greenland 1985).

The benefits of the N fixed by LCC may not be realized if the legume is harvested for food, fodder and fuel, because some of the fixed N is removed (Mafongoya et al. 2003). The N harvest index refers to the amount of N removed in the harvested products relative to the total amount in the plant. If the harvest index is high, then the harvest of legumes can result in net negative N balances. Hence, although alternative uses may encourage the adoption of legumes, this defeats the initial purpose of using LCC to add N to the soil. Production of common beans (*Phaseolus vulgaris*) for example, may not result in improved soil N status as they have a low inherent capacity

Table 7.3 Estimates of N_2 fixed by some of the grain legumes and LCC in Africa

Grain legume/LCC	Amount of N_2 fixed (kg N ha^{-1})	Duration (Days)
Arachis hypogaea (sole crop)	20–70	90–140
Cajanus cajan (sole crop)	2–90	90
Glycine max (sole crop)	159–227	97–104
Phaseolus vulgaris (sole crop)	2–58	72
Mucuna pruriens	70–130	–
Calapogonium mucunoide	126–182	1 year
M. atropurpureum	46–67	30–60

Source: (After Giller et al. 1997; Buckles et al. 1998)

to fix N and a high N harvest index (Giller and Cadisch 1995; Giller et al. 1997). Likewise if a LCC is grazed or harvested as the case with calliandra being used for fodder supply, there may also be a negative N balance unless the manure is returned to the farm. Thus the potential N replenishment by a legume is a balance between fixed N and N removed in products. As observed by Giller et al. (1994), legumes such as soybeans are very efficient at translocating their N into the grain, and even when the residues are returned to the soil there is generally a net removal of N from the field. On the other hand pigeonpeas and some varieties of cowpea (*Vigna unguiculata (L) Walp. Spp.* which lose a substantial amount of biomass in the form of roots and leaves that fall before harvest record less losses (Giller and Cadisch 1995). It is reported that as much as 280 kg N ha^{-1} $year^{-1}$ is fixed by some LCCs (Giller and Cadisch 1995). Legumes can provide sufficient N through biological nitrogen fixation to meet the needs of a subsequent crop with the exception of P hence efforts should be made to supplement P through supply of inorganic P sources (Buresh et al. 1997). Table 7.3 gives the N_2 fixing potentials of some selected legumes in Africa.

In situations where there is poor nodulation and N-fixation, legumes can exploit soil N reserves rather than contributing to them (Ssali and Keya 1984). In intercropping or agroforestry systems the N fixed mostly benefits the subsequent crop. The amount of N fixed by the legume components depends on factors such as climate, soil type, species, plant morphology, planting density of component crops, type of management and competitive abilities of the component crops (Gachene et al. 1999). Loomis and Connor (1992) and Sanchez et al. (1997) observe that symbiotic fixation in the soil and hence the overall beneficial effects from nitrogen fixation is influenced by factors such as the presence of appropriate rhizobia, deficiency of plant nutrients other than nitrogen, soil acidity and aluminium toxicity, water stress, and temperature among others. For instance, Pilbeam et al. (1995) reported that inoculated (with *Rhizobium*) and uninoculated *Phaseolus vulgaris* failed to fix N_2 in the semi arid Eastern Kenya. By contrast the cowpeas derived 50% of their N from fixation, equivalent to 197 kg N ha^{-1}. Maximization of the benefits accrued from legume is limited by widespread deficiencies of P in croplands which are mostly associated with low pH coupled by high aluminum and manganese toxicities. As observed by Irungu et al. (2002) in a study in the semi-arid lands in Kenya, application of P fertilizer at 20 kg ha^{-1} had the greatest influence on the number of nodules on cowpea roots. The amounts of N_2 fixed by the cowpeas was, however, much less than that observed by Pilbeam et al. (1995) working on an experiment at Kiboko, Kenya. Gathumbi (2000) in a study in Western Kenya using 6–8 month old fallows of *Cajanus cajan* (pigeon peas), *Sesbania sesban*, *Crotalaria grahamiana*, *Tephrosia vogelii*, *Macroptilium atropurureum* and *Arachis hypogeaea* (groundnuts) found out that most wood legume species in sole stands fixed between 57% and 74% of their total N yield. This amounted to 72–152 kg ha^{-1} of total N fixed. Tephrosia and crotalaria fixed the largest proportion of their N (68–74%). Crotalaria was the best N_2 fixer with the potential of accumulating between 113 and 152 kg ha^{-1} of fixed N in foliage. Total fallow N accumulation was highest for crotalaria and ranged between 199 and 230 kg ha^{-1}. Macroptilium fixed only 40% (71 kg ha^{-1}) of the N but extracted more N from the soil (87–105 kg ha^{-1}) compared with other species.

Nitrogen fixation is a valuable resource in any integrated soil fertility management (ISFM) system. However, because of methodological difficulties in quantifying N_2 fixation under field conditions (Sanginga et al. 1996), quantitative information on the extent of benefit that is actually realized by using legumes far from satisfactory. Furthermore it is not clearly understood what proportion of the N_2 that is fixed by a legume is actually utilized by, or potentially made available to, an associated crop during the current crop cycle and what proportion goes into the soil's N store for eventful use by subsequent crops.

7.7 Legume Inoculation

Legume inoculation is an environmental benign process through which a given leguminous crop is provided with the compatible effective bacterial strain of the genus *Rhizobium* which results in an effective symbiosis between the plant and the bacteria (Abdelgani et al. 2002). The ultimate result is higher amount of nitrogen fixed, better growth of the plant, increase in production and improvement of soil nitrogen. The need for inoculation is not, however, universal (Singleton et al. 1992) and inoculation does not always result in positive response (Olsen et al. 1996). Many factors affect the response to inoculation; these include soil nitrogen, number of native rhizobia (Karanja et al. 1995; Karanja et al. 1997), the rhizobia strain used in the inoculum (Burton 1985) and many environmental factors. Positive response to inoculation has always been attributed to low soil nitrogen content.

Further, the presence of high populations of indigenous *Bradyrhizobia spp* in tropical soils that nodulate with commonly grown legumes has also acted negatively on the response of cultivated plants to rhizobia inoculation (Karanja et al. 1995). In their study, Karanja et al. (1995) estimated the population sizes of *Bradyrhizobium spp* at between 1.25 and 2.40 $\log_{10}$ cells per gram of soil. Despite this limitation, harnessing the beneficial effects of legume-*Rhizobium* technology has led to a surplus production of inoculants. For instance, approximately 1.5 t $year^{-1}$ are produced and distributed in Kenya and Uganda, 16 t and 6 t are produced in Zambia and Zimbabwe respectively. A summary of legume inoculant production and other relevant details are presented in Table 7.4.

Table 7.4 Legume inoculant production in Eastern and Southern Africa

Country producer	Product name	Quantity (kg)	Carrier	Retail quantity (g)	Price ($)	Inoculant
Kenya: MIRCEN, University of Nairobi	Biofix	500–1,000	Filtermud	100	1.25	Lablab, soybean, beans, Alfalfa, Desmodium, Pigeonpea
South Africa Pretoria	Stimuplant	12,500	Irradiated peat	250	1.75	Soybean, groundnuts, cowpeas, Lucerne, clovers, beans, peas, Lotus, Desmodium, Medicago
Uganda: Makerere University, Kampala	Bio-N-Fixer	8,000	Peat	250	1.15	Glycine, Lucerne, beans, soybeans, calliandra
Zambia: Mt Makulu Research Station	Nitrozam	16,000	Peat	250	0.50	Soybean 90%, Lucerne, beans
Zimbabwe: DRCC, Marondera	–	6,000	Sterile bagasse	50	0.25	Soybean 90%, beans, pea Lentil, Cloves, Lucerne, Desmodium. Groundnuts, Stylosanthes, Crotalaria
Sudan: ENRRI	Okadin	2,000	Unsterile charcoal dust	500	–	Soybean, groundnut, Lucerne, Guar, fababean, Chickpea and beans

Different rhizobia inoculants have been used on different LCC. For example cowpea rhizobia has been applied to *Crotalaria spp*., Mucuna, Lablab, *P. lunatus* and *V. uinguculata*, pea/vetch rhizobia on *Vicia dasycarpa* and *Vicia benghalensis*) and *Desmodium rhizobia* inoculant on *Desmodium spp*. A survey carried out in eight countries in Eastern and Southern Africa revealed that farmers perception to the use of rhizobium technology was varied. Whereas 95% of the farmers were familiar with root nodules, only 26% considered nodules to have beneficial effects, and less than 10% had ever used legume inoculants (Woomer et al. 1997). Constraints in the agro-technology transfer to smallholder farmers ranged from pricing and marketing, farmer awareness, research and development, linkages between public institutions and industry, and policy. All inoculant production units listed in Table 7.4 are based either in National Agricultural Research Station (NARS) or in microbiology laboratories in public universities (Karanja 2000). For a long time the focus of these institutions has not been for commercial purposes, hence this non-industrial production approach has meant that per unit inoculant prices have been high leading to low farmer acceptance. Except in South Africa, Zambia, Zimbabwe where proper distribution mechanisms have been developed through commodity based co-operatives, in other countries such as Kenya and Uganda farmers are expected to contact or travel to the laboratory centers to purchase the inoculants. Coupled to this, legume inoculants are not included in the list of plant nutrient replenishing products that are recommended to farmers by the ministries of agriculture and affiliated national institutions hence are not commonly stocked in agrochemical retail shops. This has had a negative impact on the wide application of legume innoculants by the millions of smallholder farmers in Sub Saharan Africa where soil N limits food production.

7.8 Residual Effects and Interaction with Mineral Fertilizer

In general, woody fallows accumulate larger N stocks than herbaceous ones because of their larger and continuing biomass accumulation than those of herbaceous fallows as a result the residual effects of tree fallows are longer than those of herbaceous fallows (Mafongoya et al. 2003). The residual benefit does not necessarily demonstrate a contribution of N from the legume N_2-fixation but could simply be due to sparing effects of soil N. There are still relatively few studies in which the sources of N for the second crop have been separated. Giller and Cadisch (1995) reported that measurements using ^{15}N-labelled residues of grain legumes indicated that, with a few exceptions, some 10–20% of the legumes N is recovered in the first subsequent crop. The yield of subsequent maize crop after the fallows is reported to be influenced by the length of the fallow (Mafongoya and Dzowela 1999) and the fallow species (Kwesiga et al. 1999; Mafongoya and Dzowela 1999). As the length of the fallow decreases, the yield of the subsequent crop also decreases. In Western Kenya on a kaduidalfic Eutrodox soil, for example, short duration fallows (12–18 months) increased maize yield as compared with continuos cropping and natural weed fallows (Niang et al. 1996a; Jama et al. 1998b). In their experiment,

Niang et al. (1996a) reported high maize grain yields in sesbania (3.8–5.11 t ha^{-1}) tephrosia (4.1–4.8 t ha^{-1}) and cajanus (3.1 t ha^{-1}) as compared with natural fallow and continuos maize cropping which yielded 1.9 and 1.8 t ha^{-1} respectively. The sesbania fallow produced 9.7 t ha^{-1} of maize grain yield during the subsequent three seasons compared with 6.9 t ha^{-1} after grass fallow and 4.9 t ha^{-1} after continuos maize cropping. Moreover, the fallow effect at this P deficient site was substantially higher after applying P fertilizer to maize (Jama et al. 1998b).

Technologies that combine mineral fertilizers with organic nutrient sources can be considered as a better option in increasing fertilizer use efficiency, reducing risks of acidification, and providing a more balanced supply of nutrients (Donovan and Casey 1998). Combination of organic and inorganic nutrient sources has been shown to result into synergy and improved synchronization of nutrient release and uptake by crop (Palm et al. 1997) leading to higher yields; especially when the levels of inorganic used are relatively low as is the case in smallholder farms of central Kenya (Kapkiyai et al. 1998, 1999).

Leucaena biomass combined with inorganic fertilizer gave higher crop yields as compared to sole use of inorganic fertilizer or sole leucaena biomass in central Kenya (Mugendi et al. 1999; Mucheru 2003). Studies in Western Kenya by Ojiem et al. (2002) reported that there was an increment of 1 t ha^{-1} when mucuna and crotalaria were combined with inorganic fertilizer. This implies that mucuna-inorganic N combination was probably more effective in enhancing N utilization. With addition of P as triple superphosphate (TSP), at the rate of 20 kg ha^{-1}, maize yield increased considerably for crotalaria (36%), for tephrosia (52%) and for control (37.5%) compared to non-P addition (Jama et al. 1999).

7.9 Partitioning Non-N benefits of Legume-based Cropping Systems

The advantages associated with the use of cover crops include: (1) enriching the soil with biologically fixed N, (2) conserving and recycling soil mineral nutrients, (3) providing ground cover which helps minimize soil erosion, and (4) requiring little or no cash input (Franzluebbers et al. 1998). Most reports on nutrient content of tree biomass deal with N; other elements such as P and K are less commonly reported. The N content of leaf biomass will range from 2% to 5% for these N_2 fixing species and 1.4–3.5% for the non- N_2 –fixers. As for P and K, information of a general nature available in the literature shows a range of 0.15–0.29% for P and 0.9–1.52% for K in leaf biomass of common agroforestry tree species (Nair 1993; Palm 1995).

Farmers have found the green manure's labour saving effect attractive as weeding is one of their most costly operations both in terms of time and/or money (Kirungu et al. 1998). Further, studies have indicated that legumes can be used to control weeds (Mureithi et al. 2004). For example, in a study by Community Mobilization Against Desertification (CMAD) in the lower potential zone of south Nyanza evaluated the effectiveness of sunnhemp, lablab and velvet bean in the control of striga infestation, improve soil fertility and subsequently increase maize yields. A rotational system was practiced where the legumes were planted in one season and their biomass incorporated in the soil before planting maize in the following season. Striga was monitored by counting the striga seeds that had germinated. Results over the period 1996–1997 revealed that plots where the legumes were planted had on average 10 striga counts per m^2 while the control had 32. Further average maize grain yield was 2.4 t ha^{-1} in plots where a legume was planted compared to 1.4 t ha^{-1} in control plots. These preliminary results indicated that legumes had a depressing effect on striga infestation.

Results on an acid soil in Western Kenya (pH=5.1, clay=29%) indicated that rotation of maize with *S. sesban* could lessen losses in maize yield resulting from potassium deficiency (ICRAF 1996). Fischler (1999) observed that application of crotalaria green manure led to decrease in bulk density, and increase in water infiltration capacity of the soil suggesting that the yield stimulation due to crotalaria not only resulted from the increased nitrogen supply, but also from more favourable soil physical properties.

Kamidi et al. (2000) in his study in Western Kenya reported that mucuna had the highest ground cover (72%) followed by crotalaria (63%) and lablab (54%). Soyabeans and cowpeas gave the lowest ground cover (32% and 38% respectively). The groundcover offered by these green manures greatly reduced soil erosion especially during the long rain season as noted by Gachene et al. (2000). In line with improved ground cover is the issue of soil moisture conservation from legumes. Amede (2002) observed a significant difference in soil moisture conservation among legumes, and this decreased in the order of mucuna (22.8%), vetch (20.8%), stylosanthus (20.2%), crotalaria (14%), canavalia (14%) and tephrosia (11.9%).

Fuel wood is the main energy source for cooking in the rural tropics. A two-year *Sesbania sesban* fallow can produce 10 t ha^{-1} of fuel wood (Sanchez et al. 1995). Use of woody fallows can avoid the need for gathering fuel wood from adjacent forests or woodlands and the drudgery of transporting it to the household.

In favorable environments, hedgerows produced 8–12 t ha^{-1} $year^{-1}$ of prunings (dry weight) which increased SOM and ensured adequate supply of nutrients, especially N, to the alleys. Tree fallows can increase labile fractions of SOM, which supply nutrients to crops after fallows (Barrios et al. 1997).

Improved fallows of leguminous trees and shrubs accumulate N in their biomass through biological nitrogen fixation, capture of subsoil N unutilized by crops and interception of N leached beyond the crop rooting zone (Buresh and Tian 1997). Nutrients captured by trees from below the rooting zone of annual crops can become an input when transferred to surface soil in the form of leaf litter, roots and prunings of the tree leaves and branches. Deep uptake of N from the sub-soil has received considerable investigation over the years (Mekonnen et al. 1997; Buresh and Tian 1997; Kindu et al. 1997; Mugendi et al. 2003). In Western Kenya for example, an accumulation of nitrates (NO_3^-) at a depth of 0.5–2.0 m has been observed under unfertilized maize on acidic soil (Kandiudalfic Eutrodox of Kandiudalf). Nitrate concentrations of between 70 and 315 kg N ha^{-1} at a depth of 0.5–2.0 m have been observed (Hartemink et al. 1996). Mekonnen et al. (1997) showed that *Sesbania sesban* improved fallow grown in rotation with maize in this region can rapidly root into this 'chemical safety net' and take up the sorbed NO_3^- that was inaccessible to unfertilized maize. In their study, NO_3^- at a depth of 4 m was 51 kg N ha^{-1} for a 15-month *Sesbania sesban* fallow, as compared to 199 kg N ha^{-1} for unfertilized maize. The maximum rooting depth was 1.2 m for maize, whereas roots in a 15-month old *Sesbania sesban* extended below 4 m. Jama et al. (1998a) in a study on an acidic soil in Western Kenya without chemical and physical barriers to rooting in the top 4 m observed the existence of a direct relationship between the demand of trees for N and the magnitude of NO_3^- uptake from the subsoil. Fast-growing trees such as *Sesbania sesban* and *Calliandra calothyrsus* reduced soil NO_3^- in the top 2 m by about 150–200 kg N ha^{-1} by 11 months after establishment from seedlings.

In the Central highlands of Kenya, Mugendi et al. (2003) reported that treatments that received tree leafy-biomass but had no trees ± fertilizer and the inorganic fertilizer treatment recorded higher amounts of mineral-N in the 1–3 m depth averaging 300–600 kg N kg^{-1}. On the other hand, soils in treatments with *Leucaena leucocephala* and *Calliandra calothyrsus* tree hedges recorded a average of 20–60 mg N kg^{-1} in the same depth indicating that trees are capable of intercepting and recapturing the crop-inaccessible nutrients, below the roots of the annual crops by the action of their the deep roots. Indeed, results on the total root length indicated that only less than 5% of all the maize roots were located below the 90 cm soil depth while 75% leucaena and 40% calliandra roots were found below the same depth.

Mnkeni et al. (1995) studied the effects of alley cropping with *Leucaena leucocephala* and incorporation of its pruning on the phosphorus status of an Andosol from Mbeya, Tanzania, were determined. The results of soil analysis indicated plots receiving *L. leucocephala* prunings in combination with nitrogen fertilizer over a period of 11 years had a higher organic matter content than the control plots. This also caused a higher level of organic phosphorus and total P. The Olsen bicarbonate extractable P was also higher in plots receiving prunings, though to a small extent in proportion to observed increases in organic P. The results showed that incorporation of organic residues from alley cropped trees partly influences soil fertility status through its positive effects on P balance in the soil.

7.10 Biophysical and Socioeconomic Challenges of Integrating Legumes for Soil Fertility Management

In order to assess the long-term performance of hedgerow intercropping (alley cropping), Rao et al. (1997) reviewed the results of 29 trials conducted for 4 or more years, mostly with small plots, over a wide range of soils and climates across the tropics. In 28 of the trials, no N fertilizer was applied to crops, but in most trials P fertilizer was applied. The results showed both positive effects (n = 15 for cereals, n = 8 for non-cereal crops such as beans, cowpea and cassava) and negative effects (n = 13 for cereal crops and n = 1 for sweet potatoes) of hedgerow intercropping yields, indicating that the system performance is location-specific and sensitive to management, so generalizations may be difficult if not misleading. Ignoring less than 15% yield increase as unattractive to farmers, only 2 of 10 studies in semiarid sites (less than 1,000 mm rainfall) gave substantial yield increases. In subhumid environments (rainfall between 1,000 and 1,600 mm), significantly positive responses were observed in 7 of 11 studies where the soils were either inherently fertile (Kang 1993; Muriithi et al. 1994) or the acid infertile sites receiving external nutrients and lime (Matthews et al. 1992; Akyeampong 1996). In the humid tropics (rainfall greater than 2,000 mm), maize and taro did not benefit from hedgerow intercropping in four out of eight trials, but interestingly bean and cowpea yields invariably increased.

The tree species is also an important factor in different agroecological zones as far as alley cropping is concerned. For instance, Mugendi et al. (1999) reported that when alley cropped with leucaena, maize produced significantly higher yields compared to maize monoculture treatments, but when alley-cropped with calliandra, the yield of maize was less than that of the mono-cropped unfertilized control.

Whereas low yield of hedgerow prunings (2–3 t ha^{-1} $year^{-1}$) and competition of hedgerows for water were the major reasons for the negative results in water-limited areas (Mathuva et al. 1998), low yield of prunnings and competition of hedgerows for nutrients were responsible for negative results in poor soils (Matthews et al. 1992). Inadequate water limited the response of crops even though hedgerow intercropping improved soil fertility in certain sites of the semiarid tropics.

Identification of multipurpose tree and shrubs with potential for maintaining soil fertility and providing high quality fodder for livestock is a major focus for agroforestry research. Napier grass is suitable for providing the basic ration, while trees and shrubs have a significant potential as high nutrient supplements to conventional animal feeds. An example of the agroforestry trees fed on livestock is calliandra. Calliandra can be fed to dairy cattle, local cattle, goats, sheep, rabbits and poultry. For ruminants, 3 kg of fresh calliandra can replace 1 kg of concentrates. Poultry can be supplemented with dried calliandra as feed.

However, in a 15-year research on the potential benefits from herbaceous nitrogen fixing legumes in Rwanda, green manuring proved to be a risky enterprise due to highly variable biomass production and residual effects (Drechsel et al. 1996). Yield increments, from tephrosia, cajanus, crotalaria, sesbania, mucuna and mimosa planted over one or more seasons as pure green manure, in hedgerows (alley cropping), or on fields as seasonal inter-or relay-crops on-farm, of up to 74% in the first season and 46% in the second season did not compensate loss of yields and labour investments during green manuring. Even where biomass production was sufficient, residual effects were in most cases unsatisfactory, due to rapid nutrient leaching (N, K) or inappropriate foliage incorporation on-farm. In research-managed trials, residual effects were in general somewhat higher, but more than mere compensation of lost yields was not possible and farmers' adoption of these labour-intensive technologies was rather low. Due to acute land shortage, farmers were also reluctant in allocating land to fallows or hedgerows, with the exception of fields already out of

Table 7.5 Trends in informal trade in beans between Tanzania and her neighbours

	Exports		Imports	
Country	Volume (MT)	Value (USD 2000)	Volume (MT)	Value (USD 2000)
Uganda	24	8	2	1
Kenya	2,143	741	2	1
DRC	4,376	2,008	–	–
Mozambique	–	–	–	–
Malawi	327	117	7	3
Zambia	1,108	11,200	–	–

Source: Ackello-Ogutu and Echessah (1998)

production. Consequently, the concept of improving soil fertility and crop yields with the help of planted fallows or green manure in rotation failed.

Despite the many benefits of legumes in integrated soil fertility management, there are some challenges that need to be curbed so as to attain progress. In the green manure options, the pests and disease problems common to both *Phaseolus vulgaris* (common bean) and the green manure species increases the severity of these problems in both crops. Best bet green manure species' susceptibility to the root rot problem would severely constrain their productivity on farms with low soil fertility where they are needed most (Kirungu et al. 2002). According to Ndufa (2001) root knot nematodes may be a potential constriant for the use of legume species for soil fertility improvement especially when beans and maize are intercropped after a fallow. Legume fallows of Sesbania sesban, Tephrosia vogelii and *Tephrosia candida* act as host plants and are susceptible to root-knot nematodes. As observed by Desaeger and Rao (1999) root-knot nematodes (*Meloidogyne nicognita* and *M. favanica* affect sesbania and increase the nematode population in the soil. As such nematode susceptible crops such as tomatoes, carrots, eggplants, beans and tobacco may risk being increasingly parasitised after sesbania fallow and may lead to yield loss (Desaeger and Rao 2001). Although farmers often intercrop maize and beans, it is recommended that beans should not be planted after sesbania and tephrosia fallow until the second season.

As noted in Chap. 5 the issue of legumes seed/ germplasm supply and markets are critical to the sustainability of any legume based soil fertility improvement initiative. Production of pulses has traditionally been for household consumption; however, pulses are becoming a major source of income for farming households in the region through both domestic market and informal trading with neighbouring countries. Informal trade in beans in east and southern Africa, for example, is especially significant along the Kenya-Uganda border, Mozambique-Malawi border, and between Tanzania and all her neighbours (Minde and Nakhumwa 1996; Ackello-Ogutu and Echessah 1998; Bowen 1998) with research indicating that Tanzania and Mozambique are the major informal bean exporters, while Malawi, Zambia, Uganda, Kenya and Democratic Republic of Congo are the major importers of beans (Table 7.5). While cross border markets provide a niche market that can ensure income and food security to millions of smallholder bean producers in the region, it should be noted that unrestricted movement of bean varieties has implications on the quality of the varieties released by bean breeding and dissemination programs. Therefore, there is need to ensure that such programs establish mechanisms to monitor the movement of legumes varieties across the borders in order to maintain the quality of varieties being developed.

Informal cross-border trade has both positive and negative impacts on bean seed dissemination, quality and entrepreneurship. Positive aspects include availability of markets, food security and income. These should be exploited in order to benefit smallholder farmers. Negative of informal cross-border trade include seed quality concerns. Monitoring movement of varieties becomes almost impossible if legumes are increasingly traded through informal channels. This therefore calls for a continous formal and informal monitoring in order to generate sound trade statistics. Moreover, it is worth noting that informal trade thrives due to a number of factors which need to be addressed. These factors include: the tendency for traders to evade high export taxes and import duties; bureaucratic licensing, registration requirements and market failures as a result of poor policies adopted by countries and poor infrastructure.

7.11 Conclusion

Research on legume in eastern Africa has facilitated the selection of different species for different environments. The choice of these legume species has been based on characteristics such as ease of establishment, high biomass production, N-fixation, seed production, tolerance to biotic and abiotic stresses and alternative uses. Case studies on the use of legumes have shown improvements in both soil fertility and crop yields under legumes. However, lack of publication of findings hinders effective dissemination of legume technologies in this region. There is need to address in more detail, certain aspects that are likely to affect the widespread adoption of legume technologies. In particular, there is need to identify niches for these legumes in both space and time; assessing their ability to fix N under on-farm conditions and to demonstrate immediate benefits such as substantial increase of food crops, fodder and wood products and cash to the farm. Whereas a lot of research has been done on legumes in east Africa, it has focused on yields under different agro-ecological zones with scanty information on the contribution of the legumes to biological N-fixation. There is need therefore for more studies to focus on unexplored areas such as the contribution of the legumes to biological nitrogen fixation as well as other non-N benefits such as the contribution of the legume to soil organic matter and other soil physical properties. Profitability of green manure technologies, returns to labour (residue management, weed suppression), farmers' perceptions and potential for adoption need further assessment. As such there is need to carry out more detailed characterization of green manures niches inorder to understand both the spatial and temporal niches of the legumes on farmers fields. Issues of seed availability, quality and markets also need to be addressed if sustainability of the legume based soil fertility improvement technologies is to be achieved.

References

Abdelgani ME, Mohammed SS, Osman AG (2002) The use of *Rhizobium* inoculants for increasing productivity of cluster bean (*Cyamopsis tetragonoloba*) in desert-affected soils.In: Karanja NK, Kamidi JHP (eds.) Ninth Congress of the African Association for Biological Nitrogen Fixation 25–29th September 2002, Nairobi, pp 1–9

Ackello-Ogutu C, Echessah PN (1998) Unrecorded cross-border trade between Tanzania and her neighbours: Implications on food security. SD Publication Series. Technical Paper No. 89. Washington, DC. Technoserve Inc

Akyeampong E (1996) The influence of time of planting and spacing on the production of fodder and fuelwood in association of *Calliandra calothyrsus* and *Pennisetum purpureum* grown on contour bunds in the highlands of Burundi. Exp Agric 32:79–85

Amede T (2002) Integration of legumes into the farming systems of the East African highlands: a dual approach. Grainlegumes and green manures for soil fertility in southern Africa: taking stock of progress, Harare, 8–11 Oct 2002

Barrios E, Kwesiga F, Buresh RJ, Sprent JI (1997) Light fraction soil organic matter and available nitrogen following trees and maize. Soil Sci Soc Am J 61:826–831

Bowen N (1998) Trade, traders and trading. Malawi and Zimbabwean Province, Mozambique. Submitted to OXFARM, Malawi

Buckles D, Triomphe B, Sain G (1998) Cover crops in hillside agriculture. Farmer innovation with mucuna. IDRC/CIMMYT, Ottawa, Canada, p 218

Buresh RJ, Tian G (1997) Soil improvement by trees in sub Saharan Africa. Agroforest Syst 38:51–76

Buresh RJ, Sanchez PA, Calhoun F (eds.) (1997) Replenishing soil fertility in Africa. SSSA Special publication No. 51, Madison, Wisconsin, pp 193–218

Burton JC (1985) Production and use of legume inoculants in Africa. In:Ssali H, Keya SO (eds.) Proceedings of the first conference of the African association for biological nitrogen fixation, Nairobi, 23–27 July 1984, pp 119–134

Desaeger J, Rao MR (1999) The root-knot nematode problem in Sesbania fallows and scope for managing it in Western Kenya. Agroforest Syst 47:273–288

Desaeger J, Rao MR (2001) Effect of field establishment methods on root-knot nematode (Meloidogyne spp.) infection and growth of *Sesbania sesban* in Western Kenya. Crop Prot 20:31–41

Donovan G, Casey F (1998) Soil fertility management in Sub-Saharan Africa. Journal of Economic Literature, Department of Economics, University of Connecticut, USA

Drechsel P, Steiner KG, Hagedorn F (1996) A review on the potential of improved fallows and green manure in Rwanda. Agroforest Syst 33(2):109–136

Dyck E (1997) Screening legumes for potential soil productivity improvement in Kenya: Kenya legume research network. In: Poster presented at green manure cover crops conference in Santa Catarrina, Brazil, 6–12 April 1997

Fischler M (1997) Legume green manures in the management of maize-bean cropping systems in Eastern Africa with special reference to Crotalaria (C. ochroleuca). Ph.D. thesis No. 12099, ETH Zwich, Switzerland

Fischler M (1999) *Crotalaria ochroleuca* as a green manure in a maize-bean cropping systems in Uganda. Field Crop Res 61:97–107

Fischler M, Wortmann CS (1999) Green manures for maize maize-bean systems in eastern Uganda: agronomic performance and farmers' perceptions. Agroforest Syst 47:123–138

Food Agriculture Organization (2003) FAOSTAT agriculture data, FAO, Rome. http://apps.fao/cgi-binnph-db.pl?suset=agriculture

Franzluebbers K, Hossner RL, Juo ASR (1998) Integrated nutrient management for sustained crop production in sub-Saharan Africa (a review). Soil management CRSP: TropSoils/ TAMU Technical Bulletin, pp 98–03

Gachene CKK, Karanja NK, Mureithi JG (2000) Effect of soil erosion on soil productivity and subsequent restoration of productivity using green manure cover crops: data summary of a long-term field experiment. Legume Res Netw Proj Newsletter 4:2–4

Gachene CKK, Palm CA, Mureithi JG (1999) Legume cover crops for soil fertility improvement in the East African region. Report of the African highland initiative workshop TSBF, Nairobi, 18–19 Feb 1999

Gathumbi SM (2000) Nitrogen sourcing by fast growing legumes in pure and mixed species fallows in Western Kenya. PhD Thesis, Wye College, University of London

Giller KE, Cadisch G (1995) Future benefits from nitrogen fixation: an ecological approach to agriculture. Plant Soil 174:255–277

Giller KE, Cadisch G, Ehaliotis C, Adams E, Sakala WD, Mafongoya PL (1997) Building soil nitrogen capital in Africa. In: Buresh RJ, Sanchez PA, Calhoun F (eds.) Replenishing soil fertility in Africa. SSSA Special publication No. 51, Madison, Wisconsin, pp 193–218

Giller KE, McDonagh JF, Cadisch G (1994) Can biological nitrogen fixation sustain agriculture in the tropics? In: Syers JK, Rimmar DL (eds.) Soil science and sustainable land management in the tropics. CAB Int, Wallingford, pp 173–193

Gitari JN, Karumba SK, Mwaniki K (2000) Performance of green manure legumes in two contrasting sites in Embu and Mbeere districts of Mount Kenya region. Legume Res Netw Proj Newsletter 4:2–4

Greenland DJ (1985) Nitrogen and food production in the tropics: Contribution from fertilizer nitrogen and biological nitrogen fixation. In: Kang BT, van der Heide J (eds.) Nitrogen management in farming systems in humid and sub humid tropics. Institute of Soil Fertility (IB)/International Institute for Tropical Agriculture (IITA), Haren/Ibadan, pp 9–38

Hartemink AE, Buresh RJ, Jama B, Janssen BH (1996) Soil nitrate and water dynamics in sesbania fallow, weed fallows and maize. Soil Sci Soc Am J 60:568–574

ICRAF (1996) International Centre for Research in Agroforestry annual report 1995. ICRAF, Nairobi

Irungu JW, Wood M, Okalebo JR, Warren GP (2002) Enhancement of biological nitrogen fixation in smallholder farms of semi-arid Kenya through application of phosphorus. In: Karanja NK, Kamidi JHP (eds.) Ninth congress of the African Association for Biological Nitrogen Fixation, 25–29 Sept 2002, Nairobi, Kenya, pp 1–9

Jama B, Buresh RJ, Ndufa JK, Sherpherd KD (1998a) Vertical distribution of roots and soil nitrate: tree species and phosphorus effects. Soil Sci Soc Am J 62:280–286

Jama B, Amadou I, Niang I, Amadalo B, de Wolf J, Rao MR, Buresh RJ (1999) The potential of improved fallow to improve and conserve the fertility of nutrient-depleted soils of Western Kenya. Proceedings of the 6th biannual KARI scientific conference

Jama B, Buresh RL, Place FM (1998b) Sesbania tree fallows on phosphorus-deficient sites: maize yield and financial benefit. Agron J 90:7171–726

Kamidi M, Gitahi F, Osore P, Cheruiyot D (2000) Effect of green manure legume on the yield of maize and beans in Matunda Farm, Trans Nzoia District, Kenya. Legume Res Netw Proj Newsletter 4:2–4

Kang BT (1993) Alley cropping: past achievements and future directs. Agroforest Syst 23:141–155

Kapkiyai JJ, Karanja NK, Woomer PL, Qureshi JN (1998) Soil organic carbon fractions in a long-term experiment and the potential for their use as a diagnostic assays in highland farming systems of central Kenya. Afr Crop Sci J 6:19–28

Kapkiyai JJ, Karanja NK, Quresh JN, Smithson CP, Woomer PL (1999) Soil organic matter and nutrient dynamics in a Kenyan nitisol under long-term fertilizer and organic input management. Soil Biol Biochem 31:1773–1782

Karanja NK, Freire M, Gueye M, Dasilva E (2000) MIRCEN networking capacity-building and biological nitrogen fixation-technology transfer in Africa and Latin America. Ag BiotechNet 2:1–5

Karanja NK, Mwala AK, Woomer PL, Haru RW (1997) Response of Glycine max (L) Merrill and Phaseolus vulgaris L. to rhizobia inoculation, nitrogen and phosphorus fertilization in Kenya. African Crop Science Conference Proceedings 3, pp 323–330

Karanja NK, Woomer PL, Wangaruro S (1995) Indigenous Rhizobia population in East and Southern Africa. A network approach. In: Allsop D, Colwell RR, Hawksworth DL (eds.) Microbial diversity and ecosystem function. CAB International, Wallingford, pp 447–453

Kindu M, Buresh RJ, Bashir J (1997) Root and inorganic nitrogen distribution in sesbania fallow, natural fallow and maize fields. Plant Soil 188:319–327

Kirungu BAM, Kamwana S, Shiundu P, Kasiti J (1998) On farm green manure trials: 1997–1998. Legume Research Network Report

Kirungu BAM, Kasiti J, Shiundu P, Kamwana S, Mukongo M, Nderitu S, Wamalwa E (2002) The effects of legume and maize stover residue management practices on the yield of a subsequent maize crop in Kitale, Kenya.In: Proceedings of the 2nd scientific conference of the soil management and legume research network projects in Mombasa, Kenya, pp 129–143

Kwesiga F, Franzel S, Place F, Phiri D, Simwanza CP (1999) *Sesbania sesban* improved fallows in eastern Zambia: their inception, development and farmer enthusiasm. Agroforest Syst 47:49–66

Loomis RS, Connor DJ (1992) Crop ecology: productivity and management in agricultural systems. Cambridge University Press, Cambridge

Mafongoya PL, Dzowela BH (1999) Biomass production of tree fallows and their residual effect on maize in Zimbabwe. Agroforest Syst 47:139–151

Mafongoya PL, Mugendi DN, Jama B, Waswa BS (2003) Maize based cropping systems in the subhumid zone of East and Southern Africa. In: Gichuru MP, Bationo A, Bekunda MA, Goma HC, Mafongoya PL, Mugendi DN, Murwira HM, Nandwa SM, Nyathi PN and Swift MJ (eds.) Soil fertility management in Africa: a regional perspective. Academic Science Publishers, Nairobi, Kenya, pp 73–122

Maobe SN, Dyck E, Mureithi JG (1998) Screening of soil-improving herbaceous legumes for inclusion into smallholder farming systems in Kenya. In: Waddington S (ed.) Soil fertility research for maize breed farming systems in Malawi and Zimbabwe. SoilFertNet, CIMMYT, Harare, Zimbabwe, pp 105–111

Mathuva MN, Rao MR, Smithson PC, Coe R (1998) Improving maize (Zea mays) yields in semiarid highlands of Kenya: agroforestry or inorganic fertilizers? Field Crop Res 55(1–2):57–72

Matthews RB, Lungus S, Volk J, Holden ST, Solberg K (1992) The potential of alley cropping in improvement of cultivation systems in the high rainfall areas of Zambia II. Maize production. Agroforest Syst 17:241–261

Mekonnen K, Buresh RJ, Jama B (1997) Root and inorganic nitrogen distributions in sesbania fallow, natural fallow and maize. Plant Soil 188:319–327

Minde I, Nakhumwa TO (1996) Unrerecorded cross-border trade between Malawi and her neighbours: implications on food security. Technoserve Inc

Mnkeni PNS, Semoka JMR, Kamasho JA, Mwenduwa MI (1995) Effects of alley cropping with Leucaena leucocephala and incorporation of its prunings on the phosphorus status of an andosol from Uyole, Mbeya, Tanzania. Nor J Agric Sci 21(Supplement):117–123

Mucheru MW (2003) Food production improvement using locally available soil amendments in Meru South district, Kenya. MSc Thesis, Kenyatta University, Kenya

Mugendi DN, Nair PKR, Mugwe JN, O'Neill MK, Woomer PL (1999) Calliandra and leucaena alley cropped with maize. Part 1: Soil fertility changes and maize production in the sub-humid highlands of Kenya. Agroforest Syst 46:39–50

Mugendi DN, Kanyi MK, Mugwe JN, Kung'u JB, Wamicha W (2003) Mineral N movement and management in the agroforestry systems in the central highlands of Kenya. In: Mureithi JG, Macharia PN, Gichuru M, Mburu MWK, Mugendi DN, Gachene CKK (eds.) Proceedings of the 18th Soil Science Society of East Africa (SSSEA) conference held in Mombasa on 4–8 Dec 2000, pp 287–300.

Mureithi JG, Gachene CKK, Wamuongo JW (2004) Legume cover crops research in Kenya: Experiences of the legume research network project. Highlights of phase I research activities (1994–2000). KARI technical note No. 12

Muriithi FM, HJC Thijssen, DN Mugendi, JN Mwangi, MK O'Neil, OZ Nyaata (1994). Report of a survey on agroforestry technologies used for fodder production and soil fertility improvement in Meru district, Kenya. National Agroforestry Regional Project, Regional Research Centre, Embu, Kenya

Nair PKR (1993) An introduction to agroforestry. Kluwer, Dordrecht

Nandwa SM, Bekunda MA (1998) Research on nutrient flows and balances in East and Southern Africa: state-of-the-art. Agric Ecosyst Environ 71:5–18

Ndufa JK (2001) Nitrogen and soil organic matter benefits to maize by fast-growing pure and mixed species legume fallows in western Kenya. PhD Thesis, Wye College, University of London

Niang A, Amadalo B, Gathumbi S, Obonyo CO (1996) Yield response to green manure application from selected Shrubs and tree species in western Kenya: a preliminary assessment. East Afr Agric For J 62:199–207

Niang A, Githumbi S, Amadalo B (1996) The potential of short-duration improved fallow for crop productivity enhancement in the highlands of Western Kenya. East Afr Agric For J 62:103–114

Ojiem JO, Mureithi JG, Okwuosa EA (2002) Integrated management of legume green manure, farmyard manure and inorganic nitrogen for soil fertility improvement in Western Kenya. In: Proceedings of the 2nd scientific conference of the soil management and legume research network projects in Mombasa, Kenya, pp 97–102

Olsen PE, Sanda ES, Keyser HH (1996) The enumeration and identification of Rhizobial bacteria in legume inoculant quality control procedures. NifTAL Center, USA, p 96

Oswald A, Frost H, Ransom J, Kroschel JK, Shepherd KD, Saverburn J (1996) Studies on the potential for improved fallow using trees and shrubs to reduce striga infestations in Kenya. In: Proceedings of the 6th parasitic weed symposium, Cordoba, Spain

Palm CA (1995) Contribution of agroforestry trees to nutrient requirements of inter-cropped plants. Agroforest Syst 30:105–124

Palm CA, Myers RJK, Nandwa S (1997) Combined use of organic and inorganic nutrient sources for soil fertility maintenance and replenishment. In: Buresh RJ, Sanchez PA, Calhoun F (eds.) Replenishing soil fertility in Africa. SSSA special publication No. 51, Madison, Wisconsin, pp 193–218

Pilbeam CJ, Wood M, Mugane PG (1995) Nitrogen use in maize-grain legume cropping systems in semi-arid Kenya. Biol Fertil Soils 20(1):57–62

Rao MR, Mathuva MN (2000) Legumes for improving maize yields and income in semi-arid Kenya. Agr Ecosyst Environ 78:123–137

Rao MR, Nair PKR, Ong CK (1997) Biophysical interactions in tropical agroforestry systems. Agroforest Syst 38:3–50

Raquet K (1990) Green manuring with fast-growing shrub fallow in the tropical highland of Rwanda. In: Kotschi J (ed.) Ecofarming practices for smallholding. Tropical Agroecology 5:55–80

Reijntjes C, Haverkort B, Waters-Bayer A (1992) Farming for the future. An introduction to low-external input and sustainable agriculture. ILEIA, Leusden, pp 168–170

Sanchez PA, Shepherd KD, SouleMJ, Place FM, BureshRJ, Izac AMN, Mokwunje AU, Kwesiga FR, Ndiritu CG, Woomer PL (1997) Soil fertility replenishment in Africa: an investment in natural resources capital. In: Buresh, RJ, Sanchez PA, Calhoun F (eds.). Replenishing soil fertility in Africa, Soil Science Society of America (SSSA). Special publication No. 51, Madison

Sanchez PA, Valencia I, Izac AM, Pieri C (1995) Soil fertility replenishment in Africa. ICRAF, Nairobi, Kenya and World Bank, Washington DC

Sanginga N, Ibewiro B, Haungnandan P, Vanlauwe B, Okugun JA, Akubunda IO, Vesteeg M (1996) Evaluation of symbiotic properties and nitrogen contribution of mucuna to maize grown in the derived Savanna of West Africa. Plant Soil 179:119–129

Sanginga N, Vanlauwe B, Danso SKA (1995) Management of biological nitrogen fixation in alley cropping systems: estimation and contribution to nitrogen balance. Plant Soil 174:119–141

Singleton PW, Bohlool BB, Nakao PL (1992) Legume response to rhizobial inoculum in the tropics: myths and realities. In: Myths and science of the soils of the tropics, Soil Science Society of American Society of Agronomy. Special publication No. 29

Siriri D (1999) Cover crops research and dissemination in the Kigezi Highlands of South Western Uganda. ICRAF/AFRENA Project, Kabale

Ssali H, Keya SO (1984) Phosphorus and cultivar effects on nodulation, growth, dinitrogen fixation and yield of cowpea (*Vigna unguiculata*). EAAFJ 45:188–193

Stoorvogel JJ, Smaling EMA (1990) Assessment of soil nutrient depletion in sub-Saharan Africa, 1983–2000. *Report 28*. Wangeningen, Netherlands, DO Winand Starring Centre for Integrated land, soil and water research. SC-DO

Stoorvogel JJ, Smaling EMA, Jansen BH (1993) Calculating soil nutrient balances in Africa at different scales. I. Supra-national scale. Fert Res 35:227–235

Swinkels RA, Franzel S, Shepherd KD, Ohlsson E, Ndufa JK (1997) The economics of short rotation improved fallows: evidence from areas of high population density in western Kenya. Agric Syst 55:99–121

Temu AEM, Anne JB (1995) Effect of green manuring and rotations on yield in the southern highlands of Tanzania. Nor J Agric Sci Suppl 21:93–98

TSBF (1987) Tropical Soil Biology and Fertility. TSBF: inter- regional research planning workshop. Biology international special isssue No. 13. International Union of Biological Sciences, Paris

Woomer PL, Karanja NK, Mekki EI, Mwakalombe B, Tembo H, Nyika M, Silver M, Nkwine C, Ndakidemi P, Msumali G (1997) Indegenous populations of rhizobia, legume response to inoculation and farmer awareness of inoculants in East and Southern Africa. Afr Crop Sci Proc 3(1):297–308

Chapter 8
Comparative Analysis of the Current and Potential Role of Legumes in Integrated Soil Fertility Management in Southern Africa

Paul Mapfumo

Abstract Food insecurity, arising from low and non-sustainable agricultural production systems under increasing climate variability is one of the major livelihood threats for the majority of households in Southern Africa. Nitrogen (N) is the most limiting nutrient in the predominantly smallholder maize-based cropping systems, with N inputs almost invariably constituting the largest cost of crop production. Traditional methods of soil fertility management are critically short of matching the increasing food demands of a growing human population against the rising cost of living. Limited access to affordable mineral fertilizers by farmers exacerbates inherent problems of a low soil nutrient capital. This paper explores the current and potential contribution of legume biological N_2-fixation (BNF) to the N economy of cropping systems in southern African countries in the context of integrated soil fertility management (ISFM) approaches. There is potential for BNF to generate 200–300 kg N ha^{-1} under farmer management conditions, reducing N fertilizer inputs for subsequent maize by 50–100 kg ha^{-1} in a single cropping season. Legumes derive 50–90% of their N requirements from N-fixation, nearly eliminating the need for external N fertilization. However, current N inputs from BNF on smallholder farms remain as low as 5 kg N ha^{-1} $year^{-1}$ due to factors that include poor choices of legume types/varieties, small areas allocated for legume production, and poor soil fertility (particularly phosphorus [P] deficiency) and rainfall variability that lead to poor biomass accumulation. Application of P containing fertilizers was found to increase legume yields by 20–100% across dominant soils, while combinations of livestock manure and mineral fertilizer increased yields of grain legumes by 50–150% even poor sandy soils. Although legume green manures and agroforestry tree systems resulted in the highest maize yield responses of up to 400%, their lack

P. Mapfumo (✉)
Soil Fertility Consortium for Southern Africa (SOFECSA),
CIMMYT Southern Africa, P.O. Box MP163, Mt Pleasant, Harare, Zimbabwe
e-mail: p.mapfumo@cgiar.org

A. Bationo et al. (eds.), *Fighting Poverty in Sub-Saharan Africa: The Multiple Roles of Legumes in Integrated Soil Fertility Management*,
DOI 10.1007/978-94-007-1536-3_8, © Springer Science+Business Media B.V. 2011

of short-term food and income benefits severely limits their preference and adoption by farmers. In contrast, grain legumes showed the highest potential in farming systems in Malawi, Mozambique, Zambia and Zimbabwe where they are used in rotations and/or intercrops depending on rainfall conditions. Grain legumes increased yields of subsequent cereals by 60% to >200% over non fertilized controls following addition of mineral fertilizer in modest amounts. There are, however, significant technical knowledge gaps among farmers, agro-input suppliers and extension agencies on management options for optimizing BNF. The study identified lack of suitable mechanisms to deliver smallholder communities out of an existing '*maize poverty trap*' as an overaching challenge for intensification and diversification of the cereal-legume systems. Prevailing agricultural production circumstances of many smallholders favour maize monocropping. Research and development intitiatives have also tended to ignore the multiple roles of grain legumes as prioritized by communities, particularly women. Opportunities exist for multi-stakeholder platforms that generate technical and market innovations with communities to drive legume-based ISFM technologies for increased productivity and diversification of southern Africa's current farming systems.

8.1 Introduction

Nitrogen (N) deficiency is arguably an incessant problem of tropical agriculture due to high rates of soil organic matter (SOM) turnover (Nye and Greenland 1960) coupled with low levels of organic inputs and high incidences of N loss principally due to leaching. This may largely explain the invariably high demand for N inputs in the predominantly maize (*Zea mays* L.)-based cropping systems of Southern Africa, particularly in the sub-humid zones where mean annual rainfall ranges from 800 mm to ±1,600 mm. The growing human population and rising cost of living have critically diminished the scope for traditional methods of soil fertility maintenance or restoration. Consequently, food insecurity, arising from low and non-sustainable agricultural production systems under increasing climate variability, is one of the major livelihood threats for the majority of households in Southern Africa. The region is no exception to most parts of sub-Sahara Africa where per capita food production is declining (Sanchez et al. 1997). As farm sizes continue to shrink against the rising population pressure, continuous cropping has inevitably become dominant in the farming systems and there is currently a dire need for soil fertility management strategies and technologies that support intensification. For small land areas cropped every year, there are widespread reports of a downward trend in crop yields due to poor and declining soil fertility (Kumwenda et al. 1995; Sanchez et al. 1997; Waddington et al. 1998a, b). As such, scientific evidence suggests high rates of nutrient depletion in Southern Africa. Nitrogen balance estimates, for example, ranged from −31 kg N ha^{-1} for Zambia to worse than −60 kg N ha^{-1} for Malawi and Mozambique (Stoorvogel et al. 1993). These high levels of nutrient mining are largely driven by positive net nutrient exports from the fields in harvested crop products (grain, residues, fibre) since most of the farmers in the region are predominantly smallholders with limited capacity to acquire adequate external nutrient inputs for use on an annual basis.

Limited availability of nutrient resources in Southern Africa exacerbates an already existing problem of an inherently low soil nutrient capital. The dominant soil types in the whole region comprise Lixisols, Arenosols, Acrisols and Ferralsols (FAO 1990) most of which have limited capacity in holding and supplying the major plant nutrients and water. The problem cuts across the two main agro-ecologies, namely, the sub-humid and semi-arid zones. While the most apparent and obvious constraint for crop production in the semi-arid tropics is soil moisture, poor soil fertility has over the years been identified as the single most important underlying cause for low crop yields for such low rainfall areas in Southern Africa (Mapfumo and Giller 2001). Apart from the highly priced mineral fertilizers such as ammonium nitrate (AN), calcium ammonium nitrate (CAN), urea, di-ammonium phosphate (DAP) and various compound fertilizers, N_2-fixing legumes are probably the only other realistic option for supplying N and improving soil fertility in cropping systems of Southern Africa.

Leguminous plants that associate symbiotically with N_2-fixing bacteria, mainly of the genus *Rhizobium*, are able to capture atmospheric N_2 in amounts sufficient to meet their growth requirements (Peoples et al. 1989). The fixed N can significantly contribute to the N economy of agro-ecosystems (Giller 2001) and therefore constitute a renewable source of the most limiting nutrient in tropical agriculture. However, this attribute does not seem to be among the major driving factors behind the continued existence of various legume species found in the prevailing cropping systems, particularly on smallholder farms. A wide range of legumes, which include, groundnut (*Arachis hypogaea* L.), Bambara groundnut (*Vigna subterranea* [L.] Verdc), cowpea (*Vigna unguiculata* [L.] Walp), pigeonpea (*Cajanus cajan* [L.] Millsp.), common bean (*Phaseolus vulgaris L.*) and soyabean (*Glycine max* [L.] Merr.) are commonly grown by smallholder farmers in Southern Africa. Legumes are used for various purposes that include improvement of human nutrition, cash generation at local (village) and national markets, soil fertility improvement and providing high quality (protein rich) forage. Out of these multiple uses, soil fertility impact of legumes in the cropping systems is very minimal (Mapfumo et al. 2001a). Most households in Southern Africa use maize for staple food, and cereals generally rank high when compared with legumes in terms of nutritional importance. For instance, maize contributes 80% of the daily food calories and is planted to over 76% of the total arable land in Malawi (Government of Malawi 1999). With the exception of soyabean which has gained significantly in economic significance (Mpepereki et al. 1996), most of the legumes have been relegated to minor crops and are generally considered as women's crops since they are often produced by women on small land areas meant primarily for home consumption (Mapfumo et al. 2001a, b). One of the most important questions is whether, and how, legumes are serving their purpose in cropping systems. This comes in the wake of an increasing realization that legumes can play a pivotal role in sustaining crop productivity in smallholder farming systems (Giller et al. 1994; Giller 2001). Soil fertility management is considered as a means for achieving food security in Southern Africa. However, there is little documentation on the potential and realities of legumes with respect to integrated soil fertility management (ISFM). There is limited knowledge on how the other multiple functions of legumes can be turned into drivers for ISFM, particularly with reference to farming systems in Malawi, Mozambique, Zambia and Zimbabwe.

8.2 Dominant Cropping Systems in Southern Africa

In general, most of the cropping systems in Southern Africa are maize (*Zea mays* L.) based, with maize accounting for an average of 60–90% of the seasonal cropped areas and providing the staple diet for the majority of people (Kumwenda et al. 1995). With respect to soil fertility management, however, the farming systems are as diverse as the socio-cultural and environmental settings that have dictated their evolution, ranging from the *Chitemene* and *Fundikila* systems of northern Zambia to the manure- and fertilizer-based management systems of Zimbabwe. *Chitemene* is a 'lop and burn' system characteristic of shifting cultivation, while *Fundikila* is basically an *in-field* composting of predominantly low quality veld grasses in small mounds (explained in detail later in this chapter). In most of the farming systems, the maize is either monocropped (predominantly), intercropped with legumes, rotated with legumes and small grain cereals or interrupted by periods of natural fallowing. Maize is sometimes replaced with pearl millet (*Pennisetum glaucum* [L.] R.Br.) and sorghum (*Sorghum bicolor* [L.] Moench) in semi-arid zones (Chivasa 1995; Ahmed et al. 1996). Traditional intercrops involve, cowpea and Bambara groundnut with maize and sorghum. Cowpea has often been sparsely intercropped with the staple cereal crops. Dominance of any of these cropping systems partly depends on human population densities and hence land-holding, and also on availability of labour and draught power. Farm size in most areas average ~0.5 ha household in Malawi to >5 ha in Zambia. In Zimbabwe, Mozambique and southern Zambia, most households rely on animal draught power (particularly cattle) while hand-hoeing is predominantly used in Malawi (Waddington 1994; Scoones et al. 1996). Average population densities range from <45 persons km^{-2} in parts of Zimbabwe (CSO 1998) to ~250 persons km^{-2} in Malawi. With high population densities, periods of natural fallowing are reduced, and there is decreased availability of organic resources such as cattle manure. Although a wide range of nutrient resources are used, almost all are traditionally targeted towards the staple cereal crops, especially maize, with legumes and minor crops rarely receiving any external fertilization (Campbell et al. 1998; Mapfumo et al. 1999).

8.3 Legumes as a Critical Source of N in Croppings Systems

The main sources of N inputs in farming systems are mineral fertilizers, legume BNF, organic resources transferred from outside the field such as animal manures and plant biomass, and NO_3^--N reclaimed from subsoil by deep rooted shrubs

Table 8.1 A summary of N_2 fixation potential from different categories of tropical legumes

Legume system	%N derived from fixation	Amount fixed (kg N ha^{-1})	Time (days)
Grain	60–100	105–206	60–120
Pasture	45–98	115–280	120–365
Green manure	50–90	110–280	45–200
Trees	56–89	162–1,063	180–820

Adapted from Giller (2001)

and/or perennials. Of these, only mineral fertilizers and legumes (through BNF) effectively provide real N additions to the cropping systems. Nitrogen fixing legumes are, therefore, an essential driver of ISFM technologies for smallholder farming systems given the ever escalating costs of accessing mineral fertilizers. A synthesis of findings from decades of research based on controlled trials from both on-station and farmers' fields demonstrate a high potential of diverse legumes to fix large amounts of N under tropical environments (Giller 2001). Legume species commonly used for provision of grain, green manure and pastures have potential to fix between 100 and 300 kg N ha^{-1} from the atmosphere (Table 8.1).

Most annual legumes have a low C:N ratio and low levels of lignin and polyphenols, thereby satisfying the minimum chemical quality attributes required for net-mineralization in the short-term (e.g. within a single cropping season). The relatively high amounts of N fixed in legume biomass can therefore be readily available to subsequent crop, substantially reducing annual demands for chemical fertilizer use at field and farm scales. Fertilizer equivalency values between 50 and 100 kg N ha^{-1} have been reported from regional field trials from both on- station and on-farm (e.g. Bogale et al. 2001; Bezner Kerr et al. 2007). Fertilizer equivalency for legume biomass of different quality was estimated at 16% to >100% from different on-farm experiments in east and southern Africa (Murwira et al. 2002). Mineral N fertilizers constitute the largest fertilizer cost in Southern Africa's predominantly cereal-based cropping systems, further suggesting that any small N contributions from BNF can translate into significant subsidies to prevailing production costs.

Main factors that determine the amount of N fixed by legumes include: (a) legume genotype (including cultivar); (b) establishment of an effective symbiosis between the legume (macro-symbiont) and the rhizobial bacteria (micro-symbiont), from which N is fixed; (c) the amount of N accumulated in the legume plant biomass during the growth cycle as influenced by agronomic management factors including soil fertility. For purposes of enhancing the N economy in cropping systems, the area cultivated to the legume constitutes the fourth factor (d). Optimization of BNF in smallholder cropping systems will therefore hinge on management strategies that are designed to address these four factors. For instance, inoculation usually becomes necessary when a new legume is introduced into an area where the soils do not contain rhizobial strains that can effectively nodulate that legume. Inoculation may also be necessary in soils that do not support a sufficiently large populations of the rhizobia nodulating a given legume species.

8.4 Estimated Amounts of N Fixed by Different Legumes on Farmers' Fields

To date, there are limited studies quantifying N_2-fixation by different legumes in Southern Africa despite increasing awareness among research and development communities about their potential benefits to smallholder farming systems. Notably, there is a critical absence of BNF related studies from smallholder farming systems in Mozambique. However, the relatively few studies available have generated critical lessons and technical knowledge on the potential contributions of legumes to the region's farming systems. Most of the traditionally grown grain legumes that include cowpea, groundnut, Bambara groundnut and pigeonpea effectively nodulate with native soil rhizobia, and will not require inoculation across the diverse agro-ecologies (e.g. Mpepereki and Makonese 1994; Mapfumo et al. 2000). Similar patterns have been observed for most tropical legumes used for green manures such as sunnhemp species (*Crotalaria* spp.) and mucuna (*Mucuna pruriens* [L.] DC.), and a wide range of agroforestry shrubs/trees. There is therefore enormous potential for tapping the diversity of both the legumes and native rhizobia to increase BNF contributions to smallholder farming systems for increased crop productivity and enhanced nutrition of humans and livestock. Contrary to the above-mentioned legumes, soyabean generally requires inoculation in most of the region's soils, with the exception of known promiscuous varieties (Giller 2001; Musiyiwa et al. 2005). Soyabean is arguably the most researched grain legume in the region because of its economic value as a cash crop, and available national innoculant factories are basically build around production of suitable *Bradyrhizobium* strains. Thus, technically, there is a broad scope for enhancing BNF driven soyabean production for smallholders.

Field studies have shown that commonly grown grain and green manure legumes derive most of their N from N_2-fixation (Table 8.2). On granitic sandy soils with <10% clay, the legumes depend almost entirely on BNF, deriving ~90% of their N (e.g. Mapfumo et al. 1999; Kasasa 1999). Tree legume species derive about 50% of their N from fixation (Table 8.2), and it is not clear how much N_2-fixation is influenced by their ability to access subsoil N that is otherwise leached beyond the rooting zone of most annual crops. The limited studies quantifying BNF on farmers' fields seem to suggest a relatively higher dependence on N_2-fixation by annual legumes grown on poorer soils and in intercrops than those in sole cropping (e.g. Giller 2001; Giller et al. 2002). Further quantitative research is necessary to determine the implications for sustainability of some of traditional and promising (community prioritized) intercrops such as the maize/pigeopea in southern Malawi and the maize/cowpea and maize/groundnut systems in Mozambique.

Against the potential for annual fixation rates of 300 kg ha^{-1}, the amounts measured on farmer's fields are still very low (Table 8.2). The high rates of N_2-fixation of about 70–90% for most annual grain legumes do not translate into effectively high amounts N fixed at farm and field scales. Based on studies conducted mainly in Malawi, Zambia and Zimbabwe, only soyabean fixes relatively high amounts of N per unit of land, yielding between 100 and 260 kg N ha^{-1} within periods of no more than 3 months (Table 8.2). Most of the tested legumes fixed as low as 6 kg N ha^{-1} to

Table 8.2 Estimated amounts of nitrogen fixed by different legumes under smallholder farmer management conditions in Southern Africa

Legume type	%N from N_2-fixation	N fixed (kg ha^{-1})	Soil texture	Location/country of study	References
(a) Annual grain and green manure legumes					
Cowpea (*Vigna anguiculata*)	76%	73	Coarse sand	Chinamora, Zimbabwe	Mapfumo et al. (2001a, b)
	92%	79	Coarse sand	Chinyika, Zimbabwe	
	65%	54	Coarse sand	Murehwa, Zimbabwe	
	58%	28	Sandy clay loam	Domboshava, Zimbabwe	Chikowo et al. (2003)
	15–56%	21–109	Sandy loam	Matopos, Zimbabwe	Ncube et al. (2007)
Bambara groundnut (*Vigna subterranea*)	75%	47	Coarse sand	Chinamora, Zimbabwe	Mapfumo et al. (2001a, b)
	88%	54	Coarse sand	Chinyika, Zimbabwe	
	72%	57	Coarse sand	Murehwa, Zimbabwe	
	34–61%	2–72	Sandy loam	Matopos, Zimbabwe	Ncube et al. (2007)
Groundnut (*Arachis hypogaea*)	52%	28	Coarse sand	Chinamora, Zimbabwe	Mapfumo et al. (2001a, b)
	68%	46	Coarse sand	Chinyika, Zimbabwe	
	59%	50	Coarse sand	Murehwa, Zimbabwe	
	19–83%	7–86	Sandy loam	Matopos, Zimbabwe	Ncube et al. (2007)
Pigeonpea (*Cajanus cajan*)	42%	13	Sandy	Lisasadzi, Malawi	Sakala et al. (2001)
	85%	163	Sandy loam-clay	Chitedze, Malawi	
	84%	97	Sandy clay loam	Domboshava, Zimbabwe	Chikowo et al. (2003)
	81%	6	Coarse sandy	Murehwa, Zimbabwe	Mapfumo et al. (1999)
	91%	43	Coarse sandy	Murehwa, Zimbabwe	
	92%	16	Coarse sandy	Murehwa, Zimbabwe	
	28–57%	25–56	Sandy loam	Matopos, Zimbabwe	Ncube et al. (2007)
Soyabean (*Glycine max*)	92%	115	Sandy loam	Hurungwe, Zimbabwe	Kasasa (1999)
	93%	127	Sandy loam	Hurungwe, Zimbabwe	
	93%	128	Sandy loam	Hurungwe, Zimbabwe	
	88%	258	Clay	UZ farm (on-station)	

(continued)

Table 8.2 (continued)

Legume type	%N from N_2-fixation	N fixed (kg ha^{-1})	Soil texture	Location/country of study	References
Sunnhemp (*Crotalaria*	48%	16	Sandy loam – clay	Malawi	Giller (2001), Giller et al. (2002)
juncea)	49%	43		Malawi	
Mucuna (*Mucuna*	42%	58	Sandy loam	Malawi	Giller (2001)
pruriens)	73%	78	Sandy loam	Malawi	
(b) Agroforestry tree legumes (1 year)					
Tephrosia vogelii	37–78%	12–45	Clay	Malawi	Giller (2001)
Faidherbia albida	32–52%	nd	Sandy loam	Malawi	
Acacia angustisima	52%	210	Sandy loam	Kalichero, Zambia	Mafongoya et al. (2003)
Acacia angustisima	56%	122	Sandy loam	Domboshava, Zimbabwe	Chikowo et al. (2003)
Calliandra calothyrsus	44%	214	Sandy loam	Kalunga, Zambia	Mafongoya et al. (2003)
Sesbania sesban	55%	84	Sandy loam	Domboshava, Zimbabwe	Chikowo et al. (2003)

Table 8.3 Amounts of N Fixed and land areas commonly allocated to different grain legumes on smallholder farms in Zimbabwe

Legume type	N_2 fixed (kg N ha^{-1})	Area allocated to legume (ha farm^{-1})	Total N fixed per farm (kg N farm^{-1})
Groundnut	52	0.08	4.2
Cowpea	47	0.03	1.4
Bambara nut	33	0.22	7.3
Pigeonpea	39	0.34	13.3

Adapted from Mapfumo et al. (2001a, b)

a maximum that rarely exceeded 80 kg N ha^{-1}. However, medium to long-term pigeonpea fixed up to 163 kg N ha^{-1} when grown on soils with more than 20% clay. Agroforestry tree crops fixed up to 200 kg N ha^{-1}, and this can be largely attributed to their superior biomass accumulation due to the relatively long growth duration of these semi perennial and perennial crops. For both annual and perennial legumes, the actual amounts of N available to enhance soil fertility are undermined by small areas allocated to legumes at farm scales. A field survey in Zimbabwe showed that legumes were allocated to 3–15% of farm arable areas (Mapfumo et al. 2001a, b). The actual amounts of N fixed per farm are effectively reduced to an average of ~5 kg N ha^{-1} year^{-1} (Table 8.3). The practical realities of BNF contribution to cropping systems at farm scales will not therefore likely happen if constraints to legume integration are not addressed.

8.5 Opportunities for Increasing Legume Productivity

Apart from small areas allocated to legumes at farm level, some of the commonly cited biophysical reasons behind low amounts of N fixed include poor fertility of soils, rainfall related constraints, lack of improved crop varieties and limited capacity of farmers. Under favourable rainfall conditions (<750 mm year^{-1}) and relatively clayey soils (>15% clay), most of the commonly grown legumes are able to accumulate high amounts of biomass that often exceed 2 t ha^{-1}. Biomass yields of 3.5–8 t ha^{-1} have been reported for annual legumes under researcher-managed trials in Zambia and Malawi where soils are relatively fertile (e.g. Kumwenda and Gilbert 1998). Such yields would be sufficient to positively impact on N economy of the cropping systems. However, major soil parent materials in Southern Africa are well known for their limited phosphorus availability (Brown and Young 1965; FAO 1990; Nyamapfene 1991; Maria and Yost 2006). Most of the farming systems are characterised by decades of monocropping and associated nutrient mining, exacerbating an already inherent problem of a low nutrient capital, and often causing multiple nutrient deficiencies (Stoorvogel et al. 1993; Mapfumo and Giller 2001). Phosphorus deficiency is known to undermine nodulation and N_2-fixation processes, and limit biomass accumulation of the candidate legumes (Giller 2001). Responses of different legume types and varieties to P and other forms of fertilization in the dominant soils of southern Africa has, however, not been systematically investigated. Legume

Table 8.4 Response of different legumes to fertilization under smallholder farm management conditions

Legume type	Fertilizer added	Yield attained (kg ha^{-1})	% yield change over control (%)	Location of study	Reference
Sunnhemp (*Crotalaria juncea*) green manure	100 kg P_2O_5 ha^{-1} as single super phosphate (SSP)	1.7[a]	+97	Gokwe South, Zimbabwe (on-farm)	Hikwa et al. (1998)
Velvet bean	100 kg P_2O_5 ha^{-1} as single super phosphate (SSP)	2.4[a]	+24	Gokwe South, Zimbabwe (on-farm)	Hikwa et al. (1998)
Soyabean	Mono-ammonium phosphate at 60 kg P_2O_5 ha^{-1} +70 kg K ha^{-1} + 26 kg S ha^{-1} + 24 kg N ha^{-1} + 360 kg ha^{-1} of lime	4.4[a]	+92	Chibwe, Zambia (on-station)	Lungu and Munyinda (2003)
Soyabean	26 kg P ha^{-1} + cattle manure @ 6.5 t dry matter ha^{-1}	3.8[a]	+153	Chinyika, Zimbabwe	Kanonge et al. (2009)
Cowpea	Mono-ammonium phosphate at 120 kg P_2O_5 +70 kg K ha^{-1} + 26 kg S ha^{-1} + 24 kg N ha^{-1}	2.1[a]	+19	Mwanachingw-ala, Zambia (on-farm)	Lungu and Munyinda (2003)
Cowpea	Partially acidulated phosphate rock at 60 kg P_2O_5 +120 kg K ha^{-1} + 26 kg S ha^{-1} + 24 kg N ha^{-1}	2.3[a]	+35	Mwanachingwa-la (on-farm)	Lungu and Munyinda (2003)
Cowpea	26 kg P ha^{-1} + cattle manure @ 6.5 t dry matter ha^{-1}	2.2[a]	+82	Chinyika, Zimbabwe	Kanonge et al. (2009)
Groundnut	Partially acidulated phosphate rock at 60 kg P_2O_5 +70 kg K ha^{-1} + 26 kg S ha^{-1} + 24 kg N ha^{-1} +360 kg ha^{-1} lime	1.7[a]	+70	Mwanachingwala (on-farm)	Lungu and Munyinda (2003)
Groundnut	8.5 kg P ha^{-1}	0.6	39	Marange, Zimbabwe	Mupangwa and Tagwira (2005)
	34 kg P ha^{-1}	0.9	51		

[a]Biomass yield

biomass productivity on most of the smallholders' fields requiring soil fertility improvement has remained below 2 $t\,ha^{-1}$, limiting the scope for significant rotational influence to subsequent cereals as well as the much needed markatable yield incentives for farmers. The limited research findings available are nonetheless indicative of significant increases in both biomass and grain yield of different legumes following P application. Modest amounts of P, ranging from ~10 to 50 kg ha^{-1} have been shown to increase legume productivity by about 20–100% (Table 8.4). Studies from different agro-ecologies in Zimbabwe, Zambia (Table 8.4) and more recently Mozambique (Mapfumo et al. 2009a, b) underscore the challenges of addressing poor soil fertility in order to 'kick-start' a meaningful production of legumes and derive the associated benefits.

Significantly high yield improvements have been attained following application of combinations of P containing inorganic fertilizers with organic nutrient sources, and in some cases lime. On some of Southern Africa's challenging sandy soils which occur widely in Zimbabwe, application of P at 26 kg ha^{-1} and organic manure (cattle manure or woodland litter) at ~5 $t\,ha^{-1}$ increased cowpea and soyabean biomass yields by 50–150% with effective yields>2–4 $t\,ha^{-1}$ (Kanonge et al. 2009). Yields from farmers' fields often range from total crop failure to 1 $t\,ha^{-1}$. Combinations of cattle manure and a compound fertilizer which provided P and starter N, consistently gave superior legume yields in Zimbabwe (Fig. 8.1), suggesting a possibility of multiple nutrient deficiencies among other soil fertility constraints. Mupangwa and Tagwira (2005) achieved groundnut yield increases of 39–51% on similar soils when P was applied at rates between 8.5 and 34 kg ha^{-1}. However, yields increased by 57–81% when calcitic lime (200 kg ha^{-1}) was combined with P, and by 50–90% when gypsum (100 kg ha^{-1}) was used in place of lime. These findings were consistent with results from Zambia where combinations of acidulated phosphate rock and lime increased cowpea and groundnut biomass yields by 35% and 70% respectively (Table 8.4).

8.6 Rotational Benefits to Subsequent Cereal Crops

Research over the past two decades has investigated soil fertility benefits of technology options based on rotations and intercrops involving N_2-fixing green manure and grain legumes (Snapp et al. 1998; Waddington 2003) as well as agroforestry-based crop management systems (Mafongoya et al. 2006; Akinnifesi et al. 2006). Overall, there is overwhelming evidence of cereal yield benefits under these soil fertility management options, and numerous examples can be cited of grain yield increases from 50% to >400%. However, the magnitude of yield benefits, particularly for maize, vary considerably according to agro-ecology, field and farm types, farmer resource endowment, and whether the legume is used for green manuring or grain crop and in a rotation or intercrop. While the multiplicity of scenarios arising from these factors present considerable challenges for researchers to integrate legumes into the farming systems, it also broadens opportunities for

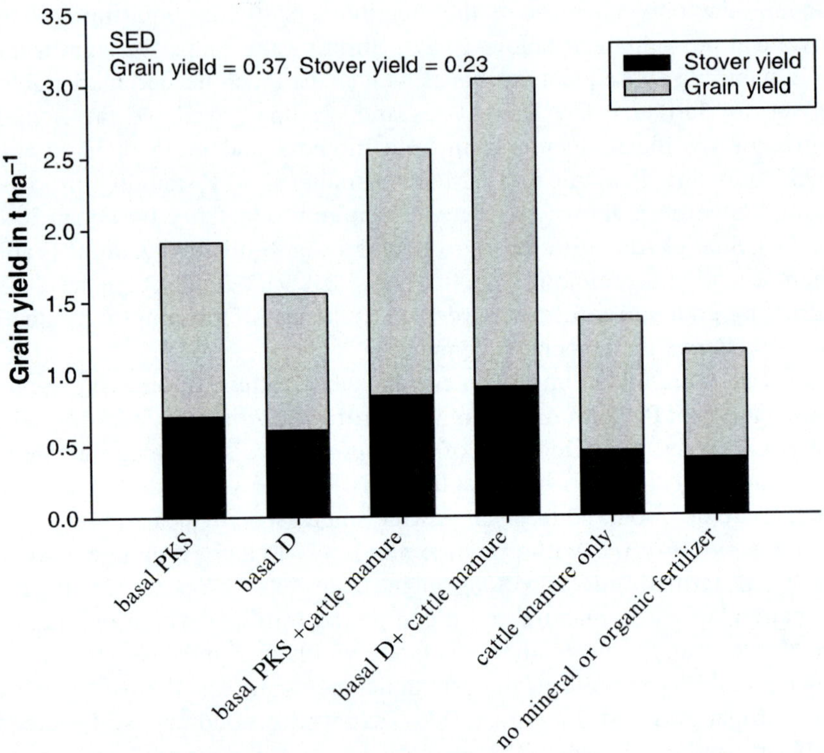

Fig. 8.1 Grain and stover yields from cowpea as affected by soil fertility management practice, in Makoni District, Zimbabwe (*SED- Standard error of the difference between means*) (Kanonge et al. 2009)

satisfying the need for the widely diverse and heterogeneous smallholder farmer circumstances.

Green manures and agroforestry biomass transfer systems have often registered the highest yield contributions to beneficiary cereals crop (Table 8.5), but have increasingly emerged as the least prioritized by smallholders farmers under their prevailing production circumstances. This can be largely attributed to their lack of short-term benefits in terms of both food and income. Critical technical lessons can, however, be drawn from research experiences in the region. In central and southern Malawi, sunnhemp (*Crotalaria juncea*), tephrosia (*Tephrosia vogelii*) and velvet bean (*Mucuna pruriens*) green manure often resulted in maize yields of 3–6 $t\,ha^{-1}$ even with no additional mineral N fertilizer added (Kumwenda et al. 1998), and were able to generate critical biomass yields of ≥2 $t\,ha^{-1}$ to influence cereal yields in rotation (Gilbert 1998). However, the same species had mixed results on granite-derived sandy soils of Zimbabwe (see Waddington et al. 1998a, b), even as they have remained the prioritized 'best-bet' green manure legumes. Higher biomass yields of green manure legumes are achieved under sole stands than when the legumes are either intercropped or relay cropped. This is often reflected in yields of the subsequent maize (Table 8.5). In trials where maize following a legume green manure was fertilized,

Table 8.5 Maize yields following different legume-based soil fertility technology options in Southern Africa

Legume technology option	Attained maize grain yield (kg ha^{-1})	% yield increase (+) or decrease (–) over controls (%)	Study area/ Country	Reference
Legume green manure – rotations				
Maize following sunnhemp/ cowpea green manure	4.3	+230[a] +148[a]	Harare research station (on-station)	Agronomy (1993)
Maize following cowpea green manure	6.2	+280[a]	Mangwende, Zimbabwe (on-farm)	Chibudu (1998)
Maize follwing mucuna green manure +100 kg P ha^{-1} +45 kg N ha^{-1}	1.6	+252[a] +140[b]	Chihota, Zimbabwe (on-farm)	Murata et al. (2000)
Maize mucuna green manure +23 kg N ha^{-1}	2.4	+170[a] +13[b]	Muswishi, Zambia (on-farm)	Mwale et al. (2003)
Sunnhemp green manure +100 kg P ha^{-1} +45 kg N ha^{-1}	2.5	+449[a] +273[b]	Chihota, Zimbabwe (on-farm)	Murata et al. (2000)
Maize following mucuna green manure	1.6	+45[a] –125[b]	Ntecheu, Malawi (on-farm)	Sakala et al. (2003a, b)
Maize following mucuna green manure	0.7	+75[a] –100[b]	Angalala, Malawi (on-farm)	Sakala et al. (2003a, b)
4th year fertilized maize following mucuna green manure	5.8	+21[b]	Chitedze, Malawi (on-station)	Kumwenda et al. (2001)
Legume/maize intercrops				
Maize relay intercropped with sunnhemp	1.3	–12[b]	Domboshawa, Zimbabwe (on-station)	Jeranyama et al. (1998)
Maize relay intercropped cowpea +120 kg N ha^{-1}	4.5	32[b]	Domboshawa, Zimbabwe (on-station)	Jeranyama et al. (1998)
Maize undersown in pigeonpea in year of improved fallow	0.4	–14[b]	Chinyika, Zimbabwe (on-farm)	Nyakanda et al. (1998)
Maize after maize/ mucuna intercrop +Compound D	2.0	+116[a] –6[b]	Muswishi, Zambia (on-farm)	Mwale et al. (2003)
Maize/ mucuna intercrop	1.0	+25[a] –110[b]	Mangwende, Zimbabwe (on-farm)	Chibudu (1998)
Maize/ cowpea intercrop	1.5	+88[a] –75[b]	Mangwende, Zimbabwe	Chibudu (1998)
Maize/*Sesbania sesban* intercrop	1.7	+244[a] –25[b]	Malosa, Zomba, Malawi (on-farm)	Phiri et al. (1999)

(continued)

Table 8.5 (continued)

Legume technology option	Attained maize grain yield (kg ha⁻¹)	% yield increase (+) or decrease (–) over controls (%)	Study area/ Country	Reference
Maize/pigeonpea intercrop	1.2	–150[b]	Chitedze, Malawi (on-station)	Kumwenda et al. (1998)
Maize/sunnhemp intercrop	1.8	–67[b]	Chitedze, Malawi (on-station)	Kumwenda et al. (1998)
Agroforestry tree crops				
S. seban improved fallow	2.1	+162[a] +17[b]	Chinyika, Zimbabwe (on-farm)	Nyakanda et al. (2002)
Tithonia diversifolia biomass transfer (without fertilizer)	3.8	+100[a] +111[b]	Bembeke, Malawi (on-farm)	Ganunga et al. (1998)
T.diversifolia biomass transfer (without fertilizer)	3.8	+100[a] +111[b]	Bembeke, Malawi (on-farm)	Ganunga et al. (1998)
1-year *S. sesban* improved fallow	3.6	+350[a] –10[b]	Eastern Zambia (on-farm)	Mafongoya et al. (2003)
1-year *Tephrosia vogelli* improved fallow	3.1	+288[a] –35[b]	Eastern Zambia (on-farm)	Mafongoya et al. (2003)

[a] % increase (+) or decrease (–) over continuous unfertilized maize
[b] % increase (+) or decrease (–) over continuous fertilized maize

grain yield gains up to four times higher than unfertilized maize were achieved, and these yields were also superior to those of maize receiving fertilizer at the prevailing recommended rates. In contrast, both intercropped grain and green manure legumes resulted in mixed yield benefits depending on quality of rainfall season. Maize grain yields following these intercrops were found to be higher than those of unfertilized maize, but often significantly lower than fertilized maize (e.g. Table 8.5). Several multilocational trials over the years have demonstrated the limited scope for legume green manuring in semi-arid areas where constraining factors of both low soil moisture and poor soil fertility are often simultaneously at play.

Most of the research studies in Malawi and Zambia demonstrated high technical feasibility of agroforestry-based soil fertility technologies, particularly with options involving biomass transfer, coppicing fallows and 2–3 year improved fallows. One-year improved fallows tended to give lower maize yields compared to continuous fertilized maize (Table 8.5). For example, in Malawi, over four consecutive cropping seasons, grain yields of maize increased by 343%, from 0.94 t ha^{-1} in unfertilized sole maize to 4.17 t ha^{-1} in gliricidia/maize intercropping (Akinnifesi et al. 2007). Despite the apparently attractive benefits of agroforestry technologies, which include improved crop yields associated with enhanced soil fertility and provision of forage and wood, there is no clear evidence to suggest sustained adoption of the technologies in the region. One underlying reason for success of agroforestry interventions in eastern Zambia and central and southern Malawi (Table 8.5) is the relatively favourable amount and distribution of rainfall received, as well as the comparatively fertile soils often with clay content >180 $g\,kg^{-1}$ soil. In contrast, Zimbabwe receives less and more erratic rainfall, with poorer soils (~100 $g\,kg^{-1}$ soil) predominating in the smallholder areas. Farmers' concerns about the high establishment costs (and often replacement) costs (including labour), general incompatibility with free ranging livestock management and lack of short-term food security benefits of agroforestry cropping systems will continue to haunt their large scale adoption in the foreseeable future.

Grain legume rotations and intercrops hold the most promise for diverse environments and farmer categories in all Southern African countries because of their potential short contribution to household food, nutrition and income security. Soil fertility benefits of grain legumes, though often only modest compared with those registered for green manures and agroforestry systems, can be substantially enhanced on the basis of prevailing knowledge gains from regional research. Grain yields for subsequent cereal are largely dependent on crop productivity during the legume phase, as well as rainfall distribution during the cereal phase. For instance, under semi-arid environments on southern Zimbabwe, sorghum yields increased by >200% following cowpea, Bambara groundnut, groundnut and pigeonpea during a relatively wet season (~650 mm $year^{-1}$), while only 30–100% yield increases were obtained during a drier year (~300 mm of rainfall). Cereal yield increases of 100–200% are commonly reported from research trials, with substantial yield gains usually realized when mineral fertilizers are additionally applied in the cereal phase. Much of the yield increases are attributable to improved N supply from the preceding legume crop, with N fertilizer equivalency values ranging from 50 to ~100 $kg\,ha^{-1}$ for most of cases where high cereal yield responses are reported (e.g. Bogale et al. 2001; Bezner Kerr et al. 2007). Improved supply of base nutrients, particularly calcium and magnesium, following legumes have has also been found to contribute to some of the non-N benefits of the rotations (Mapfumo and Mtambanengwe 2004).

Major determinants of N contribution to subsequent crops under grain legume rotations include (i) the proportion of N removed in harvested materials, (ii) the amount of plant materials/residues retained in the soil, (iii) the quality of these residues, and (iv) whether or not mineral fertilizer is used on the beneficiary crop. Legume genotypes differ in their capacity to remobilize and translocate assimilates from leaves and other plant tissue into grain. The proportion of N contained in grain relative to the rest of the plant tissue constitutes the grain N harvest index. In the commonly grown soyabean genotypes, most of the fixed N is contained in grain, with N harvest index as high as 88% (Giller 2001). In contrast, N harvest indices are notably low in most varieties of cowpea, groundnut and pigeonpea, with the latter reported to have values as low as 21% (Mapfumo 2000). Promiscuous soyabean was also found to have a low N harvest relative to other soyabean varieties (Kasasa 1999). Despite high variability in measured values of net N benefits to cropping systems by different legumes, cowpea, groundnut pigeonpea often record some of the highest maize grain yields in rotations. While a high N harvest index often implies high grain protein content, as in the case of soyabean, the plant residues may effectively contribute negligible amounts of N to subsequent crops. Reports of negative N benefits to subsequent cereals are common in soyabean and *Phaseolus* bean rotational systems (Giller 2001) as the resultant low N containing residues tend to immobilize externally added and/or native soil N during cropping. On both coarse sands and sandy clay soils in Zimbabwe, Kasasa (1999) almost doubled maize yield levels obtained from unfertilized control plots following incorporation of promiscuous soyabean residues, while no yield differences were obtained with hybrid cultivars.

Most smallholder farmers in Southern Africa either burn crop residues, incorporate then into soil at the beginning of the next cropping season or feed them to livestock by leaving them in the field for free grazing or deliberately collecting them for controlled feeding during the dry season. Farmers in Mozambique burn most of the crops residues as part of land preparation for a new cropping season, while most farmers in Zimbabwe feed crop residues to livestock. Livestock are highly valued in farming systems of Zimbabwe, southern Zambia and parts of Mozambique where they provide a variety of services including draught power, transport, manure, meat, milk and insurance against risk associated with crop failure and lack of income. In Malawi, about 55% of farmers were found to burn crop residues (Bezner Kerr et al. 2007). These management practices effectively enlarge the loop of N management beyond the field scale, increasing potential loss pathways (e.g. farmers getting residue indirectly through manure). Most grain legumes produce high quality biomass due to their high tissue N concentration, and mineralize in soils to render most of their N available for uptake by subsequent cereal crops within a relatively short period (e.g. within a single cropping season). However, high harvest indices often result in relatively low quality crop residues, causing temporary N immobilization when they are incorporated into soil. Applications of small amounts of mineral N are often adequate to offset this immobilization and increase cereal yields in rotation. Several studies have demonstrated significant increases in yields of beneficiary cereal crops following application on mineral N during the cereal phase (e.g. Mapfumo 2000; Sakala et al. 2000). In Malawi, application of small amounts of mineral fertilizer in a legume rotations resulted in maize yields 60–110% higher than non-fertilized maize (Snapp et al. 2002). In such rotations, legumes often contribute N in amounts critical for early establishment and growth of the cereal crop, eliminating the need for supply of high rates of N in basal fertilizer or early top-dressing. The non-N benefits associated with legume rotations have, however, not been adequately studied to enhance decisions on options to improve management of legume-based rotations.

8.7 Potential Entry Points for Intensification and Diversification of Legume-Cereal Systems in Southern Africa

While appropriate field, farm and socio-ecological niches exist for legume green manure and agroforestry tree crops in southern African countries, the biggest opportunity for harnessing BNF sustainably to increase productivity of the farming systems lies in grain legume – cereal systems. Emphasis should be put on increasing the capacity of farmers, local institutions and agro-service providers to support science-based approaches in boosting productivity of staple food crops, as a precursor to sustainable natural resource use and diversification into higher value crop enterprises. Work of the Soil Fertility Consortium in Southern Africa (SOFECSA) revealed the importance that farmers in Malawi, Mozambique, Zambia and Zimbabwe place on grain legumes because of their high contribution to household nutrition, food and income (Mapfumo 2009a, b). The work revealed that local level trading of legume grains, though in small quantities, is a critical source of income for women. Legumes are commonly considered as women's crops in southern Africa (Mapfumo et al. 2001a, b). Increasingly, there is improved awareness among farmers across the different countries about the potential role of legumes, not only in enhancing food security, but also in soil fertility management. For example, in a health survey in Malawi, resource-constrained farmers admitted to hospital with malnourished children clearly identified poor soil fertility as a major source of food insecurity (Bezner Kerr et al. 2007).

In recent SOFECSA participatory action research intiatives in Malawi, Mozambique, Zambia and Zimbabwe, farmers were able to prioritize legume-based cropping systems that they considered critical in the employment of ISFM technology options to enhance soil productivity and household food, nutritional and income security (Table 8.6). Among the four countries, landholding per household is lowest in Malawi (<0.5–1 ha household^{-1}) and highest in Zambia and Mozambique (>5 ha). This influences the choice of legume cropping preferred by, and technically suitable for, different farmer categories. Strategies for intensifying grain legume-cereal intercrops and rotations on small land areas are particularly critical for Malawi. However, opportunities for intensification exist given the relatively good soils with potential for building carbon and nutrient stocks. In Mozambique, current use of mineral fertilizers is still low among smallholders and this undermines productivity of legume-cereal intercops that are currently given high priority by farmers (Table 8.6). The problem is worsened by lack of access to improved seed, including poor promotion of selected traditional/local varieties whose attributes constitute farmers' preferred criteria for domestic consumption. Systematic promotion of rotations could increase general productivity and enable diversification in Mozambique. A major challenge is the current lack of empirical agricultural research information in Mozambique, particularly quantitative data on legume-based cropping systems.

In Zambia, legume integration if faced with different problems between the northern and southern regions of the country. Northern Zambia has some of SSA's most acid soils and ISFM options integrating lime application and management provide a promising entry point for increased legume productivity and diversification. In the southern regions soil acidity in less of problem, but livestock is a strong component of the farming system. Farmers in this region prioritized options for fertilization of legumes based rotations through combinations of mineral and organic fertilizers. This was the same for Zimbabwe where similar agro-ecologies are shared. Zimbabwean farmer almost exclusively prioritized rotations as opposed to intercrops (Table 8.6), primarily due to rainfall limitations and inherently low nutrient capital of soils.

Table 8.6 Legume-based entry points for integrated soil fertility management (ISFM) technology options as prioritized by smallholder farmers under SOFECSA research intiatives in Malawi, Mozambique, Zambia and Zimbabwe

Country	Prioritized technology options	Comment
Malawi	1. Grain legume – cereal rotations dominated by groundnut (*Arachis hypogaea* L.), soyabean (*Glycine max* [L.] Merr.) or cowpea (*Vigna unguiculata* [L.] Walp.) in rotation with maize (*Zea mays* L.) 2. Grain legume/cereal intercrops involving pigeonpea (*Cajanus cajan* [L.] Millsp.), cowpea and common bean (*Phaseolus vulgaris* L.) 3. Intercrops involving pigeonpea with either groudnut or soyabean (*termed double legume intercropping*) 4. Mineral fertilizers in combination with organic nutrient sources (compost) 5. Mineral fertilizer application to legumes, especially P, N and S containing formulations	There was special focus on: (i) increasing food security, nutrition and income base for households (ii) increased availability of high quality legume seed for smallholder communities through NGO driven seed pass-on schemes (iii) Increasing the diversity of grain legumes under cultivation in response to market demands
Mozambique	1. Legume/cereal intercropping, particularly cowpea/maize; groundnut/maize; pigeonpea/maize 2. Legume – cereal rotations involving soyabean (*Glycine max* [L.] Merr.), cowpea (*Vigna unguiculata* [L.] Walp.), groundnut (*Arachis hypogaea* L.), bambara nut (*Vigna subterranea* [L.] Verdc) followed by maize (*Zea mays* L.) 3. Mineral fertilizers on legumes and cereals	Key areas of focus included: (i) Increasing legume contribution to household food and nutrition security (ii) Strengthening partnerships among extension, researchers agro-service providers and farmers for improved seed and fertilizer supply as well as marketing (iii) Optimising mineral N and P fertilization for increased productivity of intercrops and rotations (iv) Increasing the use of manures (v) Home processing of legumes
Zambia	1. Liming of maize, groundnuts, cowpea, soyabean and sunflower (*Helianthus annus* L.) on acid soils 2. Intercropping of maize with either groundnut, beans or pigeonpea	Main focus was on : (i) Enabling farmers to readily access quality seed, fertilizer and lime (ii) Evaluating different liming packages promoted by NGOs, farmers' unions and research institutions for increased legume productivity on acid soils

(continued)

Table 8.6 (continued)

Country	Prioritized technology options	Comment
	3. Cover crops involving yellow sunnhemp (*Crotalaria juncea* L.) red sunnhemp (*C. ochroleuca* G. Don) and mucuna to support conservation agriculture innovations 4. Legume green manure based rotations involving yellow and red sunnhemps as well as mucuna	(iii) Inorganic x organic fertilizer combinations to support legume cereal rotations and intercrops particularly in southern region
Zimbabwe	1. Mineral fertilizer x organic (cattle manure, woodland litter, compost, crop residues) combinations for legume-based rotations 2. Grain legume – cereal rotations, involving soyabean (*Glycine max* [L.] Merr.), cowpea (*Vigna unguiculata* [L.] Walp.), groundnut (*Arachis hypogaea* L.), bambara nut (*Vigna subterranea* [L.] Verdc) and common bean (*Phaseolus vulgaris* L.) 3. Green manure legume - cereal rotations involving sunnhemp (*Crotalaria juncea* L.) and mucuna (*Mucuna pruriens* L.) for specific farm niches 4. Indigenous legumes (mixed species) for restoration of degraded (non-responsive) soils	There was a particular focus on: (i) Establishing and supporting farmer-centred co-learning and innovation platforms that enhance diversification and market linkages (ii) Developing new basal fertilizer formulations to enable legume integration (iii) increasing efficiency of nutrient use by targeting specific ISFM technology combinations to different farmer resource groups and field environments (iv) integrating water conservation techniques in semi-arid environments

Source: Adapted from Mapfumo et al. (2001a, b)

The over-arching challenge for intensification and diversification of legume-cereal farming systems in southern Africa is the lack of suitable mechanisms to deliver smallholder communities out of the '*maize poverty trap*' (Mapfumo 2009b). Experience has shown that unless there is self-sufficiency in maize at household and community levels, either through local production or sustainable marketing systems, smallholders will continue to grow maize year after year regardless of yield levels. Most of the households in southern Africa are recipients of food handouts, either through local social safety nets or food aid from donors and governments. Under prevailing intricacies of social safety net arrangements at local level, family dignity is often upheld when a household is 'seen to attempt to produce its own maize (food)' before asking for assistance, even if an outcome of total crop failure is most probable. Enhanced and sustained capacity for increased productivity of maize is therefore an apparent pre-requisite for any meaningful legume intensification and diversification. This requires broader national policy and institutional support frameworks that incorporate mechanisms for improved supply of inputs and promotion of farmer-centred co-learning processes as well as co-innovation platforms among communities and diverse stakeholders.

Other major factors constraining adoption of legumes for improved food security and livelihoods in the region include:

(i) *Lack of information, knowledge and technical support*: Education and capacity building of farmers, extension agencies and other development partners on legume-based ISFM technologies and livelihood benefits will enhance meaningful participation by farmers. For example, in their recent work to address food, nutritional and income security with smallholder communities, Mtambanengwe and Mapfumo (2009) revealed some of the prevailing misconceptions among farmers about the role of legumes in soil fertility management. Legumes are often allocated to some of the poorest soils on the farm with no external fertilization, and expected to improve soil productivity at the end of the season. Resource-constrained farmers, especially women-headed households, whose fields were poorest, could not access quality seed and the minimum levels of nutrient inputs required to boost legume productivity. More quantitative research on N_2-fixation, fertilization strategies and quantification of N and non-N rotational benefits is required, particularly under Mozambican smallholder systems. Priority should also be given to farmers' local or indigenous knowledge on sustainability of legume-based systems. For example, there is greater potential for improvement of pigeonpea/maize intercropping in Malawi than any efforts to replace it with alternative systems such as the Gliricidia/maize intercrop suggested by Chirwa et al. (2003). Following 12 years of evaluating maize-groundnut rotations, Waddington et al. (2007) found justification in farmers' current preference for maize monocropping with mineral fertilization in Zimbabwe.

(ii) *Lack of timely and ready access to quality legume seed and appropriate fertilizers by farmers*: It is apparent that currently there are limited opportunities for farmers to access legume seeds of their desired choice or characteristics. For instance, most legume seed varieties on the market tend to address industrial needs (legumes as cash crops) more than home consumption. Participatory breeding and germplasm selection initiatives should be given priority in the medium to long term strategies for legume promotion. Farmers' preferred legume genotypes tend to be ignored in current research, yet these food legumes could play a critical role in local social safety net systems in terms of enhancing food, nutritional and income security. Thus research has largely ignored the multi-functionality and gender dimension of legumes in smallholder livelihood systems.

(iii) *Increasing rainfall variability*: Although research results have shown some grain legumes (e.g. groundnut, cowpea and pigeonpea) as having high resilience in N_2-fixation and net N input across the agro-ecological and soil fertility gradients, there are concerns about increasing frequency of droughts in southern Africa. The IPCC (2007) has also projected significant reduction in rainfall amounts in the region as one of the negative impacts of climate change. The prospects of enhanced contributions of intercropping systems to smallholder farming systems are therefore somewhat limited under low and moderate rainfall areas.

(iv) *Poor soil fertility and land gedradation* : Following decades of monocropping with sub-optimal inputs, some of southern Africa's soils are now extremely poor and do not readily respond to external fertilization. Increasingly, such soils are abandoned or simply cropped as part of farmers' desparate measures to produce food, even as crop failure is often a known outcome. These 'non-responsive soils' will require non-conventional and innovative interventions, and deliberate sowing of indigenous legumes has proved a worthwhile entry point (Mapfumo et al. 2005; Nezomba et al. 2010).

(v) *Identification and development of markets for legume products*: Promotion of both local and national/international markets for legumes can potentially drive intensified production and diversification of legume-cereal systems in southern Africa by attracting the participation of agro-dealers, traders, and private agro-processors in smallholder agriculture. Currently, limited and uncertain market access, coupled to poor and highly variable prices, has remained a major disincentive for farmers to diversify (Bezner Kerr et al. 2007). There is currently a critical lack of research to determine opportunities for developing local (community and regional) markets for legumes.

References

Agronomy Institute (1993) Effect of green manure crops on the performance of a second season maize test crop. Agronomy Institute annual technical report for 1989/90, Department of research and specialist services, Ministry of Agriculture, Harare, pp 160–164

Ahmed MM, Rohrbach DD, Gono LT, Mazhangara EP, Mugwira L, Masendeke DD, Alibaba S (1996) Soil fertility management in the communal areas of Zimbabwe: current practices, constraints and opportunities for changes. Results of a diagnostic survey. Southern and Eastern Africa Regional paper no. 6, ICRISAT-Southern and Eastern Africa Region, 27 pp

Akinnifesi FK, Makumba W, Kwesiga FR (2006) Sustainable maize production using gliricidia/maize intercropping in southern Malawi. Exp Agric 42:1–17

Akinnifesi FK, Makumba W, Sileshi G, Ajayi OC, Mweta D (2007) Synergistic effect of inorganic N and P fertilizers and organic inputs from Gliricidia sepium on productivity of intercropped maize in Southern Malawi. Plant Soil 294:203–217

Bezner Kerr R, Snapp S, Chirwa M, Shumba L, Msachi R (2007) Participatory research on legume diversification with Malawian smallholder farmers for improved human nutrition and soil fertility. Exp Agric 43:437–453

Bogale T, Assenga RH, Mmbaga TE, Friesen DK, Kikafunda J, Ransom JK (2001) Legume fallow formaize-based systems in Eastern Africa: contribution of legumes to enhanced maize productivity and reduced nitrogen requirements. Seventh Eastern and Southern Africa regional maize conference. CIMMYT Int., Mexico, pp 324–329

Brown P, Young A (1965) The physical environment of Central Malawi – with special reference to soils and agriculture. Government Printers, Zomba

Campbell BM, Frost PGH, Kirchmann H, Swift MJ (1998) A survey of soil fertility management in small-scale farming systems on north-eastern Zimbabwe. J Sustain Agric 11:19–39

Chibudu C (1998) Green manuring crops in a maize based communal area, Mangwende: experiences using participatory approaches. In: Waddington SR, Murwira HK, Kumwenda JDT, Hikwa D, Tagwira F (eds.) The soil fertility network for maize-based cropping systems in Malawi and Zimbabwe. Proceedings of the soil fertility network results and planning workshop held from 7 to 11 July 1997 at Africa University, Mutare, Zimbabwe. Soil Fertility Network and CIMMYT-Zimbabwe, pp 87–90

Chikowo R, Mapfumo P, Nyamugafata P, Nyamadzawo G, Giller KE (2003) Nitrate-N dynamics following improved fallows and maize root development in a Zimbabwean sandy clay loam. Agrofor Sys 59:187–195

Chirwa PW, Black CR, Ong CK, Maghembe JA (2003) Tree and crop productivity in gliricidia/maize/pigeonpea cropping systems in southern Malawi. Agrofor Syst 59:265–277

Chivasa W (1995) Comparative performance of sunflower, sorghum and maize hybrids in Matibi II. Special study report in partial fulfilment of the B.Sc. (Hons.) Degree in Agriculture, Department of Crop Science, University of Zimbabwe, Harare 75 pp

CSO (1998) Poverty in Zimbabwe. Central Statistics Office, Harare, 129 pp

FAO (1990) Soil map of the world. FAO/UNESC/ISRIC. Food and Agricultural Organization of the United Nations, Rome, 119 pp

Ganunga T, Terokun O, Kumwenda JD (1998) Tithonia diversifolia: An organic source of nitrogen and phosphorous for maize in Malawi. In: Waddington SR (ed) Grain legumes and green manures for soil fertility in Southern Africa: taking stock of progress., Proceedings of a conference (8-11 October, 2002) Leopard Rock Hotel, Vumba, Zimbabwe. Soil Fertility Network and CIMMYT-Zimbabwe, Harare, Zimbabwe. pp 191–194

Gilbert RA (1998) Undersowing green manures for soil fertility enhancement in the maize-based cropping systems of Malawi. In: Waddington SR, Murwira HK, Kumwenda JDT, Hikwa D, Tagwira F (eds.) The soil fertility network for maize-based cropping systems in Malawi and Zimbabwe. Proceedings of the soil fertility network results and planning workshop held from 7 to 11 July 1997 at Africa University, Mutare, Zimbabwe. Soil Fertility Network and CIMMYT-Zimbabwe, Harare, pp 73–80

Giller KE (2001) Nitrogen fixation in tropical cropping systems, 2nd edn. CAB International, Wallingford, 458 pp

Giller KE, McDonagh JF, Cadisch G (1994) Can biological nitrogen fixation sustain agriculture in the tropics? In: Syers PK, Rimmer DL (eds.) Soil science and sustainable land management in the tropics. CAB International, Wallingford, pp 173–191

Giller KE, Mpepereki S, Mapfumo P, Kasasa P, Sakala WP, Pombeya H, Itimu O, Cadisch G, Gilbert R, Waddington SR (2002) Putting legume N_2-fixation to work in cropping systems of southern Africa. In: Pedrosa F, Hungria M, Yates MG, Newton WE (eds.) Nitogen fixation: from molecules to crop productivity. Kluwer, Dordrecht, pp 525–530

Government of Malawi (1999) Population housing census. National Statistics Office Zomba, Malawi

Hikwa D, Murata M, Tagwira F, Chiduza C, Murwira H, Muza L, Waddington SR (1998) Performance of green manure legumes on exhausted soils in Northern Zimbabwe: A soil fertility network trial. In: Waddington SR, Murwira HK, Kumwenda JDT, Hikwa D, Tagwira F (eds.) The soil fertility network for maize-based cropping systems in Malawi and Zimbabwe. Proceedings of the soil fertility network results and planning workshop held from 7 to 11 July 1997 at Africa University, Mutare, Zimbabwe. Soil Fertility Network and CIMMYT-Zimbabwe, Harare, pp 73–80

IPCC (2007) Climate change: impacts, adaptation and vulnerability, Inter-governmental Panel on Climate Change (IPCC) Fourth assessment report. Cambridge University Press, Cambridge, UK

Jeranyama P, Hesterman OB, Waddington SR (1998) The impact of legumes relay-intercropped into maize at Domboshawa, Zimbabwe. In: Waddington SR, Murwira HK, Kumwenda JDT, Hikwa D, Tagwira F (eds.) The soil fertility network for maize-based cropping systems in Malawi and Zimbabwe. Proceedings of the soil fertility network results and planning workshop held from 7 to 11 July 1997 at Africa University, Mutare, Zimbabwe. Soil Fertility Network and CIMMYT-Zimbabwe, Harare, pp 31–34

Kanonge G, Nezomba H, Chikowo R, Mtambanengwe F, Mapfumo P (2009) Assessing the potential benefits of organic and mineral fertilizer combinations on maize and legume productivity under smallholder management in Zimbabwe. Afr Crop Sci Conf Proc 9:63–70

Kasasa P (1999) Biological nitrogen fixation by promiscuous nodulating soyabean (*Glycine max [L.] merr.*) varieties in communal area soils of Zimbabwe. MPhil thesis submitted to the Department of Soil Science and Agricultural Engineering, University of Zimbabwe, MP 167, Mount Pleasant, pp 118

Kumwenda JDT, Gilbert R (1998) Biomass production by legume green manures on exhausted soils in Malawi: a soil fertility network trial. In: Waddington SR, Murwira HK, Kumwenda JDT, Hikwa D, Tagwira F (eds.) Soil fertility research for smallholder maize-based systems in Malawi and Zimbabwe. SoilFertNet/CIMMYT, Harare, pp 85–86

Kumwenda JDT, Waddington SR, Snapp SS, Jones RB, Blackie MJ (1995) Soil fertility management for smallholder maize-based cropping systems of Southern African: a review. Network working paper No. 1. Soil fertility network for maize-based cropping systems in countries of Southern Africa. CIMMYT, Harare, 34 pp

Kumwenda JDT, Saka AR, Snapp SS, Ganunga RP, Benson T (1998) Effects of organic legume residues and inorganic fertilizer nitrogen on maize yield in Malawi. In: Waddington SR, Murwira HK, Kumwenda JDT, Hikwa D, Tagwira F (eds.) The soil fertility network for maize-based cropping systems in Malawi and Zimbabwe. Proceedings of the soil fertility network results and planning workshop held from 7 to 11 July 1997 at Africa University, Mutare, Zimbabwe. Soil Fertility Network and CIMMYT-Zimbabwe, Harare, pp 165–171

Kumwenda JDT, Saka AR, Sakala WD, Kabambe VH, Ganunga RP (2001) Legume manure crops-maize rotation trial for the 1997/98 and 1998/99 cropping seasons. In: Kumwenda JDT, Ngwira P, Kabambe VH, Komwa MKM (eds.) Annual research project report for the 1997/98 season for the Cereals Commodity Group, Chitedze Research Station, Lilongwe: Malawi, pp 125–164

Lungu OI, Munyinda M (2003) Agronomic effectiveness of phosphate rock products, monoammonium phosphate and lime on grain legumes in some Zambian soils. In: Waddington SR (ed.) Grain legumes and green manures for soil fertility in Southern Africa: taking stock of progress. Proceedings of a conference held 8–11 Oct 2002 at the Leopard Rock Hotel, Vumba, Zimbabwe. Soil Fertility Network and CIMMYT-Zimbabwe, Harare, pp 39–44

Mafongoya PL, Kuntashula E, Kwesiga F, Chirwa T, Chintu R, Sileshi G, Matibini J (2003) Leguminous agroforestry options for replenishing soil fertility in Southern Africa. In: Waddington SR (ed.) Grain legumes and green manures for soil fertility in Southern Africa: taking stock of progress. Proceedings of a conference held 8–11 Oct 2002 at the Leopard Rock Hotel, Vumba, Zimbabwe. Soil Fertility Network and CIMMYT-Zimbabwe, Harare, pp 141–154

Mafongoya PL, Bationo A, Kihara J, Waswa BS (2006) Appropriate technologies to replenish soil fertility in southern Africa. Nutr Cycling Agroecosyst 76:137–151

Mapfumo P (2000) Potential contribution of legumes to soil fertility management in smallholder farming systems of Zimbabwe: the case of pigeonpea (Cajanus cajan *[L] Millsp.*). D.Phil thesis. Department of Soil Science & Agricultural Engineering. University of Zimbabwe, Harare, 198 pp

Mapfumo P (2009a) Developing a soil fertility consortium for Southern Africa (SOFECSA): a final project technical report prepared for the Rockefeller Foundation. SOFECSA, CIMMYT Southern Africa, Harare, 57 pp

Mapfumo P (2009b) Integrating sustainable soil fertility management innovations in staple cereal systems and other value chains to enhance livelihoods and environmental systems in Southern Africa. A SOFECSA technical annual report for the sub-Saharan Challenges Program (SSA-CP) prepared for the Forum for Agricultural Research in Africa (FARA). Soil Fertility Consortium for Southern Africa (SOFECSA), CIMMYT-Zimbabwe, Harare, 26 p

Mapfumo P, Giller KE (2001) Soil fertility management strategies and practices by smallholder farmers in semi-arid areas of Zimbabwe. In: International Crops Research Institute for the Semi-Arid Tropics (ICRISAT) with permission from the Food and Agriculture Organization of the United Nations (FAO), 60 pp

Mapfumo P, Mtambanengwe F (2004) Base nutrient dynamics and productivity of sandy soils under maize pigeonpea rotational systems in Zimbabwe. In: Bationo A (ed.) Managing nutrient cycles to sustain soil fertility in Sub-Sahara Africa. Academy Science Publishers/TSBF-CIAT, Nairobi, pp 225–238

Mapfumo P, Giller KE, Mpepereki S, Mafongoya PL (1999) Dinitrogen fixation by pigeonpea of different maturity types on granitic sandy soils in Zimbabwe. Symbiosis 27:305–318

Mapfumo P, Mpepereki S, Mafongoya P (2000) Pigeonpea-nodulating rhizobia prevalence and effect on inoculation response in smallholder-managed soils in Zimbabwe. Exp Agric 36:423–434

Mapfumo P, Campbell BM, Mpepereki S, Mafongoya P (2001a) Legumes in soil fertility management: the case of pigeonpea in smallholder farming systems of Zimbabwe. Afr Crop Sci J 9:629–644

Mapfumo P, Chikowo R, Giller K, Mugendi D, Kimetu JM, Palm C, Mutuo P (2001b) Closing the loop: identifying N sources and minimizing N losses in leguminous cropping systems. Proceedings of the Rockefeller Foundation nitrogen exploration project review workshop, 22–24 Nov 2000, Mazwikadei Resort, Banket, 70 pp

Mapfumo P, Mtambanengwe F, Giller KE, Mpepereki S (2005) Tapping indigenous herbaceous legumes for soil fertility management by resource poor farmers in Zimbabwe. Agric Ecosyst Environ 109:221–233

Maria MR, Yost R (2006) A survey of soil fertility status of four agroecological zones of Mozambique. Soil Sci 171:902–914

Mpepereki S, Makonese F (1994) Inoculation response of cowpea (Vigna unguiculata), soybean (Glycine max L. Merr) and dry bean (Phaseolus vulgaris L.) in soils with low populations of indigenous rhizobia. In: Sadiki M, Hilali A (eds.) Recent developments in biological nitrogen fixation research in Africa. IAV Hassan II, Rabat, pp 32–44

Mpepereki S, Makonese F, Giller KE (1996) Soybeans in smallholder farming systems in Zimbabwe. CIMMYT, Harare

Mtambanengwe F, Mapfumo P (2009) Combating food insecurity on sandy soils in Zimbabwe: the legume challenge. Symbiosis 48:25–36

Mupangwa WT, Tagwira F (2005) Groundnut yield response to single superphosphate, calcitic lime and gypsum on acid granitic sandy soil. Nutr Cycling Agroecosyst 73:161–169

Murata M, Waddington S, Murwira H (2000) Rehabilitation of degraded sandy soils using annual green manures in Zimbabwe. In: Target (The newsletter of the soil fertility research network for maize-based cropping systems in countries of Southern Africa), vol 21, p 4

Murwira HK, Mutuo P, Nhamo N, Marandu AE, Rabeson R, Mwale M, Palm CA (2002) Fertiliser equivalency values of organic materials of differing quality. In: Vanlauwe B, Diels J, Sanginga N, Merkx R (eds.) Integrated nutrient management in sub-Saharan Africa. CAB International, Wallingford, Oxon, UK, pp 113–122

Musiyiwa K, Mpepereki S, Giller KE (2005) Symbiotic effectiveness and host ranges of indigenous rhizobia nodulating promiscuous soyabean varieties in Zimbabwean soil. Soil Biol Biochem 37:1169–1176

Mwale M, Cassim M, Kabongo J, Phiri LK (2003) Soil fertility improvement through the use of green manure in central Zambia. In: Waddington SR (ed.) Grain legumes and green manures for soil fertility in Southern Africa: taking stock of progress. Proceedings of a conference held 8–11 Oct 2002 at the Leopard Rock Hotel, Vumba, Zimbabwe. Soil Fertility Network and CIMMYT-Zimbabwe, Harare, pp 173–177

Ncube B, Twomlow SJ, van Wijk MT, Dimes JP, Giller KE (2007) Productivity and residual benefits of grain legumes to sorghum under semi-arid conditions in southwestern Zimbabwe. Plant Soil 299:1–15

Nezomba H, Tauro TP, Mtambanengwe F, Mapfumo P (2010) Indigenous legume fallows (indifallows) as an alternative soil fertility resource in smallholder maize cropping systems. Field Crops Res 115:149–157

Nyakanda C, Maringa IK, Dzowela BH, Murwira H (1998) Biomass production and maize yield under a tree-based improved fallow of *Sesbania* and pigeonpea. In: Waddington SR, Murwira HK, Kumwenda JDT, Hikwa D, Tagwira F (eds.) Soil fertility research for the maize-based farming systems in Malawi and Zimbabwe. Proceedings of the Soil fertility Network Results and Planning Workshop, Africa University, Mutare, Zimbabwe, 7–11 July 1997. Soil Fertility Network and CIMMYT-Zimbabwe, Harare, Zimbabwe, pp 115–119

Nyakanda C, Mariga IK, Murwira HK, Nyamugafata P, Salimfu D (2002) The status of existing set aside fallows in smallholder farming: a case of two contrasting agricultural land-use systems in Zimbabwe. In: Rao MR, Kwesiga FR (eds.) Agroforestry impacts on livelihoods in Southern Africa: putting research into practice. Aventura Warmbaths, South Africa, 20–24 May 2002

Nyamapfene K (1991) Soils of Zimbabwe. Nehanda Publishers, Harare, 179 pp

Nye PH, Greenland DJ (1960) The soil under shifting cultivation. Commonwealth Agricultural Bureaux Technical Communication 51, Harpenden

Peoples MB, Faizah AW, Rerkasem B, Herridge DF (1989) Methods for evaluating nitrogen fixation by nodulated legumes in the field. ACIAR Monograph No. 11, vii + 76 p

Phiri ADK, Kanyama-Phiri GY, Snapp S (1999) Maize and Sesbania production in relay cropping at three landscape positions in Malawi. Agrofor Syst 47:153–162

Sakala WD, Cadisch G, Giller KE (2000) Interactions between residues of maize and pigeonpea and mineral fertilizers during decomposition and N mineralization. Soil Biol Biochem 32:699–706

Sakala WD, Kumwenda JDT, Saka AR, Kabambe VH (2001) The potential of legume green manured to increase soil fertility and maize yields under smallholder farm conditions. In: Phiri IMG, Saka AR, Chilembwe EHC (eds.) Agricultural Technologies for Sustainable Development in Malawi. DARTS Annual Scientific Proceedings No. 1, pp 34–40

Sakala WD, Ligowe I, Kayira D (2003a) Mucuna-maize rotations and short fallows to rehabilitate smallholder farms in Malawi. In: Waddington SR (ed.) Grain legumes and green manures for soil fertility in Southern Africa: taking stock of progress. Proceedings of a conference held 8–11 Oct 2002 at the Leopard Rock Hotel, Vumba, Zimbabwe. Soil Fertility Network and CIMMYT-Zimbabwe, Harare, pp 161–163

Sakala WD, Kumwenda JDT, Saka AR (2003b) The potential of green manures to increase soil fertility and maize yields in Malawi. Biol Agric Hortic 21:121–130

Sanchez PA, Shephard KD, Soule MJ, Place FM, Buresh RJ, Izac AN, Mokunye AU, Kwesiga FR, Ndiritu CG, Woomer PL (1997) Soil fertility replenishment in Africa: An investment in natural resource capital. In: Buresh JR, Sanchez PA, Calhoun F (eds.) Replenishing soil fertility in Africa. SSSA special publication 51, SSSA, Madison, pp 1–46

Scoones I, Chibudu C, Chikura S, Jeranyama P, Machaka D, Machanja W, Mavedzenge B, Mombeshora B, Mudhara M, Mudziwo C, Murimbarimba F, Zirereza B (1996) Hazards and opportunities. Farming livelihoods in dryland Africa: lessons from Zimbabwe. Zed Books Ltd, London/New Jersey, in association with International Institute for Environment and Development, London, 267 pp

Snapp SS, Mafongoya PL, Waddington S (1998) Organic matter technologies for integrated nutrient management in smallholder cropping systems of southern Africa. Agric Ecosyst Environ 71:185–200

Snapp SS, Rohrbach DD, Simtowe F, Freeman HA (2002) Sustainable soil management options for Malawi: can smallholder farmers grow more legumes? Agric Ecosyst Environ 19:159–174

Stoorvogel JJ, Smaling EMA, Janssen BH (1993) Calculating soil nutrient balances in Africa at different scales. I. Supranational scale. Fertil Res 35:227–235

Waddington SR (1994) Overview of the current situation and previous impact of adaptive agricultural research in southern Africa In: Craswell ET and Simpson J (eds.) Soil fertility and climatic constraints in dryland agriculture. ACIAR Proceedings No.54. Canberra, Australia: Australian council for international agricultural research (ACIAR), pp 7–14

Waddington SR (2003) Grain legumes and green manures for soil fertility in Southern Africa: taking stock of progress. Proceedings of a conference held 8–11 Oct 2002 at the Leopard Rock Hotel, Vumba, Zimbabwe. Soil Fertility Network and CIMMYT-Zimbabwe, Harare, 246 p

Waddington SR, Murwira HK, Kumwenda JDT, Hikwa D, Tagwira F (eds.) (1998a) Soil fertility research for maize-based farming systems in Malawi and Zimbabwe. Soil Fertility Network and CIMMYT-Zimbabwe, Harare, 312 p

Waddington SR, Karigwindi J, Chifamba J (1998b) Productivity and profitability of maize+groundnut rotations when compared to continuous maize under smallholder management in Zimbabwe. In: Waddington SR, Murwira HK, Kumwenda JDT, Hikwa D, Tagwira F (eds.) The soil fertility network for maize-based cropping systems in Malawi and Zimbabwe. Proceedings of the soil fertility network results and planning workshop held from 7 to 11 July 1997 at Africa University, Mutare, Zimbabwe. Soil Fertility Network and CIMMYT-Zimbabwe, Harare, pp 43–52

Waddington SR, Mekuria M, Siziba S, Karigwindi J (2007) Long-term sustainability and financial returns from grain legume-maize intercrops on a sandy soil in subhumid north central Zimbabwe. Exp Agric 43:489–503

Chapter 9
Options for Improving Plant Nutrition to Increase Common Bean Productivity in Africa

J. Jansa, A. Bationo, E. Frossard, and I.M. Rao

Abstract Common bean (*Phaseolus vulgaris* L.) is one of the most important grain legume crops in Africa. It represents an important share of food proteins and calories of poor agricultural populations in Eastern and Southern Africa. The production of bean in Africa is carried out primarily by small-scale farmers, who use little or no fertilizers or soil amendments. Average seed yields of beans in Africa are very low as compared to other production areas, mainly because of abiotic (edaphic and climatic) and biotic constraints. African soils are generally nutrient poor and application of adequate fertilizers is currently not feasible for economic and other reasons. The grain yield in the future can be increased by (i) using improved bean varieties that are better adapted to low soil fertility (ii) by application of fertilizers and managing soil fertility, and (iii) by optimizing the root symbioses to improve plant nutrient uptake under present conditions. Beans are forming two distinct types of symbiotic relations. They undergo nodulation with nitrogen-fixing bacteria and they also establish arbuscular mycorrhiza with Glomeromycota. Both of these symbioses affect in an important way the plant nutrient uptake. However, limited research efforts were made so far into studying the complex interactions between the three partners compared to the research focused either on biological nitrogen fixation or on mycorrhizal symbiosis. Thus the interactions among the symbiotic partners within the tripartite symbiosis are still poorly understood, though they might offer a potential to improve and sustain bean production in the tropics.

J. Jansa (✉) • E. Frossard
Swiss Federal Institute of Technology (ETH) Zurich, Institute of Agricultural Sciences, Eschikon 33, CH-8315 Lindau, Switzerland

A. Bationo
Alliance for a Green Revolution in Africa (AGRA), Soil Health Program, PMB KIA 114, Airport-Accra, Ghana
e-mail: abationo@agra-alliance.org

I.M. Rao
Centro International de Agriculture Tropical (CIAT), A.A. 6713 Cali, Colombia

A. Bationo et al. (eds.), *Fighting Poverty in Sub-Saharan Africa: The Multiple Roles of Legumes in Integrated Soil Fertility Management*,
DOI 10.1007/978-94-007-1536-3_9, © Springer Science+Business Media B.V. 2011

9.1 Introduction

9.1.1 *Relevance of Common Bean as Staple Crop in Africa*

The cultivation of common bean (*Phaseolus vulgaris* L.) began thousands of years ago in rather fertile lands in cool climates at medium to high altitudes in the Americas. Common beans were imported to Africa about 300 years ago by Spanish and Portuguese traders. They are now grown by both small holders and commercial producers (Mauyo et al. 2007). The Eastern, Central, and Southern Africa are the main common bean-growing regions on the continent. Africa is probably the second largest producer of common beans after Latin America with nearly 5 millions ha yielding almost 3 million tons of seeds annually (FAO 2008). Although statistics indicate higher production of dry beans in Asia than in Africa, bean production in Asia is dominated by other botanical species than *Phaseolus vulgaris* L. (Singh 1999b). The average annual yields of common bean seeds in Africa vary between 167 kg ha^{-1} in Eritrea and over 1,000 kg ha^{-1} in Madagascar and South Africa (FAO 2008). Bean (mainly in form of seeds) is a major source of protein and calories in Eastern and Southern Africa (Pachico 1993; Broughton et al. 2003).

Between one-sixth to one-third of the dietary proteins intake by humans in the Eastern and Central African highlands comes from common beans, making them the most important pulse in the region. In Rwanda, common beans cover over 60% dietary protein requirements of both the rural and urban populations (Buruchara 2009). The commercial value (producer prices) of beans ranges, depending on the variety and country of production, between US$160 per ton of seeds (Niger in 2006) and well over US$1,000 per ton in Congo, Madagascar and Zimbabwe (FAO 2008). Therefore, bean production can be both a significant source of proteins for human consumption and also a significant source of income for farmers.

However, bean production in Africa is limited, and the yields realized by farmers are far below potential yields. The yields encountered on farmers' fields are usually around 500 kg ha^{-1}, while the yield potential of the crop could be as high as 5,000 kg ha^{-1} under optimal water and nutrient supply and in the absence of pests and pathogens (Verdoodt et al. 2004). Main reasons of this discrepancy are: (i) the low fertility soils on which they are grown, (ii) lack of nutrient inputs, (iii) planting bean varieties not adapted to the environment, and (iv) lack of disease and pest controls (Thung and Rao 1999). As a consequence, the production of common beans does not yet meet the increasing demand in sub-Saharan Africa (CIAT 2009).

9.1.2 *Significance of Nitrogen, Phosphorus, and Soil Conditions for Bean Production*

The major soils found in bean-growing areas in Africa are classified as Alfisols, Ultisols, Oxisols and Inceptisols. These tend to be acidic, low in organic matter, low in exchangeable calcium (Ca), magnesium (Mg) and potassium (K), low in available phosphorus (P), high in exchangeable aluminium (Al) and/or manganese (Mn) and have often a high P fixing capacity (Wortmann et al. 1998; Thung and Rao 1999). Optimal conditions for beans are, however, light loamy soils with pH between 5.5 and 7.0, rich in organic matter, and providing an adequate supply of water, nitrogen (N) and P (Singh 1999b).

Low availabilities of N and P in agricultural soils are among the most widespread plant nutritional constraints for crop growth and yields on a global scale and more specifically in Africa. Given the acknowledged accelerated nutrient mining and lack of fertilizer use (Sanchez and Swaminathan 2005), N and P deficiencies are major and omnipresent problems in African agriculture (Drechsel et al. 2001). For example, in Ethiopia, nutrient depletion amounts to 36, 6, and 32 kg ha^{-1} $year^{-1}$ of N, P and K, respectively, while the average use of fertilizers in the country replenishes only about 10% of the nutrients removed by crops from the fields (Henao and Baanante 1999). In extreme cases such as in Rwanda and Malawi, where some of the soils are inherently quite fertile, soil nutrient depletion may reach (on a national level) up to 60, 10, and 60 kg ha^{-1} $year^{-1}$ of N, P, and K (Roy et al. 2003), with negligible nutrient inputs (Drechsel et al. 1996). Moreover, when mineral fertilizers are used, they are rather applied to crops other than common bean. Bean seeds contain large amounts of N and P (35–45 mg N g^{-1} and 4–5 mg P g^{-1}) and these concentrations do not vary largely among different growth conditions and bean cultivars (Araujo and Teixeira 2003). This means that for a bean production of 2,000 kg dry seeds ha^{-1} $year^{-1}$, almost 100 kg N and about 10 kg P will be removed each year from the system with the harvested seeds.

Common bean can obtain N from the atmosphere through biological nitrogen fixation (BNF) carried out by symbiotic rhizobia. However, the BNF rarely covers the full N demand of beans (Thung and Rao 1999). This is in contrast to soybeans and other leguminous crops, which require either little or no starter N fertilizer to establish efficient symbiosis with N fixing prokaryotes, and subsequently cover most of their N demands through BNF (Mendes et al. 2003; Campo and Hungria 2004). Beans usually respond well to N fertilization up to 200 kg ha^{-1} (Anderson 1974; Edje et al. 1975; Mayona and Kamasho 1988; Daba and Haile 2002; da Silveira et al. 2005) and rarely derive more than 50% of their N from the atmosphere (Bliss 1993a; Wortmann 2001). In spite of their relative dependency of soil mineral N, bean usually recover less than 50% of the applied N fertilizer, with highest reported values being about 70% (Thung and Rao 1999; dos Santos and Fageria 2007, 2008). This phenomenon is probably related to (i) the coarseness of bean roots (Zhu et al. 2005; Rubio and Lynch 2007), and (ii) the fact that beans develop best in soil with high organic matter content, i.e. releasing mineral N steadily through mineralization during the growing season (Thung and Rao 1999). Previous surveys estimated that over 60% of the bean production area in Central, Southern, and Eastern Africa was affected by N deficiency (Wortmann et al. 1998; Thung and Rao 1999). This caused yield losses of up to 40% compared to the N fertilized areas (Singh 1999b; Thung and Rao 1999).

In addition to N deficiency, in about 40–50% of African bean production areas, the common beans are affected by moderate to severe P deficiency (Wortmann and Allen 1994; Wortmann et al. 1998; Lunze et al. 2007). The naturally low P availability

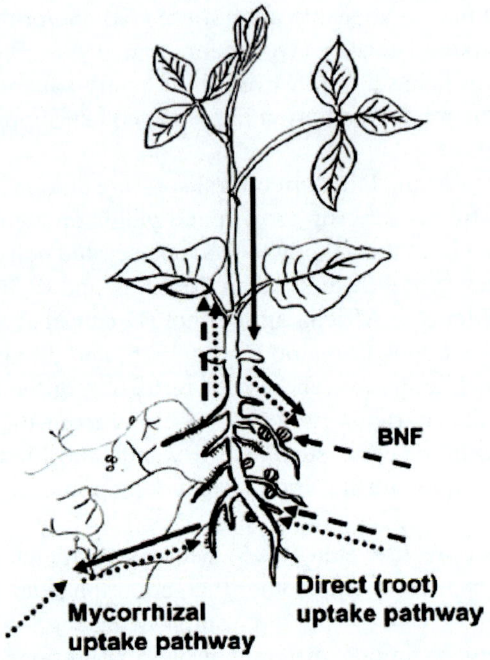

Fig. 9.1 Conceptual model of main phosphorus, nitrogen, and carbon fluxes in the tripartite symbiosis of common bean, arbuscular mycorrhizal fungus, and nitrogen-fixing prokaryote. For simplification, mycorrhizal fungus is shown only *left* to the roots and nodules are only shown on the *right* side of the root system. Phosphorus (*dotted arrows*) is derived either from the seed reserves (not shown here) or from the soil. Soil phosphorus uptake is realized either via the arbuscular mycorrhizal fungus (mycorrhizal uptake pathway) or directly through rhizodermis/root hairs (direct or root uptake pathway). Significant portion of the phosphorus transported through the vascular tissues is diverted towards the nodules due to high phosphorus demand of the biological nitrogen fixation (*BNF*). Nitrogen (*dashed arrows*) is taken up either by the roots from the soil, or obtained from the atmosphere through the BNF process mediated by the prokaryotes. Reduced carbon compounds derived from plant photosynthesis and transported belowground (*solid arrows*) are utilized by roots (not shown here), nitrogen-fixing prokaryotes, and the mycorrhizal fungi (Jansa et al.)

of African soils might be due to the low total soil P content and/or to their high P fixing capacity due to their high Al and iron (Fe) oxide contents and their low pH (Vitousek 2004; Wardle et al. 2004; Nziguheba 2007). Yield of beans can be reduced by as much as 60–75% in soils that are unable to release sufficient P levels during the growing season. P deficiency is also a major contributing factor limiting BNF in beans (Graham 1981; Thung 1992). This is because the BNF requires large amounts of P to be delivered to the rhizobia (Fig. 9.1), creating additional sink of P to the plant in the nodules, thus further depriving the plant of the P acquired from the soil (Leidi and Rodriguez Navarro 2000; Zaman–Allah et al. 2007). For these reasons, P fertilization has usually large positive effects on growth, BNF, and yields of the common beans in low P soils such as frequently encountered in Africa (Mbugua 1986; Aggarwal et al. 1997; Beebe et al. 2006). In these soils, an adequate P fertilization can thus alleviate both N and P deficiencies in beans.

About 40–50% of the bean production areas in Central, Eastern and Southern Africa are estimated to be affected by Al toxicity, which might result in 50–60% losses in yields (Thung and Rao 1999; Broughton et al. 2003). Furthermore, about 10% of the bean production areas in Africa suffer from Mn toxicity (Thung and Rao 1999; Weil 2000). Mn toxicity is typically found in soils with high rainfall and/or poor drainage, low pH, and medium to high organic matter contents. Large areas producing beans are suffering from moderate K deficiency, but this is usually not considered a major limitation for bean production (Wortmann et al. 1998; Thung and Rao 1999). Other constraints such as drought and biotic stresses also strongly limit bean production. Although these other constraints may be as important as N and P deficiencies, causing up to 100% harvest losses (Wortmann et al. 1998; Singh 1999b), they are not the subject of this review that is particularly focused on plant nutrition and root symbiotic interactions. Importance of biotic stresses for bean production and breeding efforts aiming at overcoming them have been reviewed elsewhere (Kelly 2004; Hillocks et al. 2006).

This review aims at: (i) Characterization of the components of bean cropping systems in Africa including plant genetic variability and root and root symbioses functioning; (ii) Identification of possible system changes consisting of one – or two – step modifications of the current practices, in order to improve nutrient acquisition and use by the plants and subsequently the yields, and; (iii) Proposing some further perspectives requiring deeper changes of the production systems, to increase yields and sustainability of the bean production system. Associated research needs are outlined, which are required to implement these changes. This contribution focuses on soil-microorganisms-plant relationships and leaves out social and economic aspects of the bean production, seed storage, marketing, and processing for human consumption. Although these other aspects of bean science are certainly important, they cannot be considered in this review.

9.2 Characterization of the Components of Bean Cropping Systems

9.2.1 Current Fertilization Strategies and Cropping Systems

Although frequently grown, common beans are rarely fertilized in subsistence farming in Africa. Bean production thus contributes to the widespread soil nutrient mining and eventually to soil degradation (Giller and Cadisch 1995; Wortmann et al. 1998). When mineral fertilizers are used, they tend to be applied to maize or other high-value cash crops such as banana but not to beans as reported in case studies from Malawi and Zimbabwe (Snapp et al. 1998; Nyakudya et al. 2005). Although it was repeatedly demonstrated that fertilizers would increase production, unaffordable prices and low quality of mineral fertilizers, together with the lack of local infrastructure, markets, institutions, and political will limit large scale applications of mineral fertilizers (Sanginga et al. 2007).

High levels of N and P fertilizers would indeed be necessary to overcome the low fertility of many African agricultural soils. Sanchez and Uehara (1980) and Munthali (2007) estimated that applications of 500–1,000 kg P ha^{-1} would be required to realize major and long-lasting crop yield increases in Oxisols. Similarly, large N inputs (more than 100 kg N ha^{-1} $year^{-1}$) would be required to improve bean seed yields above 2,000 kg ha^{-1} $year^{-1}$ (da Silveira et al. 2005). Since these long-term measures are currently impossible to implement, small amounts of spot – or band-applied fertilizers, mostly of local origin, have been promoted to achieve at least moderate and short-term increases in crop yields (Smithson et al. 2003; Woomer et al. 2003; Jama and Pizarro 2008). Another strategy to improve crop yields is to use locally produced green or animal manures. It has been shown that applications of plant residues (for example *Tithonia diversifolia* and *Senna hirsuta* at rates of 4 t ha^{-1}) are as effective as application of inorganic NPK fertilisers containing same amounts of N as the organic materials for increasing maize and bean yields (Kayuki and Wortmann 2001; Vanlauwe and Giller 2006). However, the plant materials usually contained less P than the corresponding inorganic fertilizers. Thus it has been recommended to use plant residues together with low rates (e.g. 10 kg ha^{-1} $year^{-1}$) of easily soluble P fertilizers (Vance 2001). On acid soils, this should be combined with liming so as to avoid strong P fixation and Al toxicity (von Uexküll and Mutert 1995; Fageria and Santos 1998). Using organic fertilizers could also limit soil organic matter depletion and thus limit soil N and bases depletion and slow down acidification (Giller et al. 1997; Thung and Rao 1999; Rutunga and Neel 2006). However, land scarcity, demand for organic materials to be used as fuel or construction materials, and generally low animal production in Africa will not allow broad application of manures in any near future (Fermont et al. 2008).

Common bean is an important component of several cropping systems on a global scale. It can be grown in monocultures (mainly in Latin America), or in different relay, strip and intercropping systems (Singh 1999b). In Africa, beans are produced either as monocultures (more bush, occasionally climbing beans) or, more often, as a component of intercropping systems (Wooley et al. 1991; Wortmann et al. 1998). Traditionally, beans are intercropped or relay-cropped with maize, banana, cassava or with other crops (Wortmann et al. 1998; Broughton et al. 2003). Climbing bean cultivars are often grown together with maize or cassava, which provide the required mechanical support instead of additional stakes or trellises. Intercropping of beans with maize seems to be the traditional way of producing these two crops, as documented by *Phaseolus* archaeology (Hastorf 1999; Zeder et al. 2006). Intercropping is first regarded as insurance against severe environmental stresses or high incidence of diseases, diversifying the farmer's production basis and decreasing production risks (Broughton et al. 2003). Further arguments for adopting intercropping systems as compared to the monocropping are the potential N benefits to the cereal component through the BNF of the legume component, reduced erosion, water conservation, and shading of the beans from direct sunlight (Clark and Francis 1985a; Vieira 1998). On the other hand, intercropping of beans and maize reduces yield of each component because of resource competition (Silwana et al. 2007). Transfer of N from bean to maize remains controversial due to generally low BNF efficiency of the common bean as compared to other legumes such as soybean or chickpea (Martin et al. 1995; Zhang and Li 2003; Ghosh et al. 2007). Although greater amount of plant biomass is sometimes produced in intercropping system as compared to the monocultures, the yields (either in terms of calories or proteins) usually do not exceed the higher yielding monocrops (Clark and Francis 1985b). In case of fodder production, intercropping maize with beans would yield the same amount of silage but with higher nutritional quality (=higher crude protein concentration) than that of the maize alone (Dawo et al. 2009). In addition to intercropping, bean may also be grown in rotation with cereals such as maize or sorghum (Wortmann 2001). Although crop rotation is known to efficiently contribute to disease and pest control, nutrient use efficiency, and system sustainability (Liebman and Dyck 1993; Kandji et al. 2003; Sangakkara et al. 2003; Vanlauwe et al. 2008), it is not widely practiced in Africa. It is usually limited to areas, where sufficient and reliable short-rains would provide sufficient moisture for the bean crop, usually sown in the short-rain season in zones with bimodal rainfall pattern (Broughton et al. 2003) or where sufficient water would be still available during the dry season such as in protected valley bottoms (Wortmann et al. 1998).

9.2.2 Genetic Variability of Beans

Phaseolus vulgaris L. belongs to a large genus, where five different botanical species have been domesticated (Debouck 1999). The genetic origin of common bean is from Ecuador and northern Peru. Wild beans dispersed from its origin northwards and southwards and formed two distinct gene pools, the Mesoamerican and the southern Andean. Post-domestication divergence resulted in three domesticated races in each of these two pools (Singh et al. 1991; Broughton et al. 2003). There are more than 30,000 domesticated and over 1,000 wild accessions of common bean housed in the germplasm collection at CIAT, Cali, Colombia (CIAT 2009). Smaller germplasm collections exist also at other institutions around the world, of which the largest is probably the one at the USDA Western Regional Plant Introduction Station at Pullman, WA. In spite of this large reservoir of genetic diversity, genetic base of commercial cultivars is rather narrow, encompassing less than 5% of the available genetic diversity of the species (Broughton et al. 2003). Out of the diversity present in the planted cultivars, only a small portion is present in Africa (Wortmann et al. 1995).

Large genetic variability exists among the common bean varieties, being manifested by a range of seed sizes and colours, tastes, growth habit etc. (CIAT 2009). For example, variation of growth habit in common bean appears to be continuous from determinate bush to indeterminate, climbing types (Singh 1982). Genetic variability of the common beans allows highly versatile use of this plant in different cropping systems, matching local culinary preferences, and local environmental constraints (Broughton et al. 2003). Indeed, large variability of traits related to mineral nutrition, tolerance to drought and other abiotic environmental stresses has been described among different bean lines (Rao 2002; Singh et al. 2003; see also a summary in Table 9.1). In brief: Low N, P and drought tolerances appear to be related to the rooting patterns, rate and duration of nutrient acquisition from the soil,

Table 9.1 Overview of variability among common bean genotypes with respect to their performance and yields under different abiotic environmental constraints

Traits	Screening efforts	Outcome	References
Tolerance to low soil fertility (N and P)	~7,000 lines	Low fertility tolerance mostly among Mesoamerican landraces (Durango, Jalisco) and improved genotypes having similar landraces in their pedigree. Tolerance to low N supply relates to rate and duration of N acquisition, root morphology, efficiency of N use in vegetative and reproductive growth, rate and duration of N accumulation in seeds and efficiency of N use in seed formation. No important variation in N remobilization and partitioning; usually more than 80% of plant N allocated to seeds. Tolerance to low P supply relates to differences in internal P use efficiency, rhizosphere acidification capacity, capacity to efficiently absorb P, shallowness of the root system, root branching and coarseness, and root hairs development. Novo Jalo and Perola are P-efficient genotypes. Bean genotypes usually well responsive to P fertilization. Limited P remobilization from roots under low P, but high under high P, probably due to P sink in the nodules. P efficiency of large seeded Andean cultivars often superior to small seeded Mesoamerican races, which are in turn usually more P responsive. Multi-stress tolerant lines identified	Wolyn et al. 1991; Lynch and White 1992; Lynch and van Beem 1993; Lynch and Rodriguez 1994; Lynch and Beebe 1995; Wortmann et al. 1995; Yan et al. 1995a; Snapp and Lynch 1996; Yan et al. 1996; Aggarwal et al. 1997; Beebe et al. 1997; Nielsen et al. 1999; Araujo and Teixeira 2000; Fageria and da Costa 2000; Ge et al. 2000; Nielsen et al. 2001; Fisher et al. 2002; Rao 2002; Singh et al. 2003; Lunze et al. 2007; Kimani and Tongoona 2008
Tolerance to Al- and Mn-toxicities	~300 lines	Different genotypes resist different levels of Al in soils; highly resistant genotypes such as Dade usually exude more organic acids (e.g. citrate) by the roots and mucigel by border cells than susceptible cultivars like Romano. Others may produce more lateral roots thus better in withstanding Al-induced damage. Large variability in Mn toxicity tolerance amongst the cultivars, but potentially confounded by other stresses if tests done only in soils.In general, Andean genotypes such as ICA Quimbaya found to be more resistant to Al than Mesoamerican genotypes	Miyasaka et al. 1991; Gonzalez and Lynch 1999; Thung and Rao 1999; Mugai et al. 2000; Miyasaka and Hawes 2001; Rao 2002; Shen et al. 2004; Rangel et al. 2005; Rangel et al. 2007, 2009

Tolerance to drought	~700 lines	Some cultivars like SEA5, SEA13, and BAT477 show superior dought tolerance. Cultivars RAB 96 and Carioca were selected as drought tolerant genotypes in some studies, but appeared more sensitive than others like Negro Argel under different field conditions. Drought tolerance is probably linked to differences in rooting patterns and water use efficiency in mobilizing photosynthates to grain. Drought tolerance may be coupled to low P tolerance, symbiotic N fixation and colonization of the roots and surrounding soil by the mycorrhizal fungi. Common bean generally drought-prone compared to very tolerant bean species like tepary bean (*P. acutifolius*). Attempts to introduce this tolerance into common bean germplasm were made at CIAT. Another strategy to cope with drought is to combine drought escape (early or late flowering cultivars) with tolerance mechanisms. Some mechanisms of drought tolerance shared with tolerance to salinity	Zaiter and Mahfouz 1993; Singh 1995; Guimaraes et al. 1996; Ramirez-Vallejo and Kelly 1998; Gomes et al. 2000; Lodeiro et al. 2000; Saadallah et al. 2001b; Singh et al. 2001; Augé et al. 2003; Bouhmouch et al. 2005; Gonzalez et al. 2006a; Munoz-Perea et al. 2007; Beebe et al. 2008; Sassi et al. 2008

and to internal utilization efficiency of the N and P. Tolerance to Al and Mn toxicities appears related to lateral root formation, root exudation and callose formation rates. The knowledge about the variability of traits among the bean lines allowed selection of adapted genotypes for particular environmental conditions and identification of genes responsible for trait expression (Kelly 2004; Blair et al. 2007). These efforts already resulted in release of dozens of improved varieties to and adoption by the farmers within the last decades (Hillocks et al. 2006). These were mainly varieties with improved disease resistance and tolerance to drought and low soil fertility. In combination with advanced breeding approaches (marker assisted selection, genomic sequencing, functional genomics, etc.), current genotype improvement efforts promise possibility of accelerated and specifically targeted breeding of common beans so as to match variety of demands (Miklas et al. 2006; Gepts et al. 2008). Although the structuring of genetic variability in beans is becoming understood, genes or gene clusters responsible for resistance to diseases, tolerance to drought, low pH and low P availability in soils are being identified and breeding strategies and molecular engineering technologies exist for insertion of the desired traits into specific genotypes, other barriers are still limiting widespread release and adoption of the improved cultivars by the farmers. Main limitations are the lack of information about availability of new lines, their benefits and required management measures, and the uncertainty about the anticipated short- and long-term changes in the production systems, which would further focus the breeding efforts. In Africa, these limitations are at least partly outweighed by participatory breeding and distribution of seeds, which have long tradition and established as viable, though less efficient alternative to the global breeding efforts (Broughton et al. 2003; Gepts 2006). For example, it was documented that even forces like civil war in Rwanda could not stop local seed exchanges and maintaining diversity of bean varieties on a local scale (Sperling 2001).

9.2.3 Biological Nitrogen Fixation

Common beans undergo symbiotic interaction with rhizobia capable of fixing atmospheric dinitrogen (N_2). This consists of formation of nodules on host roots, which harbour the bacterial symbionts, and the N_2 fixation process in the nodules, which occurs thereafter. The knowledge about signalling between the plant and the bacterial partners is well advanced, especially when compared to other kinds of symbioses like the mycorrhizas (Hungria and Stacey 1997; Cocking 2003). The molecular dialogue involves flavonoids produced by the plants, nod-factors derived from the bacterial cell walls, lectins in plant cell membranes, and an array of other molecules and signal-transfer pathways, some of which still remain to be characterized (Ferguson and Mathesius 2003; Oldroyd and Downie 2006; Shaw et al. 2006; Garg and Geetanjali 2007). Currently, we are witnessing fast developments in genome sequencing of legumes and rhizobia as well as functional genomics studies, unravelling the relationships between different genes involved in establishment and functioning of the symbiotic nitrogen fixation (Colebatch et al. 2004; Gonzalez et al. 2006b).

The common bean associates with a wide range of rhizobial strains and *Rhizobium etli* is the dominant microsymbiont at both centers of genetic diversification of the beans (Aguilar et al. 2004; Grange et al. 2007). Other rhizobial species such as *R. tropici*, *R. gallicum*, and *R. leguminosarum* were shown to associate with the beans in Colombia, Brazil, Argentina, Egypt, Tunisia, Ethiopia and elsewhere (Mhamdi et al. 1999; Beyene et al. 2004; Shamseldin et al. 2005; Kaschuk et al. 2006a; Stocco et al. 2008). As much as five different rhizobial species were shown to co-colonize common bean roots simultaneously in Spanish soil (Herrera-Cervera et al. 1999). Thus, bean can be classified as a promiscuous host, associating with a range of taxonomically rather distant prokaryotes (Martinez-Romero 2003). Genomic sequencing of *R. etli* indicated positioning of the symbiotic genes clustered within a "symbiotic island" on plasmids (Flores et al. 2005; Gonzalez et al. 2006b). Close relationship was found among symbiotic genes from different strains of rhizobia nodulating common bean, independent of the ribosomal DNA-based classification (Laguerre et al. 2001), although some differences were noted between mechanisms of nodulation by *R. tropici* and other bean-nodulating rhizobia (Broughton et al. 2003; Martinez-Romero 2003). In general, this supports the theory of frequent horizontal genetic transfer of extrachromosomal DNA across rhizobial species, and indicates co-evolution of the beans with the bacteria possessing certain symbiotic gene makeup rather than with taxonomically well-defined lineages (Vinuesa and Silva 2004).

Although there is a high level of similarity within the symbiotic genes among the rhizobial species nodulating common beans, the different strains and species of the prokaryotes may be genetically and metabolically quite different (Martinez-Romero 2003). This predisposes them to occupy different environmental niches and exert differential efficiency in N_2 fixation with different bean genotypes under different situations. In general, extreme conditions of soil pH, temperature, salinity, and moisture, and low nutrient availabilities (except N) are not favourable for the rhizobia (Graham 1981; Gonzalez et al. 2001; Saadallah et al. 2001a; Abdelmajid et al. 2008). Availability of P is recognized to be of primary importance for the nitrogen fixation in beans, because the nodules represent a strong P sink (Tsai et al. 1993; Vadez et al. 1999; Hellsten and Huss-Danell 2001; Drevon et al. 2005). Similarly, carbon costs associated with nodule development and BNF represent an important sink of reduced carbon for the plant, potentially leading to slower plant growth especially upon low light availability (Neves and Hungria 1987). It is known that the BNF activity in beans can be negatively affected by external N supply, and that this relationship is further affected by P availability. For example, upon low (1 mM) nitrate concentration in soil, increasing amount of P promoted both nodule formation and N_2 fixation in beans. However, at higher (10 mM) nitrate concentrations, nodulation and N_2 fixation did not improve with increasing P supply (Leidi and Rodriguez Navarro 2000). N_2 fixation (measured as acetylene reduction rate) and nodule weight in common beans were highest if no N fertiliser was added to the soil, and declined to very low levels at 336 kg N ha^{-1}. The N_2 fixation rate reached maximum values early in the season and declined to low values before seed filling was complete (Buttery et al.

1986). High levels of tolerance to abiotic stresses were found amongst different rhizobial strains. For example, low pH, high temperature and drought-tolerant rhizobia were found in some tropical soils, e.g. in Brazil (Hungria and Vargas 2000; Martinez-Romero 2003), and high salinity and high pH tolerance was described in *R. etli* isolated from Egyptian soils (Shamseldin and Werner 2005; Shamseldin 2008). Drought tolerance of certain rhizobial strains appear to be linked to high trehalose production by the prokaryotes within the nodules (Zacarías et al. 2004; Suarez et al. 2008), conferring drought tolerance onto the associated plants.

As already mentioned above, common bean is usually considered to fix N_2 poorly, being generally more responsive than other legumes to N fertilization (Graham 1981). The beans may also require starter N fertilizer to fully realize the BNF benefits under low soil fertility, particularly if rhizobial inoculants are used (Daba and Haile 2002). However, significant differences in the capacity to support BNF were found among bean cultivars and were shown to be heritable (StClair and Bliss 1991; Wolyn et al. 1991; Bliss 1993a). Some commercial bean lines, usually those of bushy growth habit, have the lowest N fixation efficiency among all legume crops, being far below 50% of plant N derived from the atmosphere (Isoi and Yoshida 1991; Bliss 1993a; Hardarson et al. 1993; Martinez-Romero 2003). This means that the BNF rarely provides adequate N supply to the bean crops during the whole growth cycle (Hardarson and Atkins 2003). The weak nodulation and N_2 fixation rates as compared to other grain legumes have several reasons: (i) Beans are usually grown under marginal soil conditions and high temperatures, with lacking P supply, and in presence of water stress and toxic elements. Such conditions generally suppress nodulation and/or N_2 fixation by the rhizobia (Hungria and Vargas 2000; Jebara and Drevon 2001); (ii) Beans are often grown under conditions not matching those in their areas of origin, where they would preferentially associate with *R. etli*. Since bean is generally a promiscuous legume associating with a range of rhizobial species and strains (Martinez-Romero 2003), it is likely that competition with less efficient rhizobia would impair nitrogen fixation under conditions, which were not conductive for their most efficient microsymbiont (Rao 2002; Martinez-Romero 2003); and (iii) Selection for early flowering and short growth cycle might have impaired N benefits from the BNF due to shortening the vegetative period when the N_2 fixation takes place (Graham 1981; Chaverra and Graham 1992; Thung and Rao 1999). However, it has also been recognized that certain combinations of bean and rhizobial genotypes under certain conditions considerably vary in the extent of N_2 fixation. For example, seed yields of 1–2 t ha^{-1} $year^{-1}$ could be achieved with some bean genotypes and highly efficient rhizobia without N fertilization, and in the absence of other limitations such as high temperatures, low P availability, and drought (Hungria and Neves 1987; Bliss 1993b; Thung and Rao 1999). This means, on one hand, that current bean production rates in large parts of Africa can still be substantially improved by better utilization of BNF, provided appropriate management of other constraints. On the other hand, this also means that for production rates above 2 t ha^{-1} $year^{-1}$, common bean is likely to require N fertilizer inputs anyway. It has also been demonstrated that bean genotypic variability in efficiency of BNF is at least partly linked to adaptation of the different plant genotypes to low

P conditions, particularly by higher internal P use efficiency of the plants (Araujo and Teixeira 2000; Tang et al. 2001; Vadez and Drevon 2001). Although different rhizobia may differ in their metabolic response to low P supply (Al-Niemi et al. 1997; O'Hara 2001), it is not well known whether there is any significant variation in internal P use efficiency among the different rhizobial species and strains, or whether the observed preferences in association of beans with particular rhizobia under low P conditions are mediated by the plants. All in all, satisfying P demands of the plants is of primary importance for optimizing the BNF efficiency and associated yield increases.

Rhizobial inoculation, although having the potential of improving common bean yields and production sustainability under field conditions (Vargas et al. 2000; Guene et al. 2004; Cardoso et al. 2007), is usually not practiced by farmers in Africa and elsewhere. This is somehow in contrast to other legumes like soybeans, which associate with much narrower range of rhizobia, and which generally derive greater benefits from the BNF than the common bean. For example, optimizing BNF in soybean can be achieved either by planting promiscuous varieties supporting greater rhizobial diversity, or by industrial inoculation of the fields. The latter is a multimillion business globally, though less widespread in Africa than in other parts of the world. For common beans, industrial inoculation would suffer from low economic returns (Ndakidemi et al. 2006), lack of information on competitiveness of introduced vs. indigenous rhizobia under field conditions (Hungria et al. 2000; Martinez-Romero 2003), as well as lacking technology, institutions, education, and markets for targeted production, distribution, and application of rhizobial inoculants for common bean. This last point is particularly relevant for Africa, where investments into agricultural innovations are limited because of institutional and market constraints (Sanginga et al. 2007; Jama and Pizarro 2008).

9.2.4 Arbuscular Mycorrhizal Symbiosis

Apart from the association with rhizobia, beans also establish a root association with fungi from the phylum Glomeromycota (Daft and El-Giahmi 1974). The mechanisms of establishment of this symbiosis called arbuscular mycorrhiza (AM) show several parallels to the rhizobial association. This knowledge has been established by studies on symbiotic mutants and on the gene expression and signalling events in the different symbioses involving beans and other legumes such as peas, medic, and *Lotus japonicus* (Shirtliffe and Vessey 1996; Morandi et al. 2005; Gherbi et al. 2008). Although the association with AM fungi is evolutionarily older and more widespread in the plant kingdom than the BNF (Parniske 2008), it is less understood as compared to the rhizobial association, mainly because of unusual (polynuclear) genome organization of the AM fungi, lack of transformation system allowing genetic manipulations of the AM fungi, and their resistance to grow on synthetic media in the absence of a host plant (Marsh and Schultze 2001; Sanders 2004; Jansa et al. 2006).

Under field conditions, roots of common beans are normally colonized by the AM fungi (Yan et al. 1995b), although variability was observed between the plant genotypes in the extent of root colonization as well as in the plant responses to establishment of AM symbiosis (Ibijbijen et al. 1996; Oliveira and Sanders 2000; Hacisalihoglu et al. 2005). However, great variation exists among results of different studies using slightly different approaches, which is quite typical for mycorrhizal research. For example, Mosse and Thompson (1984) identified the bean cultivar Canadian Wonder as poor host of AM fungi, with colonization levels under laboratory conditions with two AM fungal strains remaining below 10% of root length colonized, in contrast to genotype Jamapa, where values above 50% were consistently observed. However, another study on root colonization of the genotype Canadian Wonder under field conditions reported values near 100% of the root length colonized within few weeks after sowing (Oliveira and Sanders 1999). In extreme case, bean mutants have been identified, which proved completely resistant to mycorrhizal colonization (Shirtliffe and Vessey 1996; Cardenas et al. 2006).

Establishment of mycorrhizal symbiosis in beans confers multiple benefits to plant nutrition, growth, disease resistance and drought tolerance (Newsham et al. 1995; Moawad and Vlek 1998). The nutritional aspects of AM symbiosis have frequently been studied in the past in various plants including common beans. AM symbiosis usually confers improved acquisition efficiency to the plants of P and Zn, particularly under soil conditions, when the availability of these elements for plants is low (Giller and Cadisch 1995; Sharma et al. 1999; Rao 2002). These nutritional effects are generally stronger in beans than in grasses (Isobe and Tsuboki 1998), most probably due to the differences in the root system architecture. The acquisition of P and/or Zn by mycorrhizal plants is often exceeding by a factor of 2 or more the acquisition of the same elements by the nonmycorrhizal plants (Ibijbijen et al. 1996; da Silveira and Cardoso 2004; Ortas and Akpinar 2006). This is consistent with previous classification of beans as strongly mycorrhizal-dependent plants due to their better growth in non-fumigated as compared to methyl bromide fumigated soil (Plenchette et al. 1983). Next to the nutritional effects, multitude of other benefits have been attributed to AM symbiosis, such as more efficient utilization of P fertilizers by the plants, improved root development, increased tolerance of plants to drought and to heavy metal pollution (Sieverding 1991; El-Tohamy et al. 1999; Thung and Rao 1999; Augé 2004). The mechanisms behind all these effects are not completely understood, but they seem to be related to improved mineral nutrition of the plants, changes in internal hormonal balances in the plants, relief from stress through changes in soil properties and selective ion uptake from soil solution, immobilization of heavy metals in the fungal biomass as well as promotion of development of associated microbes in the mycorrhizal hyphosphere (Joner et al. 2000; Augé 2004; Joner et al. 2004; Kohlmeier et al. 2005; Hause et al. 2007; He et al. 2007). On the other hand, plants supply the AM fungi with reduced carbon originating from their photosynthesis. This means that the observed growth effects are indeed composite effects of plant performance promotion through mechanisms such as improvement of scarce nutrient uptake combined with associated carbon costs (Lynch and Ho 2005). Thus the overall greatest net benefits of AM symbiosis are usually seen under stressful environment such as low P and/or Zn availability, high Al or heavy metal availability, and drought, this means under situations, where plants would normally invest great fraction of their carbon budget into root development and exudation (Nielsen et al. 2001; Jansa et al. 2006).

In spite of numerous papers on AM symbiosis in beans with respect to mineral nutrition and growth benefits, other four important aspects of mycorrhizal symbiosis in beans received relatively little attention in the past. These are: (1) Interactions between plant, AM fungi, and pathogens; (2) Phenomenon of AM fungi interconnecting plant individuals from the same and different species; (3) Functional diversity in AM fungal communities; and (4) Mechanisms of interactions between plants, AM fungi and rhizobia within so called tripartite symbiosis. We briefly handle the first three points in this section and then we expand on the last point in the following section.

1. Colonization of bean roots by AM fungi may interfere with development of soil-borne pathogens such as *Fusarium solani* (Linderman 1992; Dar et al. 1997). The mechanisms are not precisely known, but competition for space, induced resistance, modification of root and rhizosphere physico-chemical properties and composition of associate microflora upon development of the AM fungus have all been proposed (Dar et al. 1997; Osbourn 2001; Filion et al. 2003). Of particular importance appear to be the induced systemic resistance to pathogens like *Rhizoctonia* due to AM fungal colonization of roots (Cordier et al. 1998; Guillon et al. 2002; Pozo and Azcón-Aguilar 2007). Such a resistance could potentially also be used for "immunization" of plants so as to increase their resistance against devastating foliar pathogens (Fritz et al. 2006).
2. AM fungi are colonizing soil as far as 15 cm beyond the root zone (Jansa et al. 2003a) and are very little specific with respect to the choice of their host plants (Smith and Read 2008). Therefore, mycelium networks of different plant individuals (belonging or not to the same species) may easily overlap in soils. Further, root systems of neighbouring plants can be directly interconnected by a shared mycelium network of the same AM fungal species. This situation is particularly relevant for beans not only because mycorrhizal effects on single plant and on plant stand levels are rather inconsistent and appear strongly modulated by plant density (Facelli et al. 1999; Facelli and Facelli 2002; Li et al. 2008), but also because beans are frequently grown in association with different (mycorrhizal) crop plants such as maize and sugarcane. The possibility of direct transfer of nutrients, water, and carbon between different plant species sharing a common mycorrhizal network has often been mentioned in the literature (Sierra and Nygren 2006; Wilson et al. 2006; Egerton-Warburton et al. 2007), but unequivocal experimental evidence for such transfers is virtually missing (Pfeffer et al. 2004; Whitfield 2007; Voets et al. 2008). Therefore, rather than direct transfer of compounds between the different plants, it seems that processes like diverting mycorrhiza-mediated fluxes of mineral nutrients between soil and plants acting as different sinks on a common supply line should be more carefully studied (Whitfield 2007).
3. Arbuscular mycorrhizal fungi are usually present as multispecies communities in soils and often more than one species is present in a single root system of the host

plants under field conditions (Jansa et al. 2003b). This is in contrast to most experiments, usually scrutinizing the effects of a single strain of AM fungus alone on plant growth, nutrition, and disease tolerance. We know that different AM fungal strains and species differ in their growth patterns, efficiency to acquire P from different distances from roots and/or other features (Jakobsen et al. 1992; Munkvold et al. 2004; Jansa et al. 2005). However, by simplifying the plant-AM fungal diversity of our experimental systems beyond the levels encountered in nature, we have missed possible interactions between members of a composite fungal community such as underground diversity facilitation (Maherali and Klironomos 2007) or functional complementarity (Koide 2000; Jansa et al. 2008). These phenomena deserve further attention, due to their direct relevance to the field reality. Moreover, it has been recognized that specific bacterial community does associate to the soil mycelium of the AM fungi, that these communities depend on the identity of the "carrier" AM fungus, and that these additional players may be important for soil nutrient release and other underground processes (Johansson et al. 2004; Artursson et al. 2006; Toljander et al. 2006). It would be very interesting to test whether different species compositing an AM fungal community differ not only in niche partitioning but also in functional partitioning, e.g. one species delivering P, another Zn, and another conferring disease resistance to soilborne pathogens. Since various agricultural practices such as tillage, crop rotation, and fertilization have been shown to affect the composition of the AM fungal communities in soils, it can be anticipated that these changes will also affect mycorrhizal functioning in beans and other host plants (Rosemeyer and Gliessman 1992; Jansa et al. 2006). There is a great need to quantify these changes now. Symbiotic mutants may play an important role here, especially if it would be possible in the future to obtain symbiotic mutants of beans retaining their capacity to establish association with rhizobia, but not with the AM fungi (i.e. nod^{-}myc^{+} phenotype).

So far, no particular breeding efforts have been undertaken to maximize mycorrhizal benefits to the beans, neither by the native AM fungal communities nor by industrial inoculant strains. Nevertheless, it is possible that screening for low-P tolerance might have unintentionally identified genotypes, which benefited from the AM symbiosis more than others – although evidence for this remains equivocal (Yan et al. 1995b). Positive growth responses of beans to mycorrhizal inoculation in non-sterile soil have been described (Daniels-Hylton and Ahmad 1994). However, no industrial inoculation by AM fungi is yet applied to common beans at any larger scale in the world to the best of our knowledge. This is because of lacking low-cost technology for production and application of AM fungal inoculum and uncertainties about competitiveness of the industrial AM fungal strains with native AM fungal communities.

9.2.5 Tripartite Symbiosis

The signalling pathways activated during establishment of both rhizobial and mycorrhizal symbioses by leguminous plants show high level of parallelism and interdependency (Stracke et al. 2002; Gherbi et al. 2008; Kosuta et al. 2008), indicating common evolutionary origin of the different symbioses (Parniske 2008). Since some sort of nodules occur also in podocarps, which do not accommodate any N_2 fixing prokaryotes, but the AM fungi (Baylis 1969; Russell et al. 2002), it is possible that nodulation presents an ancient feature of plants to interact with symbiotic microbes, and that this only later adapted for accommodation of the rhizobia. Further, nutrient and carbon balances and growth effects on the plants depend on identity and activity of both kinds of microsymbionts in their roots (Mortimer et al. 2008). This advocates that the interaction between plants, rhizobia, and AM fungi shall not be regarded in separation from each other, as it frequently happened in the past due to historic reasons, but that efforts shall be made to study this three-member (tripartite) symbiosis as one unit. Clearly, this means that a higher level of internal dynamics and regulatory processes will need to be considered and that additional nutrients and carbon fluxes (Fig. 9.1) will need to be quantified and understood as compared to any two-member symbiosis. However, quite remarkable amount of empirical knowledge was accumulated already about nutrient and carbon balances and mutual interactions within a tripartite symbiosis of different legumes, and some general patterns begun to emerge recently. Here we briefly review the state-of-the-art and attempt to disentangle the interdependencies between the different nutrients and carbon fluxes in bean tripartite symbiosis.

Both rhizobia and AM fungi require reduced carbon compounds originating from the plant photosynthesis to cover their energy and construction needs. Each of the microsymbionts appears to consume up to 25% of plant photosynthates (Paul and Kucey 1981; Kucey and Paul 1982; Snellgrove et al. 1982; Harris et al. 1985; Jakobsen and Rosendahl 1990) and their combined effect appears to be a sum of those figures. This means that the rhizobia and AM fungi compete for a large fraction of plant carbon. On the other hand, some evidence suggests that plant may increase its photosynthetic rates so as to compensate for the carbon supplied to its root symbionts (Harris et al. 1985; Wright et al. 1998). It has also been shown that such stimulation of photosynthetic rates negatively correlates with plant nutrient status (Jia et al. 2004), which may be the primary cause of downregulation of the symbioses upon sufficient plant nutrient status (Schulze 2004; Bittman et al. 2006; Valentine and Kleinert 2007). However, it is not known, which microsymbiont presents the dominant carbon sink, how the carbon is partitioned in the tripartite system, and how the fluxes are regulated. Recently, temporal partitioning has been suggested between AM fungal and rhizobial symbioses developing on the same bean roots, with the fungi being a major carbon sink from the early stages of plant growth, and nodules increasing their importance later in the ontogeny (Mortimer et al. 2008). Obviously, more information need to be collected on the carbon fluxes in the tripartite symbiosis, with particular attention to fine temporal and spatial resolutions, and looking at the effects of genotypic variation of plants, rhizobia, and AM fungi, as well as how the processes were influenced by changing environmental conditions (Zhang et al. 1995; Haase et al. 2007).

In exchange to carbon, the AM fungi provide to the plant P and other mineral nutrients such as Cu and Zn taken up from the soil. The efficiency of P supply via the AM fungi depends on availability of P in the soil as well as on intrinsic properties of the AM fungal species involved (see above). A significant portion of the P is immobilized in the nodules, where the P concentration may exceed several fold the P concentration in roots and shoots (Israel 1987; Al-Niemi et al. 1998; O'Hara 2001). Thus, affecting plant P acquisition, the mycorrhizal symbiosis plays an important role in enabling the rhizobial symbiosis to develop and function. Whether there are direct (not plant-mediated) interactions between the mycorrhizal fungi and the rhizobia, is not yet known – although some investigations showing unique AM fungal communities inhabiting legume nodules indicate this possibility (Scheublin et al. 2004). Nitrogen fixation by the bacteroids inhabiting root nodules is dependent almost exclusively on the nitrogen fixing capacity of the prokaryotes. Atmospheric dinitrogen is nearly always present in excess, and the fixation process is then governed by physiological state of the bacteroids, mineral nutrients and energy supply from the plants, as well as by removal of the reduced nitrogen from the nodules (Hartwig and Trommler 2001; Schulze 2004). Since the AM fungi are likely to obtain N for covering their own structural and metabolic demands from the soil (Hawkins and George 1999; Jin et al. 2005), it is not obvious, whether there could be any direct interactions between the rhizobia and the AM fungi, apart from some limited competition for root space to colonize. For example, Baird and Caruso (1994) reported about different appearance of nodules in bean plants colonised or not with AM fungi. The plants inoculated with both symbionts formed nodules in large clusters as opposed to the single or loosely clustered nodules that formed on plants inoculated only with the rhizobia. Certainly, most of the interactions between the two microsymbionts are mediated through the host plant – which has to supply its associated microbes with carbon, but which also derives multiple benefits and can perform better under nutrient limiting or otherwise stressful conditions (Izaguirre-Mayoral et al. 2000; Ruiz-Lozano et al. 2001).

Indeed, the literature is full of examples of additive or synergistic effects of establishment of tripartite symbiosis on growth and yields of different leguminous plants including common bean (Bethlenfalvay et al. 1987; Vejsadová et al. 1988; Azcón and Barea 1992; El Ghandour et al. 1996; Ibijbijen et al. 1996; Requena et al. 1997). For example, the bean plants inoculated with both rhizobia (mixture of *R. leguminosarum* and *R. tropici*) and AM fungi (*Glomus* sp.) produced significantly more biomass and greater pod yields compared to non-inoculated control and to singly inoculated plants (Aryal et al. 2003). It has also been noted that dual inoculation of the beans with rhizobia and AM fungi resulted in greater relative benefits if nutrients were supplied as organic in contrast to mineral fertilizers (Aryal et al. 2003, 2006). This has been explained by absence of inhibition of the BNF and AM fungal activities in organically fertilized plants as compared to the soils amended with mineral fertilizers, because the latter increased the immediate N and P availabilities in the soil thus mediating suppression of the symbioses by the plants (Aryal et al. 2006). On the other hand, great variability of responses was observed between different genotypes of beans combined with different strains of AM fungi and rhizobia (Daniels-Hylton and Ahmad 1994; Ibijbijen et al. 1996; Neeraj 2005). For example, bean cultivars Negro Argel and Rio Tibagi in association with rhizobia derived greater P uptake benefits from association with *Glomus manihotis* than with

G. etunicatum, whereas such differences were not evident in Carioca. However, these differences did not necessarily translate into differences in growth responses, probably due to differential carbon requirements by the different AM fungi (Ibijbijen et al. 1996). These results indicate that the best performing and most yielding combination might need to be selected for each bean genotype, probably also with regards to the target environmental conditions (Ahmad 1995). This is because different genotypes of beans as well as different genotypes of AM fungi and rhizobia do likely have different ecological optima for their performances. Moreover, the different microsymbionts will also be differentially affected by agricultural practices such as tillage and pesticide application (Zahran 1999; Jansa et al. 2006; Laguerre et al. 2006). Thus, optimizing the bean symbioses for a particular agroecosystem will certainly be a great challenge of the future research.

9.3 Improving Yield of Beans in Stressful Environments

The agriculture in Africa is confronting major challenges, facing the ongoing human population growth and its aspiration for sufficient food quantity, healthy diet and greater standard of living on one side, and stagnating crop yields, soil degradation, climatic change and raising uncertainties about future rainfall distribution on the other side (Lal and Singh 1998; de Wit and Stankiewicz 2006; Battisti and Naylor 2009). There are few possibilities how to improve agricultural production in general and the production of common bean in particular, and to increase the sustainability of agriculture in the tropics, particularly in the sub-Saharan Africa. It is possible by: (i) Improving soil management practices, e.g. using better fertilization strategies; (ii) Modifications of the cropping systems; (iii) Using improved crop cultivars; (iv) Exploitation of the potential of soil microorganisms establishing symbioses with plant roots; or (v) Combination of these different options (Singh 1999a; Broughton et al. 2003; Kelly 2004). In this section, we present potential measures requiring one- or two-step modifications of the current practices. These should be relatively easy to apply and would positively impact the production system, although their implementation may take certain time and require additional research efforts. In the following section, some deeper changes are proposed, which would generally require several modifications of the production systems or major paradigm shifts, but shall offer longer-term and more sustainable perspectives.

9.3.1 *Management of Soil Fertility*

There is a great need to increase agricultural productivity in large areas of tropical Africa, due to enormous population pressure and inherent low fertility of local soils. If the human food demands in these areas are to be met by local production, dramatic increases in agricultural productivity must be realized now (Lal 2007).

This is not possible without fertilization of the crops and without management of soil organic matter. Although fertilizers are sometimes applied in sub-Saharan Africa (Vanlauwe and Giller 2006), their quality and applied amounts rarely match the needs. Therefore, fertilizer application shall be greatly increased and go hand-in-hand with improvement (or at least slowdown of degradation) of soil organic matter status. This applies invariantly for production of any crops including the common bean. Although N demand of beans is lower than that of cereals since part of it can be covered from the BNF, bean production can still be greatly improved by application of N fertilizers (see above). Demand of beans for P fertilization is at least as high as for other crops, and low P availabilities in the soils seem to be one of the main limitations of bean production in sub-Saharan Africa. However, little knowledge is available about fertilizer demands of cropping systems others than the monocrops. This is very relevant for bean production since it is most often grown in association with other crops. Additionally, blanket fertilizer recommendations may be of limited relevance in smallholder farms, which experience a great heterogeneity in soil types and past managements (Zingore et al. 2007).

Phosphate rocks and crude oil and natural gas reserves (representing the important non-renewable resources for production of mineral P and N fertilizers) are abundant in sub-Saharan Africa, yet this does not seem to relieve the local farmers from the scarcity of high quality and financially affordable fertilizers. Additionally, many of the rock phosphates available locally show very low water solubility, which limits their direct use in agriculture (Vanlauwe and Giller 2006). Often mentioned organic fertilizers appear to be a good solution for improving soil fertility, replenishing nutrients removed from the fields with previous crops as well as enhancing soil organic matter status (Kayuki and Wortmann 2001). However, production of organic fertilizers demands both space and labour and in the end will not solve all the problems because of relocation rather than replenishment of losses of nutrients such as P and K at the farm or landscape levels (Vanlauwe and Giller 2006). Further, introducing organic materials to the soils may result in termite attraction to the fields (Bünemann et al. 2004) and in subsequent increase of the risk of crop losses due to higher pest pressure. Therefore, integrated soil fertility management is necessary to accompany system development on its way to sustainable productivity increase (Bationo et al. 2007; Frossard et al. 2009). One of the measures, which are recently being adopted to combat soil nutrient mining, is the application of small amounts of (mainly) locally produced fertilizers in bands below seeding rows or directly into the seeding holes, so called strategic fertilizer inputs. Although this usually does not have long-lasting effects on soil fertility, it increases agricultural productivity and optimizes fertilizer use efficiency at moderate costs, though with relatively high labour investments (Woomer et al. 2003; Frossard et al. 2007; Moody et al. 2008; Sokoto and Singh 2008). Application of the fertilizer in deeper horizons may confer additional benefits in drought-prone environments since the upper soil layer may be too dry for the plants to access nutrients in it during parts of the growing season (Du Toit 1997).

9.3.2 *Modifications of the Cropping Systems*

For various reasons (culture, tradition, insurance of yield, diversification of diet, control of pests, water and nutrient utilization efficiency), common bean will most probably continue to be a key component of intercropping systems rather than a sole crop in Africa (Francis and Sanders 1978; Daellenbach et al. 2005; Tamado et al. 2007). Careful selection of bean cultivars and planting times to minimize interferences between the combined crops is needed, and this requires good understanding of crop physiology and its nutrient demands (Graham and Ranalli 1997; Daellenbach et al. 2005). Efforts should be made to diversify cropping sequence by including vegetables such as cabbage or short (improved) fallows for animal feed production and soil fertility restoration. Not only this could disrupt cycles of certain pathogens, but it could also improve utilization of local phosphate rocks as shown previously (Weil 2000). Some legume fallow species may mediate significant N inputs to the system, unlike common beans (Chikowo et al. 2007), as well as they may contribute to substantial improvements of the soil organic matter content (Mando et al. 2005). However, different sites and production systems will certainly opt for different solutions. This means that understanding of each particular system is inevitable for proposing new options (Franke et al. 2008).

9.3.3 *Improvement of Plant Genotype*

Since only a small fraction of the common bean genetic variability is currently present in the commercial lines, a significant potential still exists to develop new lines or to introduce new traits into currently cultivated genotypes (Singh 2001). Furthermore, novel traits can also be introduced from closely related botanical species by traditional breeding methods (Rodriguez-Uribe and O'Connell 2006; Marquez et al. 2007) or from virtually any other organism by genetic engineering approaches (Veltcheva et al. 2005; O'Kennedy et al. 2006). These options shall be used to widen the genetic base of future bean varieties in Africa. Past breeding efforts greatly improved the resistance of beans against various pathogens. It is desirable to further enhance widespread adoption of the improved genotypes by the farmers and to improve spreading information about their availability, benefits, and requirements. Further efforts shall continue to search for and implement alternative resistance mechanisms to viral, fungal and/or bacterial diseases, since it is expected that current resistance mechanisms will be overcome by the pathogens at some point (Laine and Tellier 2008). Development of genotypes adapted to low soil fertility should also be given high priority now. However, in the long term, their importance is limited for two reasons – first, nutrient mining must be stopped in Africa to prevent nearing system collapse, and, second, it seems that high internal use efficiency of nutrients does not necessarily translate to high yields as the production of seeds requires quite a narrow range of nutrient concentrations per unit of weight of seeds.

Furthermore, it appears that tolerance to low soil fertility is not necessarily determined by a single or a few genes, as is often the case in disease resistances, and that the genetic determinants of the tolerance to low fertility are often not known. Rather, secondary markers are used such as root growth pattern and BNF efficiency for selection of lines adapted to low fertility soils (Araujo and Teixeira 2000). Genotypes adapted to high temperatures and soil acidity and Al and Mn toxicities will probably be important at both short and long-term, since they do perform better in many places in Africa suffering from these constraints. These adaptations may also support greater symbiotic activity in the roots due to higher root vitality and generally better plant performance and photosynthesis under stressful conditions (Hungria and Vargas 2000; Miyasaka and Hawes 2001).

Breeding for drought tolerance is a major challenge. Although it is clear that selection for tolerance to drought will increase yields in areas with limited rainfall, as well as it may improve performance of the beans in low P soils (Beebe et al. 2008), breeding efforts alone could not solve the problem of large environmental variations such as catastrophic droughts and extremely erratic weather. These events will, according to current climate models, become more frequent in some parts of Africa due to climatic changes, and they will probably remain largely non-predictable (Phillips and McIntyre 2000; de Wit and Stankiewicz 2006; Battisti and Naylor 2009). Moreover, pronounced changes in growth habit and/or plant water and nutrient efficiency may impose a trade-off (a penalty) on the yields and/or disease resistance (Haugen et al. 2008). Therefore, it is important to promote drought tolerant bean varieties that also yield well under a range of water and nutrient supply conditions and where other beneficial traits (such as resistance to diseases) are not sacrificed.

There is very little done with respect to breeding for greater symbiotic benefits in the common beans, in spite of the fact that a great impact of lines with enhanced BNF efficiency is anticipated for the resource poor farmers (Herridge and Rose 2000; Crouch et al. 2004). Although some variability in BNF efficiency was identified among different common bean genotypes, very few lines derived more than 50% of their N from the atmosphere (see above). Therefore, the potential to improve BNF in beans is still high. In some legume species (pea, soybean, barrel medic, *Lotus*), so called super – or hypernodulating mutants were identified, which showed very high fixation rates of atmospheric N_2 being unaffected by mineral nitrogen availability in soils (Fujikake et al. 2003; Schnabel et al. 2005; Ishikawa et al. 2008). Hypernodulating genotypes are not yet known for common bean. Compared to the BNF, even less is understood about genetic basis of variation in benefits derived from the AM symbiosis in the common bean, and to the best of our knowledge, there have been no breeding efforts to improve mycorrhizal symbiosis in beans. For any future breeding programs, it is highly recommended to check whether the rates of root colonization by AMF and nodulation as well as the symbiotic benefits are not decreased as a side effect of the breeding/genetic modifications. This appears relevant as gradual decrease in mycorrhizal responsiveness was described for modern as compared to old wheat and soybean cultivars (Khalil et al. 1994; Zhu et al. 2001), indicating that plant genotype improvement efforts may also bring along some unintended damage of the genetic base. It has to be noted here that there is a potential to speed up the progress in targeted breeding for greater symbiotic benefits now as concerted research efforts resulted in establishment of a large population of induced mutants of common beans. This resource will allow identifying genetic base of certain symbiosis-related traits (Blair et al. 2007) and establish genetic markers for the marker assisted selection.

Biofortification efforts may prove very important for the increase of nutritional quality of the beans for human consumptions. Ongoing research aims at identifying routes to enhance N, Zn, Fe, and vitamin contents in the seeds and the availability of these elements to humans (Donangelo et al. 2003; Blair et al. 2009). On the other hand, the concentration of P in the seeds shall be kept moderate to low as the major form of P in the seeds is phytic acid, which may interfere with resorption of valuable nutrients and other compounds in human gut (Frossard et al. 2000; Donangelo et al. 2003).

9.3.4 *Enhanced Symbiotic Benefits*

Apart from limited short-term perspectives in adapting the plant genotype for enhanced symbiotic benefits (e.g. super – or hypernodulating mutants) as discussed above, there are two major ways to improve the performance of the root symbioses established by the beans. One is changing the microbial communities (either by inoculation or indirectly by system management), and another is increasing the efficiency of the current symbionts by optimizing the environmental conditions. The latter may encompass simple measures such as using starter urea fertilizer to improve nodulation and BNF (Daba and Haile 2002; Mendes et al. 2003), using organic instead of mineral fertilizers to limit inhibition of the BNF and AM fungal symbioses (Aryal et al. 2006), or application of moderate P fertilizer levels to crops growing in severely P deficient soils to enhance mycorrhizal benefits (Picone 2002; Jansa et al. 2006).

No microbial inoculants are yet broadly applied in bean production. This is in spite of the facts that bean-associated rhizobia are relatively easy and cheap to produce, and also a growing number of companies is offering AM inoculum with broad plant-species specificity (Vosátka et al. 2003). Research results indicate potential of inoculation to improve plant nutrition and yields upon successful introduction of selected microbes into the rhizosphere, especially if combined in synergistic consortia (Requena et al. 1997; Lekberg and Koide 2005; Tilak et al. 2006). Apart from documented nutritional benefits (see above), introducing trehalose-overproducing rhizobia into bean rhizosphere could confer drought tolerance to the plants (Suarez et al. 2008). By enhancing development of AM fungi, soil aggregate stability could improve leading to reduction of soil erosion rates (Six et al. 2004; Cavagnaro et al. 2006). However, application of inoculants may prove to be a tricky task, and there is still only a limited mass of reliable data on survival and competitiveness of the introduced inoculants with the native ones in the soil, particularly with respect to the AM fungi (Hungria et al. 2000; Sessitsch et al. 2002; Moawad et al. 2005; Jansa et al. 2006; Bogino et al. 2008).

The consensus appears to be that symbiotic efficiency is not necessarily linked to competitiveness of the different microbial strains, as shown on example of rhizobia: The most efficient strains may not be able to survive in field soil, whereas less efficient strains (in terms of BNF efficiency) may be greatly competitive (Hafeez et al. 2001). Therefore, for industrial inoculum application, it appears imperative to select microbes, which are both highly beneficial and competitive under field conditions. Furthermore, securing inoculum quality and disentangling effects of microbes from that of inoculum additives is not an easy task (Catroux et al. 2001; Date 2001; Vosátka et al. 2003; Deaker et al. 2004). Thus the economic viability of inoculant technology in bean production shall be carefully scrutinized and compared, without preoccupations, to that of other measures such as application of mineral fertilizers (Wander et al. 2007).

Any modification to production system management is another interesting option for manipulation of indigenous microbial communities for improving the crop production. It is known that inputs of organic matter to the soil or change of tillage management may have pronounced effects on composition and activity of soil microbes including rhizobia and the AM fungi (Kiers et al. 2002; Jansa et al. 2003b, 2006; Kaschuk et al. 2006b; de Fatima et al. 2007). However, the role of the different symbiotic microbes in plant nutrition in these different agricultural management systems is very rarely quantified and separated from the other confounding effects (water availability, soil temperature, root growth, nutrient availability in soil solution etc.). This remains to be a major research gap, which shall be best closed by using well characterised bean genotypes (Jebara and Drevon 2001; Saadallah et al. 2001b) in combination with a selection of functionally diverse AM fungal and rhizobial strains (Hungria et al. 2000; Jansa et al. 2005, 2008). Such studies should also incorporate bean mutants deficient in AM and/or rhizobial symbioses formation. Using isotope labelling (either as mono – or multitracer) and spatially compartmented containers would allow direct observation of the role of AM fungi in the P, Zn, and Cu nutrition, for example (Joner et al. 2004, 2005). A second step would be identification of the diversity and activity of AM fungal and rhizobial communities in the bean-producing agroecosystems in Africa. Only then, the necessary background knowledge about the importance of root symbioses for bean production in the region would allow the decisions about potential of managing the microbial communities for improved bean production. As stressed by Vance (2001), the use of AM fungi and rhizobia in agroecosystems will be fully realised only as we: (1) Ascertain how the diversity of specific microorganisms contributes to growing crops; (2) Identify the biochemical and genetic mechanisms regulating nutrient exchange; and (3) Determine, whether there is a yield penalty for effective symbiosis.

9.4 Concluding Remarks, Challenges and Perspectives

Great challenges are approaching global agriculture in general and that in Africa in particular in the twenty-first century. Global agriculture will have to provide sufficient amounts of food for the still-growing human population on Earth and fulfil the higher food quality demands, using limited land and declining non-renewable resources.

The failure in progress to fulfil UN Millennium development goals in Africa speaks for itself (Lal 2007; UN 2008). Better utilization of bean symbioses for sustainable and high bean yields are promising, though long-term perspectives with a lot of research still to be conducted. Realization of this outlook will require, at least, moderate P inputs, best in form of organic fertilizers. It is at least theoretically possible to achieve development of highly mycorrhizal – and BNF-dependent cultivars and efficient and cost-effective inoculation strategies with superior microbial strains, both in term of symbiotic benefits and in terms of competitiveness with indigenous strains, adapted to site-specific conditions. Therefore, plant symbioses deserve much research attention, although it is premature to make any promises now. Additionally, availability of water will be critical in the next decades in many places in Africa (Phillips and McIntyre 2000), and options to irrigate cropped fields shall be carefully considered, regardless of the current economy or political situations. Livestock production in Africa shall be promoted so as to improve quality and diversification of human nutrition, provide greater income to the farmers, and increase availability of manure for soil fertilization. As a consequence, some land used these days for subsistence cropping could be converted to mixed crop-livestock systems, which would offer diverse options for management of soil fertility, plant diseases, and combating soil degradation and erosion (Rutunga et al. 2007). Furthermore, nutrients contained in human excreta shall also be consequently recycled and their losses prevented. Utilization of human excreta is already practiced in some parts of Africa such as in peri-urban agriculture in Burkina Faso (Eaton and Hilhorst 2003), but high sanitary risks are currently associated with these practices. These should be addressed so as to prevent disease outbreaks. Very long-term perspectives of bean production would encompass changing soil conditions by application of charcoal. It has been shown that charcoal additions to the soil may benefit BNF and bean yields (Rondon et al. 2007). This fascinating and ancient avenue to improve soil fertility (Glaser 2007) is currently attracting lot of attention, although its obvious limitation in the modern times is the availability of sufficient biomass quantity for charcoal production (Sunseri 2005). However, large amounts of plant biomass may potentially be available in the context of climate change remedial measures in a way the developed countries compensating for their CO_2 footprint would provide biochar for soil amendments in the tropics. If carried on a large enough scales, this would bring along benefits to both developing and developed countries.

It is not quite clear what are the main reasons for the contemporary widespread adoption of common bean in Africa in spite of the fact that it is an alien plant species to the continent. Native legumes such as peas or lentils are not produced in any comparable amounts. However, these native alternatives could be considered as potential replacement for beans shall the efforts to adapt the bean production for future environmental situations fail. Furthermore, other income sources for local populations shall be identified and supported such as tourism and diversity conservation, with accompanied financial benefits, lowering dependency of local human populations on the subsistence farming (Huston 1993; Gadd 2005). It is clear that these perspectives are closely related to political stability and significant infrastructure development. Although these are not easy to achieve in any African country, as documented by the recent riots in Kenya, Madagascar and elsewhere, they are truly within the capacity of the local governments and must be solved solely by them.

References

Abdelmajid K, Karim BH, Chedly A (2008) Symbiotic response of common bean (*Phaseolus vulgaris* L.) to iron deficiency. Acta Physiol Plant 30:27–34

Aggarwal VD, Mughogho SK, Chirwa RM et al (1997) Field-based screening methodology to improve tolerance of common bean to low-P soils. Commun Soil Sci Plan 28:1623–1632

Aguilar OM, Rival O, Peltzer E (2004) Analysis of *Rhizobium etli* and of its symbiosis with wild *Phaseolus vulgaris* supports coevolution in centers of host diversification. P Natl Acad Sci USA 101:13548–13553

Ahmad MH (1995) Compatibility and coselection of vesicular-arbuscular mycorrhizal fungi and rhizobia for tropical legumes. Crit Rev Biotechnol 15:229–239

Al-Niemi TS, Kahn ML, McDermott TR (1997) P metabolism in the bean *Rhizobium tropici* symbiosis. Plant Physiol 113:1233–1242

Al-Niemi TS, Kahn ML, McDermott TR (1998) Phosphorus uptake by bean nodules. Plant Soil 198:71–78

Anderson GD (1974) Bean responses to fertilizers on Mt. Kilimanjaro in relation to soil and climatic conditions. East Afr Agric For J 39:272–288

Araujo AP, Teixeira MG (2000) Ontogenetic variations on absorption and utilization of phosphorus in common bean cultivars under biological nitrogen fixation. Plant Soil 225:1–10

Araujo AP, Teixeira MG (2003) Nitrogen and phosphorus harvest indices of common bean cultivars: Implications for yield quantity and quality. Plant Soil 257:425–433

Artursson V, Finlay RD, Jansson JK (2006) Interactions between arbuscular mycorrhizal fungi and bacteria and their potential for stimulating plant growth. Environ Microbiol 8:1–10

Aryal UK, Xu HL, Fujita M (2003) Rhizobia and AM fungal inoculation improve growth and nutrient uptake of bean plants under organic fertilization. J Sustain Agr 21:29–41

Aryal UK, Shah SK, Xu HL et al (2006) Growth, nodulation and mycorrhizal colonization in bean plants improved by rhizobial inoculation with organic and chemical fertilization. J Sustain Agr 29:71–83

Augé RM (2004) Arbuscular mycorrhizae and soil/plant water relations. Can J Soil Sci 84:373–381

Augé RM, Moore JL, Cho KH et al (2003) Relating foliar dehydration tolerance of mycorrhizal *Phaseolus vulgaris* to soil and root colonization by hyphae. J Plant Physiol 160:1147–1156

Azcón R, Barea JM (1992) Nodulation, N_2 fixation (^{15}N) and N-nutrition relationships in mycorrhizal and phosphate-amended alfalfa plants. Symbiosis 12:33–41

Baird LM, Caruso KJ (1994) Development of root-nodules in *Phaseolus vulgaris* inoculated with rhizobium and mycorrhizal fungi. Int J Plant Sci 155:633–639

Bationo A, Waswa B, Kihara J et al (eds.) (2007) Advances in integrated soil fertility management in sub-Saharan Africa: challenges and opportunities. Springer, Berlin

Battisti DS, Naylor RL (2009) Historical warnings of future food insecurity with unprecedented seasonal heat. Science 323:240–244

Baylis GTS (1969) Mycorrhizal nodules and growth of *Podocarpus* in nitrogen-poor soils. Nature 223:1385–1386

Beebe S, Lynch J, Galwey N et al (1997) A geographical approach to identify phosphorus-efficient genotypes among landraces and wild ancestors of common bean. Euphytica 95:325–336

Beebe SE, Rojas-Pierce M, Yan XL et al (2006) Quantitative trait loci for root architecture traits correlated with phosphorus acquisition in common bean. Crop Sci 46:413–423

Beebe SE, Rao IM, Cajiao C et al (2008) Selection for drought resistance in common bean also improves yield in phosphorus limited and favorable environments. Crop Sci 48:582–592

Bethlenfalvay GJ, Brown MS, Mihara KL et al (1987) Glycine-Glomus-Rhizobium symbiosis. 5. Effects of mycorrhiza on nodule activity and transpiration in soybeans under drought stress. Plant Physiol 85:115–119

Beyene D, Kassa S, Ampy F et al (2004) Ethiopian soils harbor natural populations of rhizobia that form symbioses with common bean (*Phaseolus vulgaris* L.). Arch Microbiol 181:129–136

Bittman S, Kowalenko CG, Hunt DE et al (2006) Starter phosphorus and broadcast nutrients on corn with contrasting colonization by mycorrhizae. Agron J 98:394–401

Blair MW, Porch T, Cichy K et al (2007) Induced mutants in common bean (*Phaseolus vulgaris*), and their potential use in nutrition quality breeding and gene discovery. Isr J Plant Sci 55:191–200

Blair M, Astudillo C, Grusak M et al (2009) Inheritance of seed iron and zinc concentrations in common bean (*Phaseolus vulgaris* L.). Mol Breeding 23:197–207

Bliss FA (1993a) Breeding common bean for improved biological nitrogen-fixation. Plant Soil 152:71–79

Bliss FA (1993b) Utilizing the potential for increased nitrogen-fixation in common bean. Plant Soil 152:157–160

Bogino P, Banchio E, Bonfiglio C et al (2008) Competitiveness of a *Bradyrhizobium* sp. strain in soils containing indigenous rhizobia. Curr Microbiol 56:66–72

Bouhmouch I, Souad-Mouhsine B, Brhada F et al (2005) Influence of host cultivars and *Rhizobium* species on the growth and symbiotic performance of *Phaseolus vulgaris* under salt stress. J Plant Physiol 162:1103–1113

Broughton WJ, Hernandez G, Blair M et al (2003) Beans (*Phaseolus* spp.) – model food legumes. Plant Soil 252:55–128

Bünemann EK, Smithson PC, Jama B et al (2004) Maize productivity and nutrient dynamics in maize-fallow rotations in western Kenya. Plant Soil 264:195–208

Buruchara R (2009) Bean seed production in Rwanda: one farmer's story. Available via http://pabra.org/project07.html. Accessed 6 Jan 2009

Buttery BR, Park SJ, Findlay WI (1986) Growth and yield of white bean (*Phaseolus vulgaris* L.) in response to nitrogen, phosphorus and potassium fertilizer and to inoculation with *Rhizobium*. Can J Plant Sci 67:425–432

Campo RJ, Hungria M (2004) Sources of nitrogen to reach high soybean yields. In: Moscardi F, Hoffmann-Campo CB, Saraiva OF et al (eds.) Proceedings of the 7th world soybean research conference, 6th International Soybean Processing and Utilization Conference, and 3rd Congresso Brasileiro de Soja, Foz do Iguacu, Brazil, 29 February – 5 March 2004, pp 1275–1280

Cardenas L, Aleman E, Nava N et al (2006) Early responses to Nod factors and mycorrhizal colonization in a non-nodulating *Phaseolus vulgaris* mutant. Planta 223:746–754

Cardoso EJBN, Nogueira MA, Ferraz SMG (2007) Biological N_2 fixation and mineral N in common bean-maize intercropping or sole cropping in Southeastern Brazil. Exp Agr 43:319–330

Catroux G, Hartmann A, Revellin C (2001) Trends in rhizobial inoculant production and use. Plant Soil 230:21–30

Cavagnaro TR, Jackson LE, Six J et al (2006) Arbuscular mycorrhizas, microbial communities, nutrient availability, and soil aggregates in organic tomato production. Plant Soil 282:209–225

Chaverra MH, Graham PH (1992) Cultivar variation in traits affecting early nodulation of common bean. Crop Sci 32:1432–1436

Chikowo R, Mapfumo P, Leffelaar PA et al (2007) Integrating legumes to improve N cycling on smallholder farms in sub-humid Zimbabwe: resource quality, biophysical and environmental limitations. In: Bationo A, Waswa B, Kihara J et al (eds.) Advances in integrated soil fertility management in sub-Saharan Africa: challenges and opportunities. Springer, Berlin, pp 231–243

CIAT (2009) Common bean: the nearly perfect food. Available via www.ciat.cgiar.org/ciatinfocus/beans.htm. Accessed 6 Jan 2009

Clark EA, Francis CA (1985a) Bean maize intercrops – a comparison of bush and climbing bean growth habits. Field Crop Res 10:151–166

Clark EA, Francis CA (1985b) Transgressive yielding in bean – maize intercrops – interference in time and space. Field Crop Res 11:37–53

Cocking EC (2003) Endophytic colonization of plant roots by nitrogen-fixing bacteria. Plant Soil 252:169–175

Colebatch G, Desbrosses G, Ott T et al (2004) Global changes in transcription orchestrate metabolic differentiation during symbiotic nitrogen fixation in *Lotus japonicus*. Plant J 39:487–512

Cordier C, Pozo MJ, Barea JM et al (1998) Cell defence responses associated with localized and systemic resistance to *Phytophthora parasitica* induced in tomato by an arbuscular mycorrhizal fungus. Mol Plant Microbe Interact 11:1017–1028
Crouch JH, Buhariwalla HK, Blair M et al (2004) Biotechnology-based contributions to enhancing legume productivity in resource-poor areas. In: Serraji R (ed.) Symbiotic nitrogen fixation – prospects for enhanced application in tropical agriculture, Proceedings of the international workshop on the biological nitrogen fixation for increased crop productivity, Montpellier, France, 10–14 July 2004, pp 47–65
da Silveira APD, Cardoso E (2004) Arbuscular mycorrhiza and kinetic parameters of phosphorus absorption by bean plants. Sci Agric 61:203–209
da Silveira PM, Braz AJBP, Kliemann HJ et al (2005) Nitrogen fertilization of common bean grown under no-tillage system after several cover crops. Pesqui Agropecu Bras 40:377–381
Daba S, Haile M (2002) Effects of rhizobial inoculant and nitrogen fertilizer on yield and nodulation of common bean under intercropped conditions. J Plant Nutr 25:1443–1455
Daellenbach GC, Kerridge PC, Wolfe MS et al (2005) Plant productivity in cassava-based mixed cropping systems in Colombian hillside farms. Agr Ecosyst Environ 105:595–614
Daft MJ, El-Giahmi AA (1974) Effect of *Endogone* mycorrhiza on plant growth. VII. Influence of infection on the growth and nodulation in french bean (*Phaseolus vulgaris*). New Phytol 73:1139–1147
Daniels-Hylton KDM, Ahmad MH (1994) Inoculation response in kidney beans (*Phaseolus vulgaris* L) to vesicular-arbuscular mycorrhizal fungi and rhizobia in non- sterilized soil. Biol Fert Soils 18:95–98
Dar GH, Zargar MY, Beigh GM (1997) Biocontrol of *Fusarium* root rot in the common bean (*Phaseolus vulgaris* L) by using symbiotic *Glomus mosseae* and *Rhizobium leguminosarum*. Microb Ecol 34:74–80
Date RA (2001) Advances in inoculant technology: a brief review. Aust J Exp Agr 41:321–325
Dawo MI, Wilkinson JM, Pilbeam DJ (2009) Interactions between plants in intercropped maize and common bean. J Sci Food Agr 89:41–48
de Fatima LM, Kaschuk G, Alberton O et al (2007) Soybean [*Glycine max* (L.) Merrill] rhizobial diversity in Brazilian oxisols under various soil, cropping, and inoculation managements. Biol Fert Soils 43:665–674
de Wit M, Stankiewicz J (2006) Changes in surface water supply across Africa with predicted climate change. Science 311:1917–1921
Deaker R, Roughley RJ, Kennedy IR (2004) Legume seed inoculation technology – a review. Soil Biol Biochem 36:1275–1288
Debouck DG (1999) Diversity in *Phaseolus* species in relation to the common bean. In: Singh SP (ed.) Common bean improvement in the twenty-first century. Kluwer Academic, Dordrecht, pp 25–52
Donangelo CM, Woodhouse LR, King SM et al (2003) Iron and zinc absorption from two bean (*Phaseolus vulgaris* L.) genotypes in young women. J Agr Food Chem 51:5137–5143
dos Santos AB, Fageria NK (2007) Nitrogen fertilizer management for efficient use by dry bean in tropical lowland. Pesqui Agropecu Bras 42:1237–1248
dos Santos AB, Fageria NK (2008) Physiological characteristics of common bean in tropical varzea soils as affected by rate and nitrogen management. Ciência e Agrotecnologia 32:23–31
Drechsel P, Steiner KG, Hagedorn F (1996) A review on the potential of improved fallows and green manure in Rwanda. Agroforest Syst 33:109–136
Drechsel P, Kunze D, de Vries FP (2001) Soil nutrient depletion and population growth in sub-Saharan Africa: a Malthusian nexus? Popul Environ 22:411–423
Drevon J-J, Alkama N, Blair M et al (2005) Phosphorus use efficiency for symbiotic nitrogen fixation in common bean (*Phaseolus vulgaris*) and its consequence on soil P dynamic. In: Wang YP, Lin M, Tian ZX et al (eds.) Biological nitrogen fixation, sustainable agriculture and the environment, Proceedings of the 14th international nitrogen fixation congress, Beijing, China, 27 Oct – 1 Nov 2004, pp 277–278
Du Toit W (1997) Long-term effects of fertilizer placement on grain yield of maize. Appl Plant Sci 11:24–25

Eaton D, Hilhorst T (2003) Opportunities for managing solid waste flows in the peri-urban interface of Bamako and Ouagadougou. Environ Urban 15:53–63

Edje OT, Mughogho LK, Ayonoadu UWU (1975) Responses of dry beans to varying nitrogen levels. Agron J 67:251–255

Egerton-Warburton LM, Querejeta JI, Allen MF (2007) Common mycorrhizal networks provide a potential pathway for the transfer of hydraulically lifted water between plants. J Exp Bot 58:1473–1483

El Ghandour IA, El Sharawy MAO, Abdel-Moniem EM (1996) Impact of vesicular arbuscular mycorrhizal fungi and *Rhizobium* on the growth and P, N and Fe uptake by faba-bean. Fert Res 43:43–48

El-Tohamy W, Schnitzler WH, El-Behairy U et al (1999) Effect of VA mycorrhiza on improving drought and chilling tolerance of bean plants (*Phaseolus vulgaris* L.). J Appl Bot 73:178–183

Facelli E, Facelli JM (2002) Soil phosphorus heterogeneity and mycorrhizal symbiosis regulate plant intra-specific competition and size distribution. Oecologia 133:54–61

Facelli E, Facelli J, McLaughlin MJ et al (1999) An experimental study of the interactive effects of mycorrhizal symbiosis, intraspecific competition and resource availability using *Trifolium subterraneum* L. cv Mt Barker. New Phytol 141:535–547

Fageria NK, da Costa JGC (2000) Evaluation of common bean genotypes for phosphorus use efficiency. J Plant Nutr 23:1145–1152

Fageria NK, Santos AB (1998) Rice and common bean growth and nutrient concentration as influenced by aluminum on an acid lowland soil. J Plant Nutr 21:903–912

FAO (2008) FAOSTAT: FAO statistical database. Available via: http://faostat.fao.org. Accessed 18 Dec 2008

Ferguson BJ, Mathesius U (2003) Signaling interactions during nodule development. J Plant Growth Regul 22:47–72

Fermont AM, van Asten PJA, Giller KE (2008) Increasing land pressure in East Africa: the changing role of cassava and consequences for sustainability of farming systems. Agr Ecosys Environ 128:239–250

Filion M, St-Arnaud M, Jabaji-Hare SH (2003) Quantification of *Fusarium solani* f. sp *phaseoli* in mycorrhizal bean plants and surrounding mycorrhizosphere soil using real-time polymerase chain reaction and direct isolations on selective media. Phytopathology 93:229–235

Fisher MCT, Eissenstat DM, Lynch JP (2002) Lack of evidence for programmed root senescence in common bean (*Phaseolus vulgaris*) grown at different levels of phosphorus supply. New Phytol 153:63–71

Flores M, Morales L, Avila A et al (2005) Diversification of DNA sequences in the symbiotic genome of *Rhizobium etli*. J Bacteriol 187:7185–7192

Francis CA, Sanders JH (1978) Economic analysis of bean and maize systems: monoculture versus associated cropping. Field Crop Res 1:319–335

Franke AC, Laberge G, Oyewole BD et al (2008) A comparison between legume technologies and fallow, and their effects on maize and soil traits, in two distinct environments of the West African savannah. Nutr Cycl Agroecosys 82:117–135

Fritz M, Jakobsen I, Lyngkjaer MF et al (2006) Arbuscular mycorrhiza reduces susceptibility of tomato to *Alternaria solani*. Mycorrhiza 16:413–419

Frossard E, Bucher M, Mächler F et al (2000) Potential for increasing the content and bioavailability of Fe, Zn and Ca in plants for human nutrition. J Sci Food Agric 80:861–879

Frossard E, Bünemann EK, Carsky R et al (2007) Integrated nutrient management as a tool to combat soil degradation in sub Saharan Africa. In: Bearth T, Becker B, Kappel R et al (eds.) Afrika im Wandel. Zürcher Hochschulforum, Zurich, pp 137–146

Frossard E, Bünemann E, Jansa J et al (2009) Concepts and practices of nutrient management in agro-ecosystems: can we draw lessons from history to design future sustainable agricultural production systems? Die Bodenkultur 60:5–22

Fujikake H, Tamura Y, Ohtake N et al (2003) Photoassimilate partitioning in hypernodulation mutant of soybean (*Glycine max* (L.) Merr.) NOD1-3 and its parent Williams in relation to nitrate inhibition of nodule growth. Soil Sci Plant Nutr 49:583–590

Gadd ME (2005) Conservation outside of parks: attitudes of local people in Laikipia, Kenya. Environ Conserv 32:50–63
Garg N, Geetanjali (2007) Symbiotic nitrogen fixation in legume nodules: process and signaling. A review. Agron Sustain Dev 27:59–68
Ge ZY, Rubio G, Lynch JP (2000) The importance of root gravitropism for inter-root competition and phosphorus acquisition efficiency: results from a geometric simulation model. Plant Soil 218:159–171
Gepts P (2006) Plant genetic resources conservation and utilization: the accomplishments and future of a societal insurance policy. Crop Sci 46:2278–2292
Gepts P, Aragão FJL, de Barros E et al (2008) Genomics of *Phaseolus* beans, a major source of dietary protein and micronutrients in the tropics. In: Moore PH, Ming R (eds.) Genomics of tropical crop plants. Springer, Berlin, pp 113–142
Gherbi H, Markmann K, Svistoonoff S et al (2008) SymRK defines a common genetic basis for plant root endosymbioses with arbuscular mycorrhiza fungi, rhizobia, and *Frankia* bacteria. P Natl Acad Sci USA 105:4928–4932
Ghosh PK, Bandyopadhyay KK, Wanjari RH et al (2007) Legume effect for enhancing productivity and nutrient use-efficiency in major cropping systems – An Indian perspective: a review. J Sustain Agr 30:59–86
Giller KE, Cadisch G (1995) Future benefits from biological nitrogen fixation – an ecological approach to agriculture. Plant Soil 174:255–277
Giller KE, Cadisch G, Ehaliotis C et al (1997) Building soil nitrogen capital in Africa. In: Buresh RJ, Sanchez PA, Calhoun F (eds.) Replenishing soil fertility in Africa, SSSA Special Publication No. 51. SSSA, Madison, pp 151–192
Glaser B (2007) Prehistorically modified soils of central Amazonia: a model for sustainable agriculture in the twenty-first century. Philos T R Soc B 362:187–196
Gomes AA, Araujo AP, Rossiello ROP et al (2000) Accumulation of biomass, physiological characteristics and grain yield of bean cultivars under irrigated and dry regimes. Pesqui Agropecu Bras 35:1927–1937
Gonzalez A, Lynch J (1999) Tolerance of tropical common bean genotypes to manganese toxicity: performance under different growing conditions. J Plant Nutr 22:511–525
Gonzalez EM, Galvez L, Royuela M et al (2001) Insights into the regulation of nitrogen fixation in pea nodules: lessons from drought, abscisic acid and increased photoassimilate availability. Agronomie 21:607–613
Gonzalez AM, Monteagudo AB, Casquero PA et al (2006a) Genetic variation and environmental effects on agronomical and commercial quality traits in the main European market classes of dry bean. Field Crop Res 95:336–347
Gonzalez V, Santamaria RI, Bustos P et al (2006b) The partitioned *Rhizobium etli* genome: genetic and metabolic redundancy in seven interacting replicons. P Natl Acad Sci USA 103:3834–3839
Graham AH (1981) Some problems of nodulation and symbiotic nitrogen fixation in *Phaseolus vulgaris* L.: a review. Field Crop Res 4:93–112
Graham PH, Ranalli P (1997) Common bean (*Phaseolus vulgaris* L). Field Crop Res 53:131–146
Grange L, Hungria M, Graham PH et al (2007) New insights into the origins and evolution of rhizobia that nodulate common bean (*Phaseolus vulgaris*) in Brazil. Soil Biol Biochem 39:867–876
Guene NFD, Diouf A, Gueye M (2004) Nitrogen fixation in the common bean (*Phaseolus vulgaris*) – A multilocational inoculation trial in Senegal. In: Serraji R (ed.) Symbiotic nitrogen fixation – prospects for enhanced application in tropical agriculture, Proceedings of the International workshop on the biological nitrogen fixation for increased crop productivity, Montpellier, France, 10–14 July 2004, pp 247–252
Guillon C, St-Arnaud M, Hamel C et al (2002) Differential and systemic alteration of defence-related gene transcript levels in mycorrhizal bean plants infected with *Rhizoctonia solani*. Can J Bot 80:305–315

Guimaraes CM, Stone LF, Brunini O (1996) Adaptation of common bean (Phaseolus vulgaris L) to drought.2. Productivity and agronomic components. Pesqui Agropecu Bras 31:481–488

Haase S, Neumann G, Kania A et al (2007) Elevation of atmospheric CO_2 and N-nutritional status modify nodulation, nodule-carbon supply, and root exudation of *Phaseolus vulgaris* L. Soil Biol Biochem 39:2208–2221

Hacisalihoglu G, Duke ER, Longo LM (2005) Differential response of common bean genotypes to mycorrhizal colonization. In: Proceedings of the 118th Annual meeting of the Florida state horticultural society, Tampa, FL, 5–7 June 2005, pp 150–152

Hafeez FY, Hameed S, Ahmad T et al (2001) Competition between effective and less effective strains of *Bradyrhizobium* spp. for nodulation on *Vigna radiata*. Biol. Fert Soil 33:382–386

Hardarson G, Atkins C (2003) Optimising biological N_2 fixation by legumes in farming systems. Plant Soil 252:41–54

Hardarson G, Bliss FA, Cigalesrivero MR et al (1993) Genotypic variation in biological nitrogen fixation by common bean. Plant Soil 152:59–70

Harris D, Pacovsky RS, Paul EA (1985) Carbon economy of soybean-*Rhizobium-Glomus* associations. New Phytol 101:427–440

Hartwig UA, Trommler J (2001) Increase in the concentrations of amino acids in the vascular tissue of white clover and white lupin after defoliation: an indication of a N feedback regulation of symbiotic N_2 fixation. Agronomie 21:615–620

Hastorf CA (1999) Recent research in paleoethnobotany. J Archaeol Res 7:55–103

Haugen R, Steffes L, Wolf J et al (2008) Evolution of drought tolerance and defense: dependence of tradeoffs on mechanism, environment and defense switching. Oikos 117:231–244

Hause B, Mrosk C, Isayenkov S et al (2007) Jasmonates in arbuscular mycorrhizal interactions. Phytochemistry 68:101–110

Hawkins HJ, George E (1999) Effect of plant nitrogen status on the contribution of arbuscular mycorrhizal hyphae to plant nitrogen uptake. Physiol Plant 105:694–700

He Z-Q, He C-X, Zhang Z-B et al (2007) Mechanism of plant salt tolerance enhanced by arbuscular mycorrhizal fungi. Xibei Zhiwu Xuebao 27:414–420

Hellsten A, Huss-Danell K (2001) Interaction effects of nitrogen and phosphorus on nodulation in red clover (*Trifolium pratense* L.). Acta Agr Scand B-S P 50:135–142

Henao J, Baanante CA (1999) Estimating rates of nutrient depletion in soils of agricultural lands of Africa. Technical Bulletin T-48, International Fertilizer Development Center, Muscle Shoal

Herrera-Cervera JA, Caballero-Mellado J, Laguerre G et al (1999) At least five rhizobial species nodulate *Phaseolus vulgaris* in a Spanish soil. FEMS Microbiol Ecol 30:87–97

Herridge D, Rose I (2000) Breeding for enhanced nitrogen fixation in crop legumes. Field Crop Res 65:229–248

Hillocks RJ, Madata CS, Chirwa R et al (2006) *Phaseolus* bean improvement in Tanzania, 1959–2005. Euphytica 150:215–231

Hungria M, Neves MCP (1987) Cultivar and *Rhizobium* strain effect on nitrogen fixation and transport in *Phaseolus vulgaris* L. Plant Soil 103:111–121

Hungria M, Stacey G (1997) Molecular signals exchanged between host plants and rhizobia: basic aspects and potential application in agriculture. Soil Biol Biochem 29:819–830

Hungria M, Vargas MAT (2000) Environmental factors affecting N_2 fixation in grain legumes in the tropics, with an emphasis on Brazil. Field Crop Res 65:151–164

Hungria M, Andrade DD, Chueire LMD et al (2000) Isolation and characterization of new efficient and competitive bean (*Phaseolus vulgaris* L.) rhizobia from Brazil. Soil Biol Biochem 32:1515–1528

Huston M (1993) Biological diversity, soils, and economics. Science 262:1676–1680

Ibijbijen J, Urquiaga S, Ismaili M et al (1996) Effect of arbuscular mycorrhizal fungi on growth, mineral nutrition and nitrogen fixation of three varieties of common beans (*Phaseolus vulgaris*). New Phytol 134:353–360

Ishikawa K, Yokota K, Li YY et al (2008) Isolation of a novel root-determined hypernodulation mutant rdh1 of *Lotus japonicus*. Soil Sci Plant Nutr 54:259–263

Isobe K, Tsuboki Y (1998) The relationship between growth promotion by arbuscular mycorrhizal fungi and root morphology and phosphorus absorption in gramineous and leguminous crops. Jpn J Crop Sci 67:347–352
Isoi T, Yoshida S (1991) Low nitrogen fixation of common bean (*Phaseolus vulgaris* L). Soil Sci Plant Nutr 37:559–563
Israel DW (1987) Investigation of the role of phosphorus in symbiotic dinitrogen fixation. Plant Physiol 84:835–840
Izaguirre-Mayoral ML, Carballo O, Carreno L et al (2000) Effects of arbuscular mycorrhizal inoculation on growth, yield, nitrogen, and phosphorus nutrition of nodulating bean varieties in two soil substrates of contrasting fertility. J Plant Nutr 23:1117–1133
Jakobsen I, Rosendahl L (1990) Carbon flow into soil and external hyphae from roots of mycorrhizal cucumber plants. New Phytol 115:77–83
Jakobsen I, Abbott LK, Robson AD (1992) External hyphae of vesicular-arbuscular mycorrhizal fungi associated with *Trifolium subterraneum* L. 1. Spread of hyphae and phosphorus inflow into roots. New Phytol 120:371–380
Jama B, Pizarro G (2008) Agriculture in Africa: strategies to improve and sustain smallholder production systems. Reducing the impact of poverty on health and human development: scientific Approaches. Ann NY Acad Sci 1136:218–232
Jansa J, Mozafar A, Frossard E (2003a) Long-distance transport of P and Zn through the hyphae of an arbuscular mycorrhizal fungus in symbiosis with maize. Agronomie 23:481–488
Jansa J, Mozafar A, Kuhn G et al (2003b) Soil tillage affects the community structure of mycorrhizal fungi in maize roots. Ecol Appl 13:1164–1176
Jansa J, Mozafar A, Frossard E (2005) Phosphorus acquisition strategies within arbuscular mycorrhizal fungal community of a single field site. Plant Soil 276:163–176
Jansa J, Wiemken A, Frossard E (2006) The effects of agricultural practices on arbuscular mycorrhizal fungi. In: Frossard E, Blum W, Warkentin B (eds.) Function of soils for human societies and the environment. Geological Society, Burlington House, pp 89–115
Jansa J, Smith FA, Smith SE (2008) Are there benefits of simultaneous root colonization by different arbuscular mycorrhizal fungi? New Phytol 177:779–789
Jebara M, Drevon J-J (2001) Genotypic variation in nodule conductance to the oxygen diffusion in common bean (*Phaseolus vulgaris*). Agronomie 21:667–674
Jia Y, Gray VM, Straker CJ (2004) The influence of *Rhizobium* and arbuscular mycorrhizal fungi on nitrogen and phosphorus accumulation by *Vicia faba*. Ann Bot – London 94:251–258
Jin H, Pfeffer PE, Douds DD et al (2005) The uptake, metabolism, transport and transfer of nitrogen in an arbuscular mycorrhizal symbiosis. New Phytol 168:687–696
Johansson JF, Paul LR, Finlay RD (2004) Microbial interactions in the mycorrhizosphere and their significance for sustainable agriculture. FEMS Microbiol Ecol 48:1–13
Joner EJ, Briones R, Leyval C (2000) Metal-binding capacity of arbuscular mycorrhizal mycelium. Plant Soil 226:227–234
Joner EJ, Roos P, Jansa J et al (2004) No significant contribution of arbuscular mycorrhizal fungi to transfer of radiocesium from soil to plants. Appl Environ Microbiol 70:6512–6517
Kandji ST, Ogol CKPO, Albrecht A (2003) Crop damage by nematodes in improved-fallow fields in western Kenya. Agroforest Syst 57:49–55
Kaschuk G, Hungria M, Andrade DS et al (2006a) Genetic diversity of rhizobia associated with common bean (*Phaseolus vulgaris* L.) grown under no-tillage and conventional systems in Southern Brazil. Appl Soil Ecol 32:210–220
Kaschuk G, Hungria M, Santos JCP et al (2006b) Differences in common bean rhizobial populations associated with soil tillage management in southern Brazil. Soil Till Res 87:205–217
Kayuki KC, Wortmann CS (2001) Plant materials for soil fertility management in subhumid tropical areas. Agron J 93:929–935
Kelly JD (2004) Advances in common bean improvement: some case histories with broader applications. In: McCreight JD, Ryder EJ (eds.) Advances in vegetable breeding, Proceedings of the 26th International horticultural congress, Toronto, Canada, 11–17 August 2002, pp 99–122
Khalil S, Loynachan TE, Tabatabai MA (1994) Mycorrhizal dependency and nutrient uptake by improved and unimproved corn and soybean cultivars. Agron J 86:949–958

Kiers ET, West SA, Denison RF (2002) Mediating mutualisms: farm management practices and evolutionary changes in symbiont co-operation. J Appl Ecol 39:745–754

Kimani JM, Tongoona P (2008) The mechanism of genetic control for low soil nitrogen (N) tolerance in common beans (*Phaseolus vulgaris* L.). Euphytica 162:193–203

Kohlmeier S, Smits THM, Ford RM et al (2005) Taking the fungal highway: mobilization of pollutant-degrading bacteria by fungi. Environ Sci Technol 39:4640–4646

Koide RT (2000) Functional complementarity in the arbuscular mycorrhizal symbiosis. New Phytol 147:233–235

Kosuta S, Hazledine S, Sun J et al (2008) Differential and chaotic calcium signatures in the symbiosis signaling pathway of legumes. P Natl Acad Sci USA 105:9823–9828

Kucey R, Paul E (1982) Carbon flow, photosynthesis, and N_2 fixation in mycorrhizal and nodulated faba beans (*Vicia faba* L.). Soil Biol Biochem 14:407–412

Laguerre G, Nour SM, Macheret V et al (2001) Classification of rhizobia based on nodC and nifH gene analysis reveals a close phylogenetic relationship among *Phaseolus vulgaris* symbionts. Microbiology 147:981–993

Laguerre G, Courde L, Nouaim R et al (2006) Response of rhizobial populations to moderate copper stress applied to an agricultural soil. Microb Ecol 52:426–435

Laine AL, Tellier A (2008) Heterogeneous selection promotes maintenance of polymorphism in host-parasite interactions. Oikos 117:1281–1288

Lal R (2007) Soil science and the carbon civilization. Soil Sci Soc Am J 71:1425–1437

Lal R, Singh BR (1998) Effects of soil degradation on crop productivity in East Africa. J Sustain Agr 13:15–36

Leidi EO, Rodriguez Navarro DN (2000) Nitrogen and phosphorus availability limit N_2 fixation in bean. New Phytol 147:337–346

Lekberg Y, Koide RT (2005) Arbuscular mycorrhizal fungi, rhizobia, available soil P and nodulation of groundnut (*Arachis hypogaea*) in Zimbabwe. Agr Ecosys Environ 110:143–148

Li HY, Smith FA, Dickson S et al (2008) Plant growth depressions in arbuscular mycorrhizal symbioses: not just caused by carbon drain? New Phytol 178:852–862

Liebman M, Dyck E (1993) Crop rotation and intercropping strategies for weed management. Ecol Appl 3:93–122

Linderman RG (1992) Vesicular-arbuscular mycorrhizae and soil microbial interactions. In: Bethlenfalvay GJ, Linderman RG (eds.) Mycorrhzae in sustainable agriculture, ASA Special Publication 54. ASA, Madison, pp 45–70

Lodeiro AR, Gonzalez P, Hernandez A et al (2000) Comparison of drought tolerance in nitrogen-fixing and inorganic nitrogen-grown common beans. Plant Sci 154:31–41

Lunze L, Kimani PM, Ngatoluwa R et al (2007) Bean improvement for low soil fertility adaptation in Eastern and Central Africa. In: Bationo A, Waswa B, Kihara J et al (eds.) Advances in integrated soil fertility management in sub-Saharan Africa: challenges and opportunities. Springer, Berlin, pp 325–332

Lynch J, Beebe SE (1995) Adaptation of beans (*Phaseolus vulgaris* L.) to low phosphorus availability. HortScience 30:1165–1171

Lynch JP, Ho MD (2005) Rhizoeconomics: carbon costs of phosphorus acquisition. Plant Soil 269:45–56

Lynch J, Rodriguez NS (1994) Photosynthetic nitrogen use efficiency in relation to leaf longevity in common bean. Crop Sci 34:1284–1290

Lynch J, van Beem JJ (1993) Growth and architecture of seedling roots of common bean genotypes. Crop Sci 33:1253–1257

Lynch J, White JW (1992) Shoot nitrogen dynamics in tropical common bean. Crop Sci 32:392–397

Maherali H, Klironomos JN (2007) Influence of phylogeny on fungal community assembly and ecosystem functioning. Science 316:1746–1748

Mando A, Bonzi M, Wopereis MCS et al (2005) Long-term effects of mineral and organic fertilization on soil organic matter fractions and sorghum yield under Sudano-Sahelian conditions. Soil Use Manage 21:396–401

Marquez ML, Teran H, Singh SP (2007) Selecting common bean with genes of different evolutionary origins for resistance to *Xanthomonas camplestris* pv. *phaseoli*. Crop Sci 47:1367–1374

Marsh JF, Schultze M (2001) Analysis of arbuscular mycorrhizas using symbiosis-defective plant mutants. New Phytol 150:525–532

Martin RC, Eaglesham ARJ, Voldeng HD et al (1995) Factors affecting nitrogen benefit from soybean [*Glycine max* (L) Merr cv Lee] to interplanted corn (*Zea mays* L cv Co-op S259). Environ Exp Bot 35:497–505

Martinez-Romero E (2003) Diversity of *Rhizobium–Phaseolus vulgaris* symbiosis: overview and perspectives. Plant Soil 252:11–23

Mauyo LW, Okalebo JR, Kirkby RA et al (2007) Spatial pricing efficiency and regional market integration of cross-border bean (*Phaseolus vulgaris* L.) marketing in East Africa: The case of Western Kenya and Eastern Uganda. In: Bationo A, Waswa B, Kihara J et al (eds.) Advances in integrated soil fertility management in sub-Saharan Africa: challenges and opportunities. Springer, Berlin, pp 1027–1033

Mayona CM, Kamasho J (1988) Research experiences with inorganic and organic fertilizers in the southern highlands of Tanzania. In: Wortmann CS (ed.) Proceedings of a workshop on soil fertility research for bean cropping systems in Africa, CIAT African Workshop Series No 3. CIAT, Cali

Mbugua GW (1986) Effects of plant density and phosphate levels on growth, yield and yield components of field beans (*Phaseolus vugaris* L.). Phaseolus Newsl East Afr 15:15–16

Mendes IC, Hungria M, Vargas MAT (2003) Soybean response to starter nitrogen and *Bradyrhizobium* inoculation on a Cerrado oxisol under no-tillage and conventional tillage systems. Rev Bras Cienc Solo 27:81–87

Mhamdi R, Jebara M, Aouani ME et al (1999) Genotypic diversity and symbiotic effectiveness of rhizobia isolated from root nodules of *Phaseolus vulgaris* L. grown in Tunisian soils. Biol Fert Soil 28:313–320

Miklas PN, Kelly JD, Beebe SE et al (2006) Common bean breeding for resistance against biotic and abiotic stresses: from classical to MAS breeding. Euphytica 147:105–131

Miyasaka SC, Hawes MC (2001) Possible role of root border cells in detection and avoidance of aluminum toxicity. Plant Physiol 125:1978–1987

Miyasaka SC, Buta JG, Howell RK et al (1991) Mechanism of aluminum tolerance in snapbeans: root exudation of citric acid. Plant Physiol 96:737–743

Moawad AM, Vlek PLG (1998) Potential contribution of (vesicular-) arbuscular mycorrhiza to nutrient efficient crops. In: Proceedings of the BTIG workshop on oil palm improvement through biotechnology, Bogor Indonesia, 16–17 April 1997, pp 48–58

Moawad H, El-Rahim WMA, El-Aleem DA et al (2005) Persistence of two *Rhizobium etli* inoculant strains in clay and silty loam soils. J Basic Microb 45:438–446

Moody PW, Cong PT, Legrand J et al (2008) A decision support framework for identifying soil constraints to the agricultural productivity of tropical upland soils. Soil Use Manage 24:148–155

Morandi D, Prado E, Sagan M et al (2005) Characterisation of new symbiotic *Medicago truncatula* (Gaertn.) mutants, and phenotypic or genotypic complementary information on previously described mutants. Mycorrhiza 15:283–289

Mortimer PE, Perez-Fernandez MA, Valentine AJ (2008) The role of arbuscular mycorrhizal colonization in the carbon and nutrient economy of the tripartite symbiosis with nodulated *Phaseolus vulgaris*. Soil Biol Biochem 40:1019–1027

Mosse B, Thompson JP (1984) Vesicular-arbuscular endomycorrhizal inoculum production. 1. Exploratory experiments with beans (*Phaseolus vulgaris*) in nutrient flow culture. Can J Bot 62:1523–1530

Mugai EN, Agong SG, Matsumoto H (2000) Aluminium tolerance mechanisms in *Phaseolus vulgaris* L.: citrate synthase activity and TTC reduction are well correlated with citrate secretion. Soil Sci Plant Nutr 46:939–950

Munkvold L, Kjøller R, Vestberg M et al (2004) High functional diversity within species of arbuscular mycorrhizal fungi. New Phytol 164:357–364

Munoz-Perea CG, Allen RG, Westermann DT et al (2007) Water use efficiency among dry bean landraces and cultivars in drought-stressed and non-stressed environments. Euphytica 155:393–402

Ndakidemi PA, Dakora FD, Nkonya EM et al (2006) Yield and economic benefits of common bean (*Phaseolus vulgaris*) and soybean (*Glycine max*) inoculation in northern Tanzania. Aust J Exp Agric 46:571–577

Neeraj SK (2005) Impact of VA-mycorrhiza, *Rhizobium* and phosphorus on growth and yield of *Phaseolus vulgaris* L. J Phytol Res 18:59–63

Neves MCP, Hungria M (1987) The physiology of nitrogen fixation in tropical grain legumes. CRC Crit Rev Plant Sci 6:267–321

Newsham K, Fitter A, Watkinson A (1995) Multi-functionality and biodiversity in arbuscular mycorrhizas. Trends Ecol Evol 10:407–411

Nielsen KL, Miller CR, Beck D et al (1999) Fractal geometry of root systems: field observations of contrasting genotypes of common bean (*Phaseolus vulgaris* L.) grown under different phosphorus regimes. Plant Soil 206:181–190

Nielsen KL, Eshel A, Lynch JP (2001) The effect of phosphorus availability on the carbon economy of contrasting common bean (*Phaseolus vulgaris* L.) genotypes. J Exp Bot 52:329–339

Nyakudya IW, Katsvanga CAT, Mugayi EC et al (2005) Working with communities to improve soil productivity: Sadziwa Ward, Zimbabwe. J Sustain Dev Africa 7:183–191

Nziguheba G (2007) Overcoming phosphorus deficiency in soils of Eastern Africa: recent advances and challenges. In: Bationo A, Waswa B, Kihara J et al (eds.) Advances in integrated soil fertility management in sub-Saharan Africa: challenges and opportunities. Springer, Berlin, pp 149–160

O'Hara GW (2001) Nutritional constraints on root nodule bacteria affecting symbiotic nitrogen fixation: a review. Aust J Exp Agric 41:417–433

O'Kennedy MM, Grootboom A, Shewry PR (2006) Harnessing sorghum and millet biotechnology for food and health. J Cereal Sci 44:224–235

Oldroyd GED, Downie JA (2006) Nuclear calcium changes at the core of symbiosis signalling. Curr Opin Plant Biol 9:351–357

Oliveira AAR, Sanders FE (1999) Effect of management practices on mycorrhizal infection, growth and dry matter partitioning in field-grown bean. Pesqui Agropecu Bras 34:1247–1254

Oliveira AAR, Sanders FE (2000) Effect of inoculum placement of indigenous and introduced arbuscular mycorrhizal fungi on mycorrhizal infection, growth, and dry matter in *Phaseolus vulgaris*. Trop Agr 77:220–225

Ortas I, Akpinar C (2006) Response of kidney bean to arbuscular mycorrhizal inoculation and mycorrhizal dependency in P and Zn deficient soils. Acta Agr Scand B-S P 56:101–109

Osbourn AE (2001) Plant mechanisms that give defence against soilborne diseases. Australas Plant Path 30:99–102

Pachico D (1993) The demand for bean technology. In: Henry G, Henry G (eds.) Trends in CIAT Commodities 1993, Working Document No. 128. CIAT, Cali, pp 60–74

Parniske M (2008) Arbuscular mycorrhiza: the mother of plant root endosymbioses. Nat Rev Microbiol 6:763–775

Paul EA, Kucey RMN (1981) Carbon flow in plant microbial associations. Science 213:473–474

Pfeffer PE, Douds DD, Bucking H et al (2004) The fungus does not transfer carbon to or between roots in an arbuscular mycorrhizal symbiosis. New Phytol 163:617–627

Phillips J, McIntyre B (2000) ENSO and interannual rainfall variability in Uganda: implications for agricultural management. Int J Climatol 20:171–182

Picone C (2002) Managing mycorrhizae for sustainable agriculture in the tropics. In: Vandermeer JH (ed.) Tropical agroecosystems. CRC Press, Boca Raton, pp 95–132

Plenchette C, Fortin JA, Furlan V (1983) Growth responses of several plant species to mycorrhizae in a soil of moderate P-fertility.1. Mycorrhizal dependency under field conditions. Plant Soil 70:199–209

Pozo MJ, Azcón-Aguilar C (2007) Unraveling mycorrhiza-induced resistance. Curr Opin Plant Biol 10:393–398

Ramirez-Vallejo P, Kelly JD (1998) Traits related to drought resistance in common bean. Euphytica 99:127–136
Rangel AF, Mobin M, Rao IM et al (2005) Proton toxicity interferes with the screening of common bean (*Phaseolus vulgaris* L.) genotypes for aluminium resistance in nutrient solution. J Plant Nutr Soil Sci 168:607–616
Rangel AF, Rao IM, Horst WJ (2007) Spatial aluminium sensitivity of root apices of two common bean (*Phaseolus vulgaris* L.) genotypes with contrasting aluminium resistance. J Exp Bot 58:3895–3904
Rangel AF, Rao IM, Horst WJ (2009) Intracellular distribution and binding state of aluminum in root apices of two common bean (*Phaseolus vulgaris*) genotypes in relation to Al toxicity. Physiol Plant 135:162–173
Rao IM (2002) Role of physiology in improving crop adaptation to abiotic stresses in the tropics: The case of common bean and tropical forages. In: Pessarakli M (ed.) Handbook of plant and crop physiology. Marcel Dekker, New York, pp 583–613
Requena N, Jimenez I, Toro M et al (1997) Interactions between plant-growth-promoting rhizobacteria (PGPR), arbuscular mycorrhizal fungi and *Rhizobium* spp. in the rhizosphere of *Anthyllis cytisoides*, a model legume for revegetation in mediterranean semi-arid ecosystems. New Phytol 136:667–677
Rodriguez-Uribe L, O'Connell MA (2006) A root-specific bZIP transcription factor is responsive to water deficit stress in tepary bean (*Phaseolus acutifolius*) and common bean (*P. vulgaris*). J Exp Bot 57:1391–1398
Rondon MA, Lehmann J, Ramirez J et al (2007) Biological nitrogen fixation by common beans (*Phaseolus vulgaris* L.) increases with bio-char additions. Biol Fert Soils 43:699–708
Rosemeyer ME, Gliessman SR (1992) Modifying traditional and high-input agroecosystems for optimization of microbial symbioses – a case study of dry beans in Costa Rica. Agr Ecosyst Environ 40:61–70
Roy RN, Misra RV, Lesschen JP (2003) Assessment of soil nutrient balance: approaches and methodologies, FAO fertilizer and plant nutrition bulletin 14. FAO, Rome
Rubio G, Lynch JP (2007) Compensation among root classes in *Phaseolus vulgaris* L. Plant Soil 290:307–321
Ruiz-Lozano JM, Collados C, Barea JM et al (2001) Arbuscular mycorrhizal symbiosis can alleviate drought-induced nodule senescence in soybean plants. New Phytol 151:493–502
Russell AJ, Bidartondo MI, Butterfield BG (2002) The root nodules of the Podocarpaceae harbour arbuscular mycorrhizal fungi. New Phytol 156:283–295
Rutunga V, Neel H (2006) Yield trends in the long-term crop rotation with organic and inorganic fertilisers on Alisols in Mata (Rwanda). Biotechnologie, Agronomie, Société et Environnement 10:217–228
Rutunga V, Janssen BH, Mantel S et al (2007) Soil use and management strategy for raising food and cash output in Rwanda. J Food Agric Environ 5:434–441
Saadallah K, Drevon J-J, Abdelly C (2001a) Nodulation and growth of nodules in the common bean (*Phaseolus vulgaris*) under salt stress. Agronomie 21:627–634
Saadallah K, Drevon J-J, Hajji M et al (2001b) Genotypic variability for tolerance to salinity of N_2-fixing common bean (*Phaseolus vulgaris*). Agronomie 21:675–682
Sanchez PA, Swaminathan MS (2005) Hunger in Africa: the link between unhealthy people and unhealthy soils. Lancet 365:442–444
Sanders IR (2004) Intraspecific genetic variation in arbuscular mycorrhizal fungi and its consequences for molecular biology, ecology, and development of inoculum. Can J Bot 82:1057–1062
Sangakkara UR, Richner W, Steinebrunner F et al (2003) Impact of the cropping systems of a minor dry season on the growth, yields and nitrogen uptake of maize (*Zea mays* L) grown in the humid tropics during the major rainy season. J Agron Crop Sci 189:361–366
Sanginga PC, Kaaria S, Muzira R et al (2007) The resources-to-consumption system: A framework for linking soil fertility management innovations to market opportunities. In: Bationo A, Waswa B, Kihara J et al (eds.) Advances in integrated soil fertility management in sub-Saharan Africa: challenges and opportunities. Springer, Berlin, pp 979–992

Sassi S, Gonzalez EM, Aydi S et al (2008) Tolerance of common bean to long-term osmotic stress is related to nodule carbon flux and antioxidant defenses: evidence from two cultivars with contrasting tolerance. Plant Soil 312:39–48

Scheublin TR, Ridgway KP, Young JPW et al (2004) Non-legumes, legumes, and root nodules harbor different arbuscular mycorrhizal fungal communities. Appl Environ Microb 70:6240–6246

Schnabel E, Journet EP, de Carvalho-Niebel F et al (2005) The *Medicago truncatula* SUNN gene encodes a CLV1-like leucine-rich repeat receptor kinase that regulates nodule number and root length. Plant Mol Biol 58:809–822

Schulze J (2004) How are nitrogen fixation rates regulated in legumes? J Plant Nutr Soil Sci 167:125–137

Sessitsch A, Howieson JG, Perret X et al (2002) Advances in *Rhizobium* research. Crit Rev Plant Sci 21:323–378

Shamseldin A (2008) Plasmid content of salt stress-tolerant *Rhizobium* strains from Egyptian soils nodulating common bean (*Phaseolus vulgaris* L.). World J Microb Biot 24:1603–1606

Shamseldin A, Werner D (2005) High salt and high pH tolerance of new isolated *Rhizobium etli* strains from Egyptian soils. Curr Microbiol 50:11–16

Shamseldin AAY, Vinuesa P, Thierfelder H et al (2005) *Rhizobium etli* and *Rhizobium gallicum* nodulate *Phaseolus vulgaris* in Egyptian soils and display cultivar-dependent symbiotic efficiency. Symbiosis 38:145–161

Sharma AK, Srivastava PC, Johri BN (1999) Multiphasic zinc uptake system in mycorrhizal and nonmycorrhizal roots of French bean (*Phaseolus vulgaris* L.). Curr Sci 76:228–230

Shaw LJ, Morris P, Hooker JE (2006) Perception and modification of plant flavonoid signals by rhizosphere microorganisms. Environ Microbiol 8:1867–1880

Shen H, Yan XL, Cai KZ et al (2004) Differential Al resistance and citrate secretion in the tap and basal roots of common bean seedlings. Physiol Plant 121:595–603

Shirtliffe SJ, Vessey JK (1996) A nodulation (Nod(+)/Fix-) mutant of *Phaseolus vulgaris* L has nodule-like structures lacking peripheral vascular bundles (Pvb(−)) and is resistant to mycorrhizal infection (Myc(−)). Plant Sci 118:209–220

Sierra J, Nygren P (2006) Transfer of N fixed by a legume tree to the associated grass in a tropical silvopastoral system. Soil Biol Biochem 38:1893–1903

Sieverding E (1991) Vesicular-arbuscular mycorrhiza management in tropical agrosystems. Deutsche Gesellschaft fur Technische Zusammenarbeit (GTZ) GmbH, Eschborn

Silwana TT, Lucas EO, Olaniyan AB (2007) The effects of inorganic and organic fertilizers on the growth and development of component crops in maize/bean intercrop in Eastern Cape of South Africa. J Food Agric Environ 5:267–272

Singh SP (1982) A key for identification of different growth habits of frijol Phaseolus vulgaris L. Annual Reports of Bean Improvement Cooperation Nr 25, pp 92–95

Singh SP (1995) Selection for water stress tolerance in interracial populations of common bean. Crop Sci 35:118–124

Singh SP (1999a) Integrated genetic improvement. In: Singh SP (ed.) Common bean improvement in the twenty-first century. Kluwer, Dordrecht, pp 133–166

Singh SP (1999b) Production and utilization. In: Singh SP (ed.) Common bean improvement in the twenty-first century. Kluwer, Dordrecht, pp 1–24

Singh SP (2001) Broadening the genetic base of common bean cultivars: a review. Crop Sci 41:1659–1675

Singh SP, Gepts P, Debouck DG (1991) Races of common bean (*Phaseolus vulgaris*, Fabaceae). Econ Bot 45:379–396

Singh SP, Teran H, Gutierrez JA (2001) Registration of SEA 5 & SEA 13 drought tolerant dry bean germplasm. Crop Sci 41:276–277

Singh SP, Teran H, Munoz CG et al (2003) Low soil fertility tolerance in landraces and improved common bean genotypes. Crop Sci 43:110–119

Six J, Bossuyt H, Degryze S et al (2004) A history of research on the link between (micro)aggregates, soil biota, and soil organic matter dynamics. Soil Till Res 79:7–31

Smith S, Read D (2008) Mycorrhizal symbiosis. Academic, New York

Smithson P, Jama B, Delve R et al (2003) East African phosphate resources and their agronomic performance. In: Rajan SSS, Chien SH (eds.) Proceedings of the international meeting on direct application of phosphate rock and related appropriate technology-latest developments and practical experience, Kuala Lumpur, Malaysia, 16–20 July 2001, pp 123–133

Snapp SS, Lynch JP (1996) Phosphorus distribution and remobilization in bean plants as influenced by phosphorus nutrition. Crop Sci 36:929–935

Snapp S, Aggarwal V, Chirwa R (1998) Note on phosphorus and cultivar enhancement of biological nitrogen fixation and productivity of maize/bean intercrops in Malawi. Field Crop Res 58:205–212

Snellgrove RC, Splittstoesser WE, Stribley DP et al (1982) The distribution of carbon and the demand of the fungal symbiont in leek plants with vesicular-arbuscular mycorrhizas. New Phytol 92:75–87

Sokoto AL, Singh A (2008) Yield and yield components of cowpea (*Vigna unguiculata* (L.) Walp.) as influenced by Sokoto phosphate rock and placement methods in the semi-arid zone of Nigeria. Nutr Cycl Agroecosys 81:255–265

Sperling L (2001) The effect of the civil war on Rwanda's bean seed systems and unusual bean diversity. Biodivers Conserv 10:989–1009

StClair DA, Bliss FA (1991) Intrapopulation recombination for ^{15}N-determined dinitrogen fixation ability in common bean. Plant Breeding 106:215–225

Stocco P, do Santos JCP, Vargas VP et al (2008) Assessment of biodiversity in rhizobia symbionts of common bean (*Phaseolus vulgaris* L.) in Santa Catarina, Brazil. Rev Bras Cienc Solo 32:1107–1120

Stracke S, Kistner C, Yoshida S et al (2002) A plant receptor-like kinase required for both bacterial and fungal symbiosis. Nature 417:959–962

Suarez R, Wong A, Ramirez M et al (2008) Improvement of drought tolerance and grain yield in common bean by overexpressing trehalose-6-phosphate synthase in rhizobia. Mol Plant Microbe In 21:958–966

Sunseri T (2005) 'Something else to burn': forest squatters, conservationists, and the state in modern Tanzania. J Mod Afr Stud 43:609–640

Tamado T, Fininsa C, Worku W (2007) Agronomic performance and productivity of common bean (*Phaseolus vulgaris* L.) varieties in double intercropping with maize (*Zea mays* L.) in eastern Ethiopia. Asian J Plant Sci 6:749–756

Tang CX, Hinsinger P, Jaillard B et al (2001) Effect of phosphorus deficiency on the growth, symbiotic N_2 fixation and proton release by two bean (*Phaseolus vulgaris*) genotypes. Agronomie 21:683–689

Thung M (1992) Phosphorus: a limiting nutrient in bean production in Latin America and field. Screening for efficiency and response. In: El Bassam N (ed.) Genetic aspects of plant mineral nutrition. Kluwer Academic, Dordrecht, pp 501–521

Thung M, Rao IM (1999) Integrated management of abiotic stresses. In: Singh SP (ed.) Common bean improvement in the twenty-first century. Kluwer, Dordrecht, pp 331–370

Tilak KVBR, Ranganayaki N, Manoharachari C (2006) Synergistic effects of plant-growth promoting rhizobacteria and *Rhizobium* on nodulation and nitrogen fixation by pigeonpea (*Cajanus cajan*). Eur J Soil Sci 57:67–71

Toljander JF, Artursson V, Paul LR et al (2006) Attachment of different soil bacteria to arbuscular mycorrhizal fungal extraradical hyphae is determined by hyphal vitality and fungal species. FEMS Microbiol Lett 254:34–40

Tsai SM, Bonetti R, Agbola SM et al (1993) Minimizing the effect of mineral nitrogen on biological nitrogen fixation in common bean by increasing nutrient levels. Plant Soil 152:131–138

UN (2008) The Millennium development goals report 2008. United Nations, New York

Vadez V, Drevon J-J (2001) Genotypic variability in phosphorus use efficiency for symbiotic N_2 fixation in common bean (*Phaseolus vulgaris*). Agronomie 21:691–699

Vadez V, Lasso JH, Beck DP et al (1999) Variability of N_2-fixation in common bean (*Phaseolus vulgaris* L.) under P deficiency is related to P use efficiency. Euphytica 106:231–242

Valentine AJ, Kleinert A (2007) Respiratory responses of arbuscular mycorrhizal roots to short-term alleviation of P deficiency. Mycorrhiza 17:137–143
Vance CP (2001) Symbiotic nitrogen fixation and phosphorus acquisition: plant nutrition in a world of declining renewable resources. Plant Physiol 127:390–397
Vanlauwe B, Giller KE (2006) Popular myths around soil fertility management in sub-Saharan Africa. Agric Ecosyst Environ 116:34–46
Vanlauwe B, Kanampiu F, Odhiambo GD et al (2008) Integrated management of *Striga hermonthica*, stemborers, and declining soil fertility in western Kenya. Field Crop Res 107:102–115
Vargas MAT, Mendes IC, Hungria M (2000) Response of field-grown bean (*Phaseolus vulgaris* L.) to *Rhizobium* inoculation and nitrogen fertilization in two Cerrados soils. Biol Fert Soils 32:228–233
Vejsadová H, Hršelová H, Přikryl Z et al (1988) Interrelationships between vesicular-arbuscular mycorrhizal fungi, *Bradyrhizobium japonicum* and soybean plants. Dev Soil Sci 18:115–123
Veltcheva M, Svetleva D, Petkova S et al (2005) *In vitro* regeneration and genetic transformation of common bean (*Phaseolus vulgaris* L.) – Problems and progress. Sci Hortic 107:2–10
Verdoodt A, Van Ranst E, Ye LM (2004) Daily simulation of potential dry matter production of annual field crops in tropical environments. Agron J 96:1739–1753
Vieira C (1998) Growing beans in favorable environments and in association with crops of high social and economic value. In: Voysest O (ed.) An ecoregional framework for bean germplasm development and natural resources research. Working Document. CIAT, Cali, pp 128–154
Vinuesa P, Silva C (2004) Species delineation and biogeography of symbiotic bacteria associated with cultivated and wild legumes. In: Werner D (ed.) Biological resources and migration. Springer, Berlin, pp 143–155
Vitousek PM (2004) Nutrient cycling and limitation: Hawai'i as a model system. Princeton University Press, Princeton
Voets L, Goubau I, Olsson PA et al (2008) Absence of carbon transfer between *Medicago truncatula* plants linked by a mycorrhizal network, demonstrated in an experimental microcosm. FEMS Microbiol Ecol 65:350–360
von Uexküll HR, Mutert E (1995) Global extent, development and economic impact of acid soils. Plant Soil 171:1–15
Vosátka M, Dodd JC, Patten R et al (2003) A joint initiative for the use of mycorrhizal fungi in plant production (the establishment of the Federation of European mycorrhizal fungi producers – FEMFiP). Folia Geobot 38:235–237
Wander AE, Didonet AD, Moreira JAA et al (2007) Economic viability of small scale organic production of rice, common bean and maize in Goias state, Brazil. J Agr Rural Dev Trop 108:51–58
Wardle DA, Walker LR, Bardgett RD (2004) Ecosystem properties and forest decline in contrasting long-term chronosequences. Science 305:509–513
Weil RR (2000) Soil and plant influences on crop response to two African phosphate rocks. Agron J 92:1167–1175
Whitfield J (2007) Fungal roles in soil ecology: underground networking. Nature 449:136–138
Wilson GWT, Hartnett DC, Rice CW (2006) Mycorrhizal-mediated phosphorus transfer between tallgrass prairie plants *Sorghastrum nutans* and *Artemisia ludoviciana*. Funct Ecol 20:427–435
Wolyn DJ, StClair DA, Dubois J et al (1991) Distribution of nitrogen in common bean (*Phaseolus vulgaris* L) genotypes selected for differences in nitrogen fixation ability. Plant Soil 138:303–311
Wooley J, Lepiz R, Aquinas-Portesy CT et al (1991) Bean cropping systems in the tropics and subtropics and their determinants. In: van Schoonhoven A, Voysest O (eds.) Common beans: research for crop improvement. CAB International and CIAT, Wallingford, pp 679–706
Woomer PL, Okalebo JR, Maritim HK et al (2003) PREP-PAC: a nutrient replenishment product designed for smallholders in western Kenya. Agr Ecosyst Environ 100:295–303
Wortmann CS (2001) Nutrient dynamics in a climbing bean and sorghum crop rotation in the Central Africa highlands. Nutr Cycl Agroecosys 61:267–272

Wortmann CS, Allen DJ (1994) African bean production environments: their definition, characteristics and constraints. Network on bean research in Africa Occasional Publication Series, no. 11. CIAT, Kampala

Wortmann CS, Lunze L, Ochwoh VA et al (1995) Bean improvement for low fertility soils in Africa. Afr Crop Sci J 3:469–477

Wortmann C, Kirgby R, Eledu C et al (1998) Atlas of common bean (*Phaseolus vulgaris* L.) production in Africa. CIAT, Cali

Wright DP, Read DJ, Scholes JD (1998) Mycorrhizal sink strength influences whole plant carbon balance of *Trifolium repens* L. Plant Cell Environ 21:881–891

Yan XL, Beebe SE, Lynch JP (1995a) Genetic variation for phosphorus efficiency of common bean in contrasting soil types. 2.Yield response. Crop Sci 35:1094–1099

Yan XL, Lynch JP, Beebe SE (1995b) Genetic variation for phosphorus efficiency of common bean in contrasting soil types. 1.Vegetative response. Crop Sci 35:1086–1093

Yan XL, Lynch JP, Beebe SE (1996) Utilization of phosphorus substrates by contrasting common bean genotypes. Crop Sci 36:936–941

Zacarías JJJ, Altamirano-Hernández J, Cabriales JJP (2004) Nitrogenase activity and trehalose content of nodules of drought-stressed common beans infected with effective (Fix^+) and ineffective (Fix^-) rhizobia. Soil Biol Biochem 36:1975–1981

Zahran HH (1999) *Rhizobium*-legume symbiosis and nitrogen fixation under severe conditions and in an arid climate. Microbiol Mol Biol Rev 63:968–989

Zaiter HZ, Mahfouz B (1993) Salinity effect on root and shoot characteristics of common and tepary beans evaluated under hydroponic solution and sand culture. J Plant Nutr 16:1569–1592

Zaman-Allah M, Sifi B, L'Taief B et al (2007) Symbiotic response to low phosphorus supply in two common bean (*Phaseolus vulgaris* L.) genotypes. Symbiosis 44:109–113

Zeder MA, Emshwiller E, Smith BD et al (2006) Documenting domestication: the intersection of genetics and archaeology. Trends Genet 22:139–155

Zhang FS, Li L (2003) Using competitive and facilitative interactions in intercropping systems enhances crop productivity and nutrient-use efficiency. Plant Soil 248:305–312

Zhang F, Hamel C, Kianmehr H et al (1995) Root-zone temperature and soybean [*Glycine max* (L) Merr] vesicular-arbuscular mycorrhizae: development and interactions with the nitrogen fixing symbiosis. Environ Exp Bot 35:287–298

Zhu YG, Smith SE, Barritt AR et al (2001) Phosphorus (P) efficiencies and mycorrhizal responsiveness of old and modern wheat cultivars. Plant Soil 237:249–255

Zhu JM, Kaeppler SM, Lynch JP (2005) Topsoil foraging and phosphorus acquisition efficiency in maize (*Zea mays*). Funct Plant Biol 32:749–762

Zingore S, Murwira HK, Delve RJ et al (2007) Soil type, management history and current resource allocation: three dimensions regulating variability in crop productivity on African smallholder farms. Field Crop Res 101:296–305

Conclusion

The discussions presented in this book have highlighted the important role played by legumes in cropping systems. Legumes can be incorporated into cropping systems as green manure, intercropped or rotated with cereals and as leguminous shrubs in improved fallow systems. Legumes are an important source of nutrition to both humans and livestock by providing the much needed protein, minerals, fiber and vitamins. Legumes are a major source of income through the sale to of grain, leaves and fiber. The case studies reviewed show that legumes play an important role in maintaining soil fertility and sustain crop production especially in resource scarce systems such as those present in SSA. This is mainly due to the special relationship between leguminous plants with nitrogen-fixing bacteria called Rhizobium. By biologically fixing nitrogen (BNF) in the soil, legumes provide a relatively low-cost method of replacing otherwise expensive inorganic nitrogen in the soil. Legumes also improve other soil physical properties, provide ground cover and reduce soil erosion, increase soil organic matter, cation exchange capacity, microbial activity and lowers soil temperature and suppress weeds and pests. All these beneficial effects result in enhanced soil fertility and boosts subsequent cereal crop yields. Legumes therefore play an important role of improving the livelihoods of smallholder farmers around the world.

Improving legume production will require investment in addressing the key challenges identified by the various studies. There is need to improve general soil fertility and in particular phosphorus availability which is essential for proper legume establishment and biological nitrogen fixation. Variety selection and breeding is needed to adapt to biophysical stresses such as low soil pH, high salinity, drought and flooding. Beyond the biophysical constraints is the need to address the challenges posed by diseases, pests and weeds. Strategies are also needed to improve farmers access to inputs (improved seed and fertilizers), markets and access to relevant production information.

Strategies to scale out adoption of legume-based technologies should therefore focus at developing appropriate agroecological and social niches within which

A. Bationo et al. (eds.), *Fighting Poverty in Sub-Saharan Africa: The Multiple Roles of Legumes in Integrated Soil Fertility Management*,
DOI 10.1007/978-94-007-1536-3, © Springer Science+Business Media B.V. 2011

potential of the different legume crops can be fully realized. Potential niches for the "best bet" legumes should aim at 1) defining the sequential or simultaneous occurrence of food crops and legumes so as to reduce competition and maximize on productivity while at the same time spreading risk, and 2) defining the best place to grow legumes on the farms depending on the farmers priorities and preferences, soil fertility, farm size, climate and other factors.

Index

A. Bationo et al. (eds.), *Fighting Poverty in Sub-Saharan Africa: The Multiple Roles of Legumes in Integrated Soil Fertility Management*, DOI 10.1007/978-94-007-1536-3,

© Springer Science+Business Media B.V. 2011